Technology Market Transactions

For Elfriede Margarethe Emma Dora Tietze

Technology Market Transactions

Auctions, Intermediaries and Innovation

Frank Tietze

Institute for Technology and Innovation Management
Hamburg University of Technology
Germany

Edward Elgar
Cheltenham, UK • Northampton, MA, USA

Published by
Edward Elgar Publishing Limited
The Lypiatts
15 Lansdown Road
Cheltenham
Glos GL50 2JA
UK

Edward Elgar Publishing, Inc.
William Pratt House
9 Dewey Court
Northampton
Massachusetts 01060
USA

A catalogue record for this book
is available from the British Library

Library of Congress Control Number: 2012935311

FSC
www.fsc.org
MIX
Paper from
responsible sources
FSC® C004839

ISBN 978 1 78100 009 0

Printed and bound by MPG Books Group, UK

Contents

List of Figures vi
List of Tables vii
Foreword ix
Acknowledgements xii

PART I SETTING THE SCENE

1. Introduction 3
2. Research Methodology 21

PART II TECHNOLOGY TRANSACTIONS AND AUCTIONS

3. A Firm Perspective on Technology Transactions 47
4. Technology Market Intermediaries 66
5. Auctions for Technology Transactions 76
6. Technology Properties 97
7. Transaction Cost Theory 122

PART III AN EMPIRICAL STUDY OF TECHNOLOGY AUCTIONS

8. Methodological Approaches 139
9. Auction Governance Structures 168
10. Analysis of Auctioned Technologies 227
11. Discussion of Results 294
12. Conclusions, Implications, and Research Recommendations 318

Annexes 334
References 337
Index 378

Figures

1.1	Inter-firm technology transactions on the MfTI	7
1.2	Scope of the study and related research fields	13
2.1	Conceptual framework	22
2.2	Research strategy	30
3.1	Transactions along the innovation process involving TMI	48
3.2	Systematization of the forms of technology exploitation	53
3.3	Distribution channels for technologies	54
3.4	Extended framework incorporating exploitation channels	55
3.5	External technology deployment process model	56
3.6	External technology commercialization process model	58
4.1	The transaction nature without / with TMI involvement	74
5.1	Schematic view of a generic auction process	84
6.1	Types of carriers for technological information	99
6.2	Patent portfolio value distribution and recommended actions	112
6.3	The value of European patents across macro-technology classes	114
7.1	Continuum of transaction governance structure types	127
7.2	Matching governance structures with commercial transactions	131
7.3	Transaction costs and asset specificity of governance structures	133
8.1	Conceptual framework of the quantitative study	149
8.2	Sensitivity analysis of value ranges	162
9.1	Technology auction process structure	176
9.2	Phases and distinguishing events in the auction process	181
9.3	Six standardization elements of technology auctions	207
9.4	Profit and cost comparison of bargaining negotiations / auctions	221
10.1	Transactions closed at different occasions of the auction process	229
10.2	Alternative measures for technological uncertainty	235
10.3	Conceptual model for regression analyses	261
11.1	Suspected u-shaped relationship of technological uncertainty and sales price	296
11.2	Firms' decision sequence in technology market transactions	303
11.3	Transaction costs of auctions and bargaining negotiations	306
11.4	Dynamic TC as reason for disintegration / reintegration phases	308
11.5	Alternation of disintegration and reintegration phases	309

Tables

3.1	Selected empirical ETE studies since the mid-1980s	49
5.1	Four common bidding procedure types	88
6.1	Average and median values of patents by industry	115
6.2	Patent use across different invention sources	117
8.1	Overview of auctions covered in the dataset	150
8.2	Data compiled from different sources	150
8.3	Classification of application areas	152
8.4	Examples of consolidated assignee names	154
8.5	Missing data overview	156
8.6	Normality of original / transformed variables	159
9.1	Listing fees for different asset types of an OT auction	169
9.2	Bidding processes and MRPs of OT and IPA auctions	177
9.3	Fee structures of OT and IPA auctions	178
9.4	Duration of due diligence period	180
9.5	Overview of case study firms and technologies	184
9.6	Comparative properties of technologies	197
9.7	Perceived technology values	198
9.8	Internal management of technology transactions	200
9.9	Perceived problems of transactions	201
9.10	Perceived advantages of technology auction transactions	202
9.11	Role of auction firm in the transaction	203
9.12	Overall success of transactions	204
10.1	Overview of variables	228
10.2	Example of inventor team composition	238
10.3	Technologies offered and sold at the auctions	242
10.4	Sources of auctioned technologies	243
10.5	Distribution of auctioned technologies across application areas	244
10.6	Sales prices across application areas	245
10.7	Overview of sales prices across the technology auctions	246
10.8	Sources of technology and technology sales prices	246
10.9	Comparison of expected value and sales prices	247
10.10	Descriptive statistics for technology properties	248
10.11	Technological complexity across auctions	250
10.12	Technological complexity for different sources	251

10.13 Technological uncertainty across auctions 252
10.14 Technological uncertainty for different sources 253
10.15 Technological impact across auctions 254
10.16 Technological impact for different sources 255
10.17 Technology quality across auctions 256
10.18 Technology quality for different sources 256
10.19 Value perception of sellers across auctions 257
10.20 Value perceptions of sellers for different sources 258
10.21 Distributions of technology properties and sales prices 260
10.22 Multivariate logistic regression models 264
10.23 Nagelkerke R^2 change statistics for logistic regression 265
10.24 Logistic regression models across value ranges - right tail 268
10.25 Logistic regression models across value ranges - left tail 271
10.26 Logistic regression models for auction sub-samples 273
10.27 OLS regression models 275
10.28 Change statistics for OLS regression models 276
10.29 Log-linear OLS regression models for sub-samples of young / old technologies 279
10.30 Log-linear OLS regression models across value ranges - right tail 281
10.31 Log-linear OLS regression models across value ranges - left tail 283
10.32 Log-linear OLS regression models for auction sub-sets 285
10.33 Summary of logistic and log-linear OLS regression results 287

Foreword

This book is a product of the 21st century. For a number of reasons it could hardly have been produced in or before the 1990s. First, the by now substantial interest in patents and IP management among economists and management scholars had not by and large taken root, at least not in Europe. Second, the business model of auctioning had not yet been applied to patents, despite the many centuries of history of both auctions and patents. Rapid developments in auction technology and patenting in the 1980s and 1990s paved the way for an encounter in the 2000s, however. Third, platforms for inter-university research collaboration in technology and innovation management hardly then existed in Europe.

Various entrepreneurial activities are behind the current situation. Industrial entrepreneurs have increasingly invested in and commercialized patents, business entrepreneurs have created patent auctions, academic entrepreneurs have created collaborative platforms and entrepreneurial researches have entered the IP field.

This study is the result of a joint research project between the Hamburg University of Technology (Germany) and Chalmers University of Technology (Sweden). Both participating institutes are founding members of the European Institute for Technology and Innovation Management (EITIM).

With this book Frank Tietze makes a substantial contribution to the research and literature on transactions of patents and technologies as well as the effects of intermediaries acting on the markets for technology and innovation (MfTI). His work clearly helps to better understanding auctions, their functioning and role in technology trade, innovation and exploitation through conceptual, empirical, theoretical and practical contributions.

This study comes at the right time since technology-based firms have in recent years widely realized the importance of exploring different strategies to appropriate the benefits from investing in technology development and innovation. Today firms increasingly innovate openly, sourcing and exploiting technologies outside the boundaries of their own firm. For this purpose they eagerly need to learn how to successfully market technologies and ideas on the markets. In this context the MfTI are becoming increasingly attractive alternatives.

This affects management skills and practices and the implicit learning needs to address the challenges to manage transactions, specifically when it comes to the acquisition and exploitation of technologies for the benefit of the firm.

But many practical hurdles still complicate efficient clearing on the MfTIs causing high management and transaction costs. These costs in turn offer opportunities for intermediaries to enter the MfTIs, offering innovative services and novel transaction models. In this realm among others, technology auctions have recently become popular. But are these suitable for all types of technology and patents? Probably not and firms need to (as always) decide on a good if not the best suitable model with a favorable transaction governance structure adapted to the specific characteristics of the technologies and patents to be auctioned. This again is likely very context dependent; but how to optimally decide?

Working together with experts Frank Tietze developed a set of variables for designing technology transaction models, based on transaction cost theory. Building on these insights he further developed and tested them with multiple case studies and analyzed over 390 patented technologies from six technology auctions using inferential statistical techniques.

His results show that technology auctions appear suitable as a transaction model primarily for technologies with moderate market values. Such technologies are difficult to transact efficiently via "classical" bargaining negotiations commonly characterized by high search and transaction costs. On the other hand, the highly standardized transaction governance structure of technology auctions - particularly the perceptual, non-limited commission fees - imposes constraints on high value technologies. These can anyhow be transacted profitably via costly bargaining negotiations that also allow for a higher degree of flexibility in contract design. The governance structure of the investigated technology auctions should further not be considered preferable for transactions of highly complex technologies that need additional transfer of tacit knowledge complementing the ownership rights of the related intellectual property assets. Rather auctions can be seen as spot market exchanges of technologies that can simply be transacted without the need to structure and negotiate complex royalty schemes.

The results of this study are especially relevant to firms, universities, independent inventors and their support institutions through offering a much better understanding of technology transaction processes in general and auctions in particular. The results further support intermediaries and particularly auction firms to help optimize the design of transaction models. Furthermore, the results of this study support innovation and intellectual property policy makers when designing schemes to further develop the technology markets. This study's contribution to the current debate must not least be considered highly relevant because Frank Tietze has integrated the micro level perspective of innovating firms (reflected well by the expertise at the partner institute in Hamburg) with the market perspective (reflected through the industrial economics orientation of the Swedish partner institute).

As an entrepreneurial researcher Frank Tietze has produced this book with its underlying study as one of the first of its kind. As such he charters new territory, not crafting complete maps but rather finding fertile new areas and directions for further investigations, be they theoretical, statistical or managerial. It is with great pleasure that we can recommend this book to a broad range of readers. All in all, it is highly relevant to both academia and industry: Comprehensive, interdisciplinary, containing almost all elements to be expected, and as such very well structured and readable.

Hamburg/ Gothenburg
November 2011

Cornelius Herstatt
Ove Granstrand

Acknowledgements

Although this book is an outcome of my own pursuit, it is the result of a collaborative endeavor of many people without whom I would not have been able to do what I was finally able to achieve. I owe a great deal to Cornelius Herstatt. Your permanent belief in my work, the flexibility to pursue this research literally independently at your institute as well as your continuous support and guidance in all matters, whether related to this study or of a private nature, have been invaluable. Your patience, combined with deliberate endorsements created the endurance so indispensable for this intellectual expedition. Not less I owe my deepest thanks to Ove Granstrand. Your sound opinion, critical but inspiring advice has been instrumental opening space for original avenues of thought. The number of ample debates with you have been intellectually more than rewarding. Following you when expressing any complex thought in a distinctively clear manner, has inspired me to enjoy research.

I thank my current and previous colleagues for engaging into discussions with me about my research, particularly Tim Schweisfurth, Dominic de Vries, Rajnish Tiwari, Florian Skiba, Viktor Braun, and Christoph Stockstrom. I appreciate the comments on my manuscript from Christoph Mirow, at that time based at the Berlin Institute of Technology. I also thank my colleagues from the MELT project at Chalmers University of Technology, particularly Marcus Finlöf Holgersson for stimulating ideas, and Thomas Ewing for important comments from a practitioner's perspective. I further thank Wolfgang Kersten for his support throughout the years and Frerich Keil for leading my examination committee.

My former students Antoneta Hristova, Andre Rocha Barreto, Henning Krenz, Sascha Exner and Sascha Havestadt contributed to this book. Various intense discussions helped me to maintain the spirit to advance this research. I particularly appreciate the thorough language proof reading from Ilze Hugo and the support of Ozan Mahmutluoglu for preparing the final manuscript. I owe further thanks to the interviewees, who helped me phrasing the topic and the research questions during the early stage of this study.

I wish to express my warmest gratitude to my mother Annemarie, my father Günther as well as my grandfather Günter for your steady belief in me and my own way. My deepest thanks go to Alexandra. It feels incredibly difficult to express how profoundly you have contributed to this accomplishment.

I owe further thanks to Jan-Peter Meier, Dominic de Vries, Ulla von Lühmann, Julius Meinecke, Arist von Harpe, Laleh Pakzad, Sönke Bruhns and Stig Kleiven. Your accompaniment often made demanding times much easier. I am grateful that I can consider you my friends.

Throughout this endeavor I had the pleasure to enjoy inspiring discussions and relaxing moments with EITIM Doc fellows from across Europe. I hope we continue our joint journey. I thank the EITIM professors for enriching my insights into the academic life, encouraging me to continue this venture.

Finally, this research would not have been possible without financial support. The Awapatent Foundation for the Promotion of Scientific Research in the Area of Intellectual Property Rights kindly granted me funding twice in 2007 and 2010. Also the Hans-Sauer Foundation which facilitates ecologically responsible innovation development, as well as the Ernst Geber Jubileumsfonds, and the EU Regional Development Fund (ERDF) via the RIS Hamburg Project contributed financially to the completion of this study.

To all of you and many others who cannot be mentioned here, who have shown interest and indulgence, I express my heartfelt thanks. Although I have tried to the best possible extent, I have not been able to follow all your advice and the final responsibility is, of course, mine.

Hamburg
November 2011

Frank Tietze

Research conducted with kind support from:

AWAPATENT

HANS SAUER STIFTUNG

PART I

Setting the Scene

1. Introduction

1.1 INNOVATION, TRANSACTIONS AND MARKET INTERMEDIARIES

To create and sustain a competitive advantage and subsequently ensure continuous growth, many firms focus on the creation of innovations. These firms often find themselves in an innovation race against competitors, thus being forced to accelerate their processes from 'simply being ideas' to their actual launch on the markets (Cooper, 2008). While firms' tendency to innovate depends primarily on their internal resources in their research and development (R&D) departments, this has become difficult mainly due to the increasing 'complexity of modern technology' resulting from the cumulative nature of many technologies[1] and technical products (Hall, 2004: 4). According to Granstrand (2000b: 9), 'products and services are not only becoming increasingly based on new technologies, but increasingly based on many different technologies. That is, products and services become more multi-technological'.[2]

Nowadays, for instance, automobiles can hardly be regarded as discrete products but must be seen as complex technical systems that include a wide range of electronics and software components that were not built into automobiles in the 1970s (Miyazaki and Kijima, 2000). The global system for mobile communications' (GSM) standard for mobile telephony is another example. According to Bekkers et al. (2002), GSM includes at least 140 essential patent families with the major share being scattered across large multinational competitors (that is Motorola owns 18 per cent; Nokia 13 per cent; Alcatel 10 per cent; Philips 9 per cent, and Telia 7 per cent).[3] There are various other examples. For instance, the DVD media technology consists of 500 patents from 28 countries owned by nine patent holders, the MPEG four technology is made up of 196 patents from 21 countries owned by 22 patent holders, the Ethernet technology comprises 70 patents from four countries owned by 65 patent holders, and Wifi technology (802.11 wireless) holds 100 patents from seven countries and 91 patent holders (Aggarwal and Walden, 2009).

The growing complexity of technical products is well reflected in empirical statistics. Kash and Rycroft (2002) prove that complex technologies comprised 43 per cent of the 30 most valuable world goods exports in 1970, however by 1996, complex technologies represented 84 per cent of those goods. von Graevenitz et al. (2008) show that there was a strong increase in complex technology patents between 1978 and 2003. This was calculated by determining whether the annual patent applications filed at the EPO and USPTO are complex or discrete according to dichotomic categorization proposed by Cohen et al. (2000).

The growing complexity of technical products and systems has increased firms' uncertainty with regard to freedom-to-operate (that is whether they possess all the necessary intellectual property (IP) assets to be allowed to manufacture a certain product). Consequently, the risk of litigation has increased, especially for large multinational firms. Particularly in the US, but also to some extent in Europe, non-practicing entities (often labeled 'patent trolls') have recently filed enormous infringement cases against large firms pressing for damages and licensing royalties. For instance, in 2001, Research in Motion (RIM), the Blackberry manufacturer, was sued by the 'non-manufacturing entity' NTP for infringing on its patents 'covering the use of radio frequency wireless communications in e-mail systems.' In 2006, the case was settled with RIM paying US$612.5 million to NTP (Tietze and Herstatt, 2010: 8). In another high profile patent case, Intel reached a US$525 million settlement of a suit alleging that Intel's Pentium family of microprocessors infringed Intergraph's patents (Gilbert and Katz, 2007). In Europe, one of the first high-number infringement cases started in 2008 when IP-Com sued Nokia at the German patent court in Mannheim for infringing about a thousand patents of some 150 patent families. IPCom, who bought the patents from Robert Bosch GmbH in 2006, claims €12 billion in damages (Nokia, 2009: 124).

Behnken (2005: 1) concluded this trend, arguing that because 'products and technologies are becoming increasingly complex... supporting activities to conceptualize, develop and promote the product become increasingly detailed and comprehensive... [and] the competencies to develop new technologies or products are functionally and spatially dislocated... [therefore, firms require] various competencies over organizational and geographical boundaries to develop a complex innovation.'[4] In addition, Granstrand (2003a: 233f) summarized this trend arguing that nowadays new technologies are 'interacting with each other and with old technologies in more complex and interdependent ways ... [meaning] that patents and businesses become more cross-linked and interdependent with each new business becoming reliant on an increasing number of patents and each new patent having an impact on an increasing range of businesses on average… Thus, in a new technology as well as in a product market, there will be not only more agents on average, but increasingly interdependent agents in a mixture of cooperation and competition ('coopetition' or 'competeration').'

Opening up the innovation process to cope with growing complexity
The increasing complexity of many technical products and systems makes it economically impossible for a single firm to internally develop all the technologies necessary for an innovation and particularly the IP assets required to enable freedom-to-operate in order to prevent costly litigation. Specifically with regard to IP assets firms – particularly those in highly competitive industries – often face an IP assembly problem that might be substantial, especially if important strategic patents are scattered across competing firms.[5] Hence, such an IP assembly problem can create hold-up problems, slowing down development processes, for instance 'when a patent owner refuses to allow a technology developer the use of a patent after the technology has already been developed' and it is difficult and costly to invent complementary parts for that technology (Aggarwal and Walden, 2009: 24).[6] In addition, where an innovation requires several complementary patents, fragmented property rights can limit firms' willingness to invest in R&D (Clark and Konrad (2008), Ziedonis (2008)). In contrast, firms often undertake redundant research and develop technologies that are valuable to other firms as well. Even if they are actively used by their inventors, technologies could also be embedded into other firms' products (for example, because they have been established as an industry standard). Moreover, technologies that are not used by their inventors might still be valuable to other firms that could then avoid own R&D efforts.[7]

Thus, mainly due to the growing complexity of many technical products and systems, firms not only need to manage the innovation process (that is their R&D activities), but also draw increasing attention to the management of their IP assets.[8] While firms should at least manage their IP portfolio to eliminate the risk of litigation, they can furthermore realize benefits for the more efficient creation of innovations through successful IP management. While a firm can conduct internal R&D efforts to develop its own solution to a technical problem, taking a license from another firm that has already developed a solution, might not only save resources but also accelerate the firm's innovation activities.

As a result, firms have recently started to innovate more openly in (international) networks, often even together with competitors that own technologies or IP assets relevant for industry standards.[9] Technology trade on the markets for technologies and ideas (MfTI)[10], for example, through licensing and particularly cross licensing, becomes increasingly necessary as a means to solve the IP assembly problem and holds advantages for both the technology owner and the firm willing to acquire or license it (Granstrand, 2003a). Firms that own a technology can generate additional revenues, thereby increasing their own R&D return ratio through the external exploitation of those technologies.[11] According to the European Commission (2008: 32), nowadays, 'firms increasingly diffuse intellectual property beyond firm and even country boundaries, as firms innovate more openly.' Lichtenthaler (2005), referring to Cesaroni et al. (2004), Granstrand (2004a), Chesbrough (2003b) and Grindley and Teece (1997) summarizes that the increasing technological content of products accompanied by shorter product and technology lifecycles and more intense competition have encouraged stronger external knowledge exploitation. Lichtenthaler (2005) argues that this 'knowledge push effect' has been intensified by a growing knowledge convergence and fusion, which has led to higher numbers of knowledge components from different areas being incorporated into a single product. In fact, there is great exploitation potential. Having investigated 867 patents (applied for between 1994 and 1997), Braunerhjelm and Svensson (2010) find that, on average, 39 per cent of all patents are 'fallow patents.' Nevertheless, a firm's 'willingness-to-exploit' (for example, license or sale) a technology depends mainly on the purpose of the technology in the portfolio (for example, strategic, fencing, blocking, evergreening).[12] However, firms on the supply side as well as on the demand side turn beyond the own firm boundaries to source technologies from the MfTI. Investigating 133 technology licensees and an equal number of matched non-licensees, Reichstein (2009) proves that licensed technology indeed acts as a catalyst for accelerating innovation speed. Moreover, Laursen and Salter (2006) documented the positive effects of externally sourced knowledge on innovation outcomes in an empirical study.

Hence, in order to maintain competitiveness in the innovation race (for example, by accelerating its innovation speed) and generate sustainable firm growth, an increasing number of firms extend their innovation processes beyond their own firm boundaries (Gassmann (2006), Rosenkopf et al. (2001)) – a trend labeled by Chesbrough (2003c) as open innovation.[13] Following successful firms like Procter and Gamble, more and more firms have started innovating in networks, trying to exploit and source ideas, technologies and IP assets on the MfTI. Consequently, 'with a shift towards "open innovation" … contributors to innovations are more likely to come from different types of organizations' including universities, other firms (for example, start-ups), leading clients and users (Murray and O'Mahony (2007: 1008), Owen-Smith and Powell (2003)).[14]

With the paradigm changing from 'closed' to the 'open' innovation, the 'classical' (closed) innovation process (for example, according to Cooper (2008)), is now being split up and divided among different actors and innovation is carried out cooperatively (for example, in joint ventures, alliances, and R&D cooperations) where technology transactions among the involved actors play an important role. Figure 1.1 illustrates a technology transaction[15] on the MfTI within open innovation processes of two firms (Firm_A and Firm_B). Firm_A is exploiting a technology in an advanced stage of the innovation process to Firm_B who acquires the technology in an early innovation process stage. While transactions can take place directly between Firm_A and Firm_B, they can also involve technology market intermediaries (TMIs)[16] supporting the transactions.

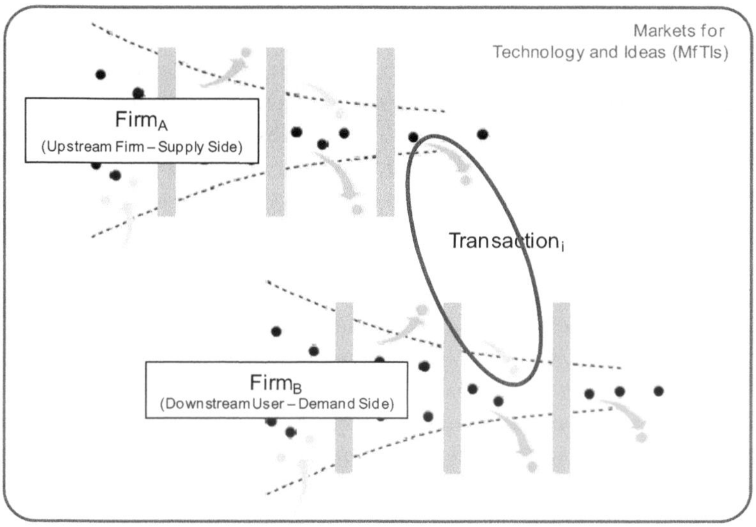

Notes: Own illustration with depiction of open innovation processes adapted from Chesbrough (2003c)

Figure 1.1 - Inter-firm technology transactions on the MfTI

Empirical studies have proven the trend towards open innovation. For instance, Sheehan et al. (2004) show that there has been an increase in the importance of out-licensing in 51.4 per cent of their surveyed firms in the preceding 10 years. Furthermore, 63 per cent of the respondents in their study expected this trend to continue and anticipated that the out-licensing of patented inventions would become more important for their firms in the following five years. This trend was also confirmed in a survey by PWC (2007) who reported that 54 per cent or their respondents expected the importance of out-licensing to increase in the following three to five years. There are also various firm examples. Among the most cited cases is IBM. While IBM is among the top patent holders in the world, its licensing revenues until 1993 amounted to approximately US$300 million a year. This changed drastically in 1993 when, under the newly appointed CEO Lou Gerstner, IBM adopted an active licensing program that was expected to yield US$1.4 billion to US$1.5 billion in 2000 according to Salomon Smith Barney (1999). Owing to the substantially higher gross margin on licensing revenues than on other IBM revenues, the patent royalty contribution to IBM's bottom line can be regarded as considerably larger than of those from other revenue sources. According to Shulman (2003) and Lang (2001), IBM's licensing revenues accounted for 20 per cent of their total profits in 1999 and in the last decade a total of almost €8.2 billion. Dow Chemicals is another example often cited. According to Roos et al. (1997), the company set up an 'Intellectual Asset Function' in 1993 and obtained licensing revenues of €110 million in 2000 compared to €22 million in 1994. Another example is the Denmark-based healthcare firm, Novo Nordisk A/S. According to Reitzig (2004), the firm gained a dominant market position in Europe with a diabetes drug since it had a license to a technology for manufacturing insulin from animal sources. From 197 survey responses and 30 interviews with senior executives in the five principal regions across the world, PWC (2007) reported that above 80 per cent of all surveyed top managers agreed or strongly agreed with the statement that IP management is important for the success of a firm.

The increased adoption of open innovation approaches is also well reflected by the growth in the MfTI. Although MfTI have existed for decades,[17] empirical data from various sources indicate that they have grown sizably since the 1990s[18], especially in some high-technology areas.[19] Arora et al. (2001: 40) compare estimates at an aggregated level from three different data sources that were 'subject to numerous caveats' but led to rather consistent results. Limiting their analysis to technological knowledge, Arora et al. (2001: 40) estimates indicate that the annual worldwide 'markets for technology ' were worth between US$35 billion and US$50 billion in 2000. In addition, Elton et al. (2002) and Kline (2003) estimated that the overall US patenting licensing revenues have skyrocketed from below US$15 billion per year in the early 1990s to around US$100 billion per year in 2002. Results from a survey by Sheehan et al. (2004) indicate that a majority of 81% of the responding firms expected an increasing number of out-licensing transactions from 2005 to 2010, while 54% of the respondents had experienced a growth of out-licensing since 1995. A study by the Institut der deutschen Wirtschaft Köln (2006) estimates that the German MfTI has a potential size of €8 billion. Gambardella et al. (2006) estimated the market for the EU-8 countries at €9.4 billion from 1994 to 1996, €12.7 billion from 1997 to1999 and €15.6 billion from 2000 to 2002, which corresponds to 0.16 per cent, 0.19 per cent and 0.20 per cent of the GDP and a total growth of 65 per cent between the third and the first period. According to Gambardella et al. (2006: V), 'the total value of patents licensed has increased considerably in the... industries [they included in their study] in the 1990s, suggesting that the markets for technology are growing in these sectors at a significant pace.' Athreye and Cantwell (2007) have compared the growth of non-US patents and worldwide licensing receipts and came to conclude that the growth of patenting coincided with the growth of MfTI after the 1980s. Moreover, the growth is expected to continue in the future. Gambardella et al. (2006) estimate that the MfTI in Europe could increase by 50 per cent, that is the potential has grown from US$14.8 billion to US$24.4 billion. As noted earlier, the market potential suggests that there are notable untapped opportunities for enhancing the MfTI in Europe and subsequently for using this means in order to increase the utilization rate of patented technologies.

Obstacles preventing efficient transactions

As a result, when a growing number of firms started expanding their innovation processes, they needed to acquire dedicated knowledge to establish competences and capabilities for an effective management of transactions on the MfTI. While firms that source technologies externally need to establish acquisition competences, those firms who own technologies and are willing to exploit them to those sourcing them (for example, to generate additional R&D returns through outlicensing or technology sales) need to establish competences to efficiently manage the exploitation process.[20]

However, according to Lichtenthaler and Ernst (2006a), the management approaches developed so far are still in their infancy. Firms face various obstacles when managing transactions on the MfTI including the not-invented-here-syndrome (NIH)[21], valuation difficulties to assess market prices, identifying buyers, etc.[22] In addition to internal obstacles related to management processes and firms' innovation cultures, further obstacles relate to the interfirm relations or the institutional structures of the MfTI in which the transactions take place. Most of the studies that were conducted to better understand MfTI transactions on firm level either from the seller's perspective (for example, Lichtenthaler (2006a), Escher (2005)), the buyer's perspective (for example, Granstrand et al. (1992)) or on a national level (for example, Gambardella (2002); Granstrand (2004a)) came to similar conclusions. The market design and the institutional structure are far from optimal. There are many obstacles that inhibit MfTI from clearing efficiently. According to Troy and Werle (2008: 3), the 'well-functioning market for patented new technological knowledge is confronted with several obstacles... [and the]...markets are far from functioning smoothly.' Teece (1998b: 545), referring to his early work in Teece (1981), noted that he had already recognized the 'first signs of an emerging market for know-how' nearly twenty years prior to his 1998 research. However, at the same time, he had stated that 'much technology does not enter it ... either because the firm is unwilling to sell or because of difficulties in transacting in the market for know-how.' Teece (1998b: 62) further noted that 'the market for knowledge is riddled with imperfections...' and 'one class of assets that is especially difficult, although not impossible, to trade involves knowledge assets.' Owing to various obstacles on the MfTI, transaction difficulties and high transaction costs ultimately prevent efficient market clearing, which leads to market failure or constitutes a transaction challenge for firms.[23] Consequently, firms facing the transaction challenge have lower incentives to engage in technology transactions that could actually facilitate the more efficient creation of innovations.

Technology market intermediaries

Recently specialized market actors started to enter the MfTI using transaction obstacles as business opportunities.[24] These technology market intermediaries (TMI) develop novel transaction models offering them to firms willing to acquire or exploit technologies and IP assets of the MfTI.

OECD et al. (2005: 10) point out that 'market intermediaries have become more numerous and diverse as the demand for technology transfer and patent valuation has grown.' According to the EPO et al. (2006: 1) the 'marketplace is nowadays in a probe and learn period in which the number of intermediaries is rising.' EPO et al. (2006) draw further attention to the rise of new models that intermediaries apply. Examples of such models include partnerships, technology pools, special purpose investments vehicles, auctions, publicly traded IP indexes, as well as patent value funds which aim at taking care of IP logistics issues (for example, finding and negotiating with potential licensees). According to EPO et al. (2006: 1) these new models 'make one step towards the development of a market... [and]... contribute to the maturation of the IP market.'

Aside from governmental organizations, scholars from this field have also recognized the emergence of TMI. Koruna (2004) observes that, with new services and instruments on the market, the process of externally exploiting technologies is getting easier and thus will probably also gain more acceptance among firms. Chesbrough (2006: 3) reported 'that a small number of intermediary firms has arisen in recent years to assist in the process of identification, negotiation and the transfer of patents from one firm to another.' Troy and Werle (2008: 20) note that 'the number of intermediaries is growing, as is the propensity of firms to employ specialized intellectual property professionals. These and other actors potentially involved in patent transactions gain trading experience, experiment with different modes of trade and invent ways to cope with uncertainty.'[25] However, for the purpose of this study only few statistics documenting this trend were available. Subsequently, I collected own data by means of a pre-study done in December 2006. Based on interviews with a range of industry experts, about 70 TMI were identified. The growth rate of those firms was calculated since the year of their foundation. The results of this pre-study confirm the trend. An approximated exponential curve fit indicates an annual growth rate of 8 per cent since 1980.[26]

Today, TMI have developed a number of novel transaction models that are characterized by different transaction governance structures (TGS). However, it remains insufficiently understood whether the different models are equally suitable for every technology transaction or which factors determine the choice of preferred model. Among the various novel transaction models, technology auctions have gained particular interest since the first, widely recognized public technology auction was organized in San Francisco, US in spring 2006. Since then, subsequent technology auctions were held not only in the US but also in Europe and Asia. However, whether all or only certain technologies can and should be auctioned remains an open question in the literature and motivated this study.

To summarize, firms have nowadays widely realized the importance of innovation, while recent market developments force them to continuously innovate (for example, by accelerating their innovation speed). However, to maintain competitiveness, particularly for the development of (increasingly) complex technical products, firms need to combine internally developed technologies with externally acquired ones as innovation is often a cumulative process. Hence, firms no longer only use internal resources to innovate but there is an increasing tendency to innovate openly, sourcing from and exploiting technologies outside of the own firm's boundaries on the MfTI. Hence, these firms become more dependent on rigorous management of the interfirm innovation process as well as on the management of transactions to acquire or exploit technologies. These firms need to acquire knowledge to establish competences and capabilities for the efficient management of technology and IP asset transactions on the MfTI.

However, various obstacles prohibit efficient clearing on the MfTI (for example, through high transaction costs). Those obstacles present a transaction challenge for firms that engage in technology transactions or open innovation. TMI have started entering the MfTI (that is as organizational innovation in the institutional structure). They develop and offer novel transaction models (service innovations) to firms in order to facilitate the more efficient creation of innovations.

However, each novel transaction model is characterized by a different TGS and is – most likely – not equally suitable for any technology transaction. Among these various novel transaction models, firms have to choose the best suited model with the most favorable TGS for each transaction. Thus, firms need to understand how they can best utilize the currently emerging TMI or the transaction models they offer. Technology auctions are among these new models. The question addressed in this study is whether and, if so, for which technologies auctions represent a suitable TGS for conducting efficient transactions.

1.2 FIELD OF RESEARCH AND FOCUS OF THE STUDY

This study is of an interdisciplinary nature and combines concepts from both management science and industrial organization literature. Figure 1.2 illustrates the main research fields that are related to this study and which are further discussed in the following section. The research fields that are closely related to the present research are highlighted with bold frames.

In terms of management science, this study is primarily grounded in the innovation management literature with a focus on the firm.[27] In this study, innovation is considered a result of an iterative, interacting and cumulative process, in which various pieces of complementary knowledge (tacit and explicit)[28] as well as related IP assets[29] are assembled from different sources. These sources are either inside the own firm (including other business units) or, to an increasing extent, from outside the own firm's boundaries. This definition builds on the notions of various scholars. For instance, Murray and O'Mahony (2007: 1008) argue that 'for innovation to occur … innovators must have the ability to actually combine or accumulate knowledge.' Aghion et al. (2001: 470), subscribing to endogenous growth theory, argue along the lines of Harris and Vickers (1987) and Budd et al. (1993) that technological progress is 'emerging from a dynamic process of "step-by-step" innovation.' Green and Scotchmer (1995: 20) note that 'knowledge and technical progress are cumulative in the sense that products are often the result of several steps of invention, modification and improvement.' Nelson (1994: 50), subscribing to evolutionary growth theory, also observes that 'in many fields, technological advance … [is]… "cumulative" in the sense that today's technological advances tend to proceed from yesterday's, building on and from what had previously been achieved and improving it in various directions. In many cumulative technologies … "natural trajectories" tend to appear, with the cumulative improvements proceeding along particular lines of advance that reflect both what technologists understand they likely can achieve and what entrepreneurs believe customers will buy.'[30]

In management science, following the introduction of the innovation concept by Schumpeter (1912), systematic research on innovation management began in the 1960s and has since then exploded (Hauschildt and Salomo, 2007: X).[31] Innovation management has nowadays grown into a complex research stream combining a wide range of topics including, for instance, different generations of innovation processes (for example, distributed, open, linear, and interactive) with its different phases / stages,[32] innovation strategies, project management, organizational structures (for example, patent and licensing departments), incentive systems, sources of innovation,[33] actors in innovation projects, knowledge management approaches, corporate entrepreneurship, product lifecycle models, diffusion process models, various management tools, innovation types (for example, radical, incremental, cumulative), and system innovations.[34]

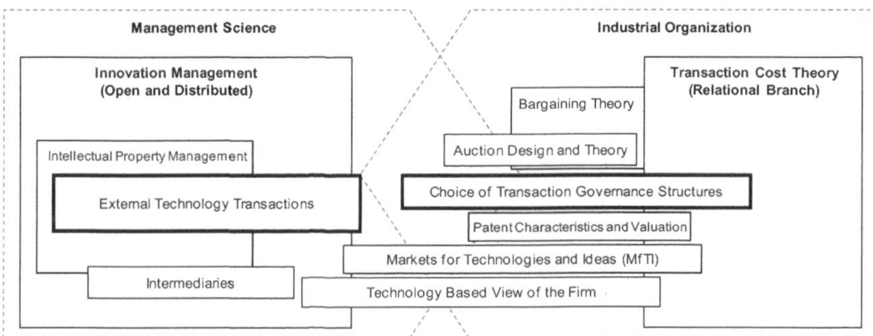

Figure 1.2 - Scope of the study and related research fields[35]

In terms of innovation literature, this study analyses the transactions of the firm's (patented) technologies. Similarly, this study looks into the management (exploitation and acquisition) of a firm's 'technology base'[36] supporting the creation of innovations. Firms that transact and recombine technologies for the creation of innovations are considered to be 'technology-based' in this study. Technology-based firms (TBF) are the primary actors that engage in transactions with their innovation and IP (IIP) managers who manage these transactions. Although this study focuses on transactions in general I apply the perspective of firms exploiting technologies rather than of those acquiring them. More specifically, this study focuses on the organizational structures and the management of technology transactions that take place beyond the boundaries of firms on the MfTI.

Each technology transaction is assumed to happen within the overarching frame of an innovation process. It has often been assumed that the firm who develops an invention and eventually files for patent protection will be the same actor to complement the patented invention with all necessary resources to carry it all the way through the different stages of the innovation process up to the market launch, that is turn the patented invention into an innovation. However, a significant share of patents is not used directly by its inventors. To facilitate innovation, it is thus necessary to ensure that patented inventions are owned by the market actors who can and are willing to exploit them. The interacting innovation process across firm boundaries has recently been labeled as open or distributed innovation. Those distributed, inter-firm innovation processes, particularly when TMI are involved, are related to this study.[37]

As this study is primarily concerned with patented technologies as the transaction object, it is closely linked to the IP management literature. While – in light of the above mentioned developments – previous studies of patents have been mainly addressed by the legal discipline and only to a minor extent by economists[38], the traditional view on patents has changed gradually since the early 1990s, which is reflected by the increasing attention scholars have drawn to this field.[39] Based on some early contributions (for example, as often cited Teece (1998a), Grindley and Teece (1997), Nonaka and Takeuchi (1995)), patent management and more broadly IP management is currently widely accepted and integrated into the innovation management literature, particularly with importance to open and distributed innovation. Within IP management research, scholars have been concerned with processes, tools, strategies, and structures of technology acquisition and exploitation[40] transactions and the marketing of technologies.[41] However, the role of the increasing number of TMI influencing transactions has so far been neglected to a large extent.[42] This study focuses particularly on how TMI influence transactions (particularly those conducted via technology auctions). To enhance the readers' understanding of TMI, this research borrows different concepts from the literature on brokers and intermediaries in innovation systems (see Chapter 3.2). A large stock of the IP management literature focuses on licensing. Although this topic is closely related to this study, I rather take an organizational perspective, focusing on the structure of transactions, and thus hardly refer to the licensing literature which often rather focuses on, for example, contract design and royalties.[43]

As previously mentioned, to achieve an economically efficient resource allocation along the (open) innovation processes from inventors of technologies to those best equipped to exploit them, MfTI represent the institutional framework as part of the innovation system[44] in which transactions take place (see Figure 2.1). Thus, applying this outside, inter-firm perspective involving different market actors and their relationships during transactions within this institutional framework, this study is also related to the industrial organization literature (for example, Carlton and Perloff (2008), Tirole (2008)). In the industrial organization literature, a particular stream is concerned with the patent valuation.[45] Those studies investigate properties of patents and particularly the determinants of patent value. For the quantitative analysis of factors that influence sales probability and sales prices, measures are applied to operationalize the variables developed primarily in this research stream.

This study adopts transaction cost theory's overarching theoretical concept, which is reflected in the main research question (see Chapter 2) and combines the choice of governance structures of transactions with properties of the traded asset. As this study is concerned with transactions between a technology owner and an acquirer, that is with the relation between these actors, it is related to the relational branch of transaction cost theory as opposed to 'organizational governance.' The relational branch of transaction cost theory provides this study's theoretical basis. According to Shelanski and Klein (1995), the relational branch of transaction cost theory aims to explain how trading partners choose from a set of feasible institutional alternatives, that is this theory is concerned with make-or-buy as well as keep-or-sell decisions. In the open innovation literature, a number of empirical studies have applied the transaction cost theory in a similar way to this study in that they investigate 'relational governance' structures. In terms of technology acquisition, these studies include, for instance, (Van de Vrande et al. (2006), Delmas (1999)), make-or-buy decisions (Klein (2005)), licensing contracts (Bessy et al. (2008), Brousseau et al. (2007)), and alliances (Colombo (2003), Oxley (1999)). According to some authors, for instance Leiblein (2003), closely related concepts also include contracting for innovation, contractual relations, vertical disintegration, and interfirm collaboration.

Investigating technology auctions as a specific transaction model, this study is clearly related to the theoretical and empirical stream of the auction literature (see Chapter 5), which in turn is part of game and bargaining theory. This study borrows a theoretical auction process model from the auction literature. For my discussion on the choice of technology auctions, particularly when and under which circumstances they should be preferred to bargaining negotiations, I also refer to the fundamental principles of bargaining theory. Furthermore, to some extent, this study also draws on the market design literature.

NOTES

[1] Rycroft and Kash (1999) provide a comprehensive examination of the complexity phenomenon. Schmookler (1966: x) regards the cumulative nature of technical knowledge for inventions as 'the "bits" that are added to the existing stock of knowledge.' According to Pénin (2008), similar concepts discussed in the literature are 'step-by-step innovation,' 'multi-invention products,' 'collective learning process,' 'collective mode of innovation,' 'combinative capabilities and replication,' 'evolutionary innovation,' 'sequential innovation,' and 'multi-stage innovation.' However, sequential innovation does not appear to be a suitable term, since it can be confused with the rather early generations of sequential innovation processes. Moreover, cumulative innovation should not to be confused with collective innovation. For a discussion comparing open, user, and cumulative innovation, see West and Bogers (2010).

[2] Related concepts in the literature are multi-invention products (Somaya and Teece, 2008) and modularity (Campagnolo and Camuffo, 2010).

[3] Also see Ernst and Unctad (2005) and Lindmark (2002).

[4] Nevertheless, determining the boundaries of a firm is not trivial. Within industrial organization a related stream of literature investigates what determines the firm's size (firm scope). A review of the related literature can be found in Holmström and Roberts (1998).

[5] The notions of the 'IP assembly problem' and the hold-up problem further relate to the theory of the anti-commons originally referring to the difficulties of assembling lease rights for buildings in Russia (for example, Granstrand (2003c) and Aggarwal and Walden (2009)) or in biomedical research (Heller and Eisenberg, 1998).

[6] Nevertheless, patents do not necessarily create hold-up problems in cumulative innovation processes as demonstrated by Granstrand (2006) in mathematics as an extreme case of cumulative innovation.

[7] It is a common assumption that, along the innovation process, the market actor who files for the patent protection of a technical invention is the same one who ultimately exploits the patented invention on the market, turning it into an innovation. Empirical evidence however proves that this is not the case. A considerable share of patents is not used directly by its inventors. Gambardella et al. (2006) reported that 36 per cent of the patents in their 'huge' sample of EU patents are not used internally or for licensing. While about half of these patents (18.7 per cent) may even assume a potentially high value as they help block competitors, the other half (17.4 per cent), labeled 'sleeping patents,' are left virtually unexploited. The Institut der deutschen Wirtschaft Köln (2006) proved that in Germany a quarter of patents (24.6 per cent) is not used at all. Chesbrough (2006: 5) reported that in the 'US over 95 per cent of issued patents are unlicensed and over 97 per cent never generate any royalties.'

[8] Granstrand (2004b) provides further arguments for the shift from R&D and technology management towards IP and intellectual capital management.

[9] Parker et al. (1996) provides further arguments explaining this trend. They point out that, through marketing, firms who intentionally or unintentionally find that their internal R&D efforts are limited to line extensions, can gain access to the breakthrough ideas created by inventors who are not confined to the corporate context. Moreover, Somaya and Teece (2000: 1) argue that 'inventions may be combined into … multi-invention products using three alternative organizational modes - viz., licensing of inventions, trade in components that embody inventions or by integrated production.'

[10] Various notions are applied in the literature. Troy and Werle (2008) and Gambardella et al. (2007) use the term 'market for patents' as a narrow definition. Chesbrough (2006) uses the term 'markets for IP,' Lichtenthaler and Ernst (2006b) use 'market for knowledge,' Arora et al. (2001) use the term 'markets for technology,' and Gu and Lev (2000) use 'markets in intangibles.' Recently, however, a special issue of ICC (Volume 19 Issue 3) summarizes the

different notions as markets for technologies and ideas (MfTI) that cover not only technologies but also ideas that are traded in the rather early stages of innovation processes (for example, Innocentive.com).

11 According to March (1991: 71), the relation 'between the exploration of new possibilities and the exploitation of old certainties ... is of central concern' in economic studies. While exploration includes aspects captured by terms such as search, variation, risk taking, experimentation, play, flexibility, discovery, and innovation in contrast, exploitation includes aspects such as refinement, choice, production, efficiency, selection, implementation, and execution. Bessy and Brousseau (1998: 452) argue that 'when an innovator is allowed to license his technology, he spreads it in the economic system, while at the same time he increases his return on innovation investments and efforts.'

12 While technologies directly related to a firm's core competences and competitive advantage might not be suitable for external exploitation – or at least not to direct competitors – other technologies might not be critical to the firm's competitive advantage and may be very well suited for external exploitation. However, looking into the different strategic purposes of the technologies within technology portfolios is beyond the scope of this study. The reader might rather refer to, for example, Tschirky and Koruna (1998, chapter 4.2.10) who present different typologies and approaches to classify technologies. Granstrand (2000a) also provides a discussion of different strategic purposes of patents.

13 According to Chesbrough (2003a), open innovation is the process of cultivating and internalizing value from opportunities external to the firm, as well as the skillful deployment of internal discoveries to external complements. Although this phenomenon has only recently been observed in certain industries, this innovation principle is not particularly new. Teece (1989: 35) argues that 'the institutional structure of innovation in capitalist economies is extremely variegated and involves a complex network of backward, forward, horizontal, lateral relationships and linkages within, among and between firms and other organizations.' Furthermore, based on an original idea published in Granstrand (1982), Granstrand and Sjölander (1990) propose a typology for technology acquisition and exploitation strategies that would nowadays be subsumed under the notion of an open innovation.

14 Firms also actively engage and sponsor online communities (for example, Janzik et al., forthcoming) where users develop software (for example, Raasch and Herstatt, forthcoming) as well as tangible products (for example, Raasch et al. (2009)).

15 The label 'technology trade' is avoided as it is often used in the literature for inter-country technology transactions commonly associated with the transaction of technical know-how towards developing countries for instance by Yang and Maskus (2009), Hoekman et al. (2005) and Chen et al. (1994).

16 For a detailed definition of this concept, see Chapter 4, which refers to Schumpeter (1912). Such intermediaries can be seen as ventures started by entrepreneurs and thus as a source of change (North, 1996).

17 Even at the beginning of the 20th century, there have been organized MfTI (Lamoreaux and Sokoloff (1998)).

18 Explanations for this growth effect are currently under scrutiny by various scholars. However, explaining this phenomenon is beyond the scope of this study.

19 For instance, pharma firms rely extensively on outside knowledge for their products (Ceccagnoli et al., 2009). However, it is interesting to note that the size of the MfTI, respectively for intangible assets, patents and licensing remains difficult to determine. Besides the absence of solid measures to systematically collect data, few studies have tried to approximate the market size and so far only few official statistics have been regularly collected by international authorities, for instance, EUROSTAT.

20 In the literature, various terms are used for essentially similar processes. These include deployment, for example, by Escher (2005) and commercialization, for example, by Lichten-

thaler (2006a). For a typology of external technology sourcing strategies and a detailed discussion of them, see Chapter 3.

21 For further readings on the NIH syndrome, see, for example, Lichtenthaler and Ernst (2006a) and Katz and Allen (1982).

22 For details and a discussion on obstacles, see Chapter 3.

23 For instance, Escher (2005: 75) argues that firms 'often fail to initiate such an exploitation program due to market imperfections and high initial financial commitments.'

24 Certain specialized intermediaries, such as brokers and dealers, have been present for a long time (for example, reported in Arora et al. (2001) as 'specialized engineering firms'). The recent emergence of TMIs rather refers to those firms included in the rather narrow definition provided in Chapter 4.

25 It is interesting to note that the emergence of intermediaries is not significant to markets for technology when markets do not clear efficiently. For example, in agricultural markets, Klerkx and Leeuwis (2008: 260) report that 'due to market and systemic failures, both supply side and demand side parties in this market have experienced constraints in effecting transactions and establishing the necessary relationships to engage in demand-driven innovation processes. To mitigate these constraints, a field of intermediary organizations has emerged to assist agricultural entrepreneurs to articulate demand, forge linkages with those that can provide innovation support services and manage innovation processes.'

26 Founding dates could be identified for only 60 TMIs of the sample. Counting for 80 per cent of the TMIs, by far the majority of the TMIs is based in the US clustering around two centers at the west and east coasts. While a considerable number of them are concentrated around Silicon Valley at the west coast, another cluster can be found at the east coast, including New York and Massachusetts. The non-US based TMIs are mainly European and Canadian firms. In Europe, the majority of TMIs are British and German. Several TMIs hold regional offices in Europe, Japan, China and East Asia.

27 Schumpeter (1912: chapter 2) introduces the innovation concept to the economic literature and defines it as the 'introduction of new goods …, new methods of production …, the opening of new markets …, the conquest of new sources of supply …, and the carrying out of a new organization of any industry.' Since then, various definitions have been developed. For instance, Drucker (1954: xi) defines innovation as 'change that creates a new dimension of performance' and Nelson and Winter (1982: 130) define innovation as the 'the creation of any sort of novelty in art, science or practical life that consists – to a substantial extent – of a recombination of conceptual and physical materials that were previously in existence.' Nowadays, it is widely accepted that innovations comprise a variety of domains (for example, organizational, financial, social or service) and nowadays different typologies have been proposed (for example, Granstrand, 2000). The OECD (2005: 46) defines innovation as 'the implementation of a new or significantly improved product (good or service) or process, a new marketing method or a new organizational method in business practices, workplace organization or external relations.'

28 Polanyi (1966) introduced the dichotomy of tacit and explicit knowledge. At the same time 'tacit' also relates to the notion of 'sticky knowledge' as introduced by von Hippel (1994).

29 In this study, following Pearce (1992: 18), an asset is defined as an 'entity possessing market of exchange value and forming part of the wealth or property of the owner. In economics, an important distinction is made between real assets, which are tangible resources like plant, building and land yielding services in terms of production or directly to consumers; and financial assets which include money, bonds and equities and which are claims or titles to receive income or to receive value from others.'

30 The cumulative nature of innovation has further been on the research agenda in relation to the policy debate, for example, by Furman and Stern (2006), O'Donoghue (1998), Mazzoleni and Nelson (1998), Chang (1995), and Scotchmer (1991).

31 In addition, see the comprehensive review of innovation research from its early inception at the beginning of the 20th century as conducted by Fagerberg and Verspagen (2009: 220).

Accordingly, 'before 1960, scholarly publications on innovation were few and far between. The main exception to this rule was the work of the Austrian-American social scientist Joseph Schumpeter. However, at least until the 1990s, most research was conducted in the economic field rather than by management scholars. For a review of the development of research on innovation and particularly on innovation processes during the last decades, see, for example, Xu et al. (2007), Teece (2006), Herstatt and Verworn (2004), and Rothwell (1994).

[32] For instance, Herstatt (2007) focuses in particular on the early stages of innovation processes, that is the 'fuzzy front end'.

[33] See von Hippel (1988) as a prominent example.

[34] For an overview of the wide range of literature within innovation management, see, for example, Hauschildt and Salomo (2007). Gerybadze (2004) and Tschirky and Koruna (1998) also illustrate the various sub-themes and interconnections of topics within technology and innovation management.

[35] Applying to the JEL classification system, this study relates to the following categories: Primarily, that is, closely connected, to O31 (Innovation and Invention: Processes and Incentives), D45 (Rationing; Licensing), M21 (Business Economics), O3 (Technological Change; Research and Development) and D44 (Auctions). To a broader extent, concepts from the following classes are embedded in this study: K11 (Property Law), D23 (Organizational Behavior; Transaction Costs; Property Rights), D43 (Oligopoly and Other Forms of Market Imperfection), L1 (Market Structure, Firm Strategy and Market Performance), L14 (Transactional Relationships; Contracts and Reputation; Networks), and L21 (Business Objectives of the Firm).

[36] Granstrand (2000a) understands technology as a resource and subsequently developed the concept of the technology base, that is the portfolio of technologies that are 'continually being acquired, combined, and exploited in various ways...' Referring to Granstrand and Sjölander (1990), Escher (2005: 44) points out that 'the main task of the technology-based enterprise is to optimize its resource base through acquisition and exploitation processes.'

[37] Among other ideas, the use of the concept of distributed innovation processes is an outcome of various discussions with colleagues from the MELT project (which synthesizes Management, Economics, Law and Technology) at Chalmers University of Technology, Sweden.

[38] According to Penrose (1951: xi), 'although the patent system was developed primarily to promote economic ends, economists have devoted very little attention to it and none at all to the international patent system.' Granstrand (2000a, ch.2.5) provides further arguments that there was no change in this attitude between the 1950s until the 1990s.

[39] Despite this 'old observation that the value of a firm as a going concern considerably exceeds the value of its physical assets' (Arrow, 1996: 126), a growing number of IP-related articles reflects the increased importance of IP management since the mid-1980s (Ziedonis, 2008). Moreover, Granstrand (2000a: ch.2.3.2) argues that the growing awareness and importance of IP is linked to the emergence of the 'pro-patent era' in the 1980s that was followed by some substantial legislative changes in the US patent jurisdiction.

[40] In the literature, various terms are used for essentially similar processes. These include deployment (for example, Escher (2005)) and commercialization (for example, Lichtenthaler (2006a)).

[41] For instance, Escher (2005: 68) provides a well-structured overview of various studies.

[42] There are a few exceptions, for instance, Escher (2005) briefly discusses the use of auctions to establish market prices.

[43] See, for instance, Anand and Khanna (2000) for an empirically-based discussion on licensing features, Lafontaine and Slade (forthcoming) for an analysis of licensing contracts, and Braun and Herstatt (2007) for a discussion on the relation between licensing activities and innovation.

[44] According to Nelson and Nelson (2002: 265), 'the development of the conception of an innovation system … has largely been the work of economists and other scholars of technological advance who adhere to an evolutionary theory of economic growth.' Within innovation systems, 'scholars of technological advance have always understood the important role of institutional structures in supporting and molding efforts to advance technology'.

[45] For instance, Giuri et al. (2007), Hall et al. (2005) and Harhoff et al. (2003).

2. Research Methodology

2.1 CONCEPTUAL FRAMEWORK AND RESEARCH QUESTIONS

To guide this research, a conceptual framework was employed, reflecting the generic institutional structure of technology transactions. The framework was inspired by the 'contingent effectiveness model of technology transfer' which was applied by Bozeman (2000) to order the literature on technology transfer that also relates to the standard microeconomic supply and demand model.[1] Based on the assumption that participants in a technology transfer have multiple goals, the model includes five broad dimensions. The 'transfer agent' is the actor interested in transferring a technology. The 'transferred medium' is the means by which the technology is to be transferred (for example, license, sales). The 'transfer object' is the content and form of what is to be transferred (for example, technical knowledge, IP assets). The 'transfer recipient' is the actor who will be using the technology and finally the 'demand environment' is characterized by the market conditions relevant to the technology to be transferred. However, Bozeman's (2000) model does not include intermediaries that can play an essential role in technology transactions. Subsequently, for the purpose of this study, I added such intermediaries to it.

Figure 2.1 illustrates the framework applied in this study. The framework comprises of six elements, which can be further ordered into four categories: (i) the actors (sellers, buyers and TMI), (ii) the transaction, (iii) the traded asset, and (iv) the institutional (market) environment. Furthermore, the framework can be divided into a supply and a demand side. The supply side is concerned with owners offering technologies and IP assets on the MfTI (that is external technology exploitation (ETE)) and the demand side is concerned with prospective buyers that externally acquire technologies.

Three different actor groups are involved in transactions. These actors are connected by the transaction and particularly its governance structure through which the asset, that is a technology, is transferred from the seller to the buyer either directly or indirectly with the support of an intermediary.

This process is subject to certain restrictions and rules of the market and the particular industrial environment that are set by superordinate actors (for example, federal and local governments, WTO, patent and trademark offices).

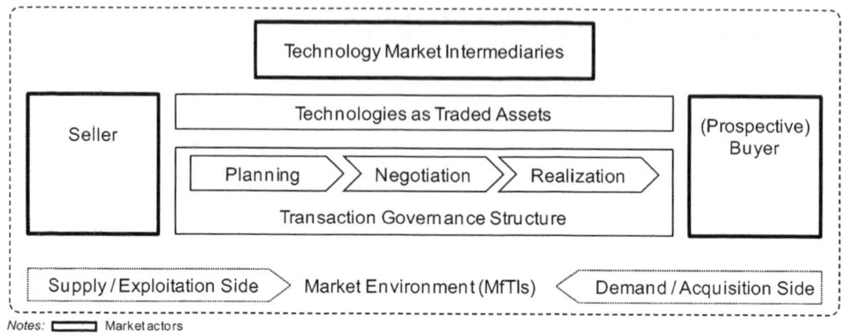

Figure 2.1 - Conceptual framework

Having decided to engage in a transaction, a current technology owner acts as a *seller* on the supply side of the market, who is willing to exploit a technology externally. There are various types of sellers. One can generally distinguish between five types that are similar to certain types of patent applicants at the patent offices. These include individuals, universities or research institutes, small and medium sized enterprises (SME) and large firms. Aside from the nature of the seller, the technology offered by the seller might come from various sources. The seller could have developed the technology for sale either internally within the corporate research institute or in a business subsidiary or the seller could have acquired the technology from another party.[2] In addition, non-practicing entities might also own technologies without having developed them, by rather acquiring them from the MfTI. Furthermore, the technology owner could hire an intermediary, for instance, a patent law firm (PLF), to offer a technology on its behalf, for example, to remain anonymous. When exploiting technologies externally, sellers have to choose the most economical governance structure for each transaction.

On the demand side of the market, actors want to acquire technologies and act as (prospective) *buyers* who are interested in becoming future owners and users of a technology. In general, sellers can also act as buyers. However independent inventors as well as universities and research institutes rather act as sources for developing technologies offering these on the market, and seldom act as buyers. Thus, the types of actors acquiring technologies are rather limited to SME and large firms or an intermediary, for example, a law firm acting on the behalf of another firm. Buyers acquire technologies from different sources and have to choose the appropriate governance structure for each transaction.[3]

Service firms represent the third actor group. They offer a range of services to facilitate transactions. In this study, I refer to those firms that directly impact the transaction as *technology market intermediaries (TMI)*. In Chapter 4, I provide a definition for TMI and discuss their roles in innovation systems.

Actors are connected through the transaction and particularly its *governance structure*. If successfully completed, a technology (the traded asset) will be successfully transferred to the new owner. Applying a process perspective for the technology transaction has become generally accepted in the literature according to Lichtenthaler (2006a). In this study, I refer primarily to two transaction process models that were proposed by Lichtenthaler (2006a) and Escher (2005). A more detailed discussion of these is presented in Chapter 3.

The object of the transaction is the *traded asset*, which can range from a single patent that is sold for monetary compensation only to a complex technology with all its accompanying explicit and tacit knowledge even including R&D employees. In an extreme case, even firms' separate business (for example, start-ups) might be transacted for innovation purposes. A more detailed explanation of the traded asset is presented in Chapter 6.

Finally, transactions are embedded in an institutional *environment* that sets the rules and regulations for IP transactions. In this study, I adopt the common notion of markets for technologies and ideas (MfTI). This environment includes the institutions of society (for example, patent systems and patent courts), industry unions, standardization and certification bodies, scientific societies, trade associations, etc. According to Bessy and Brousseau (1998), MfTI play an important role because they set up (sometimes informal) rules and provide agents with supervision and enforcement means (both for these collective rules and for their private agreements). These rules and enforcement mechanisms complete the institutional framework. Since this study is situated in management science, it does not directly consider the institutional environment.[4]

Although attempting to take a holistic perspective on technology transactions in this study, whenever necessary, this study focuses on the supply side, primarily applying the seller perspective. Within the conceptual framework illustrated in Figure 2.1, the focus of this study is primarily on three of the six dimensions: TMI, the transaction, and the traded asset.

TMI have recently started to develop novel transaction models[5] which they offer to TBF willing to exchange technologies externally and more economically on the MfTI. Hence, with an increasing number of alternative transaction models – each characterized by a specific transaction governance structure – technology-based firms face an increasingly difficult choice between the alternatives. For IIP managers, it thus becomes essential to decide which transaction model is suitable for which technology type. This issue is of major concern in this study, with its focus particularly on technology auctions as a specific novel transaction model. Assuming that not all technologies are equally suitable for technology auctions and that the best suited transaction model has to be determined for each technology type led to this study's general research question:

> Which technologies can be auctioned economically profitable?

This general research question was used as a conceptual guideline throughout this study, that is it was used to organize the research project and to give it direction and coherence (Punch, 2005: 33ff). However, this research question encompasses a large number of facets. For instance, one might question: What are the determinants for the choice of the preferred transaction model? To what extent does the governance structure of technology auctions reduce the costs of negotiation, contract design, supervision, conflict settlement, etc.? To what extent do technology auctions increase the efficiency of technology transactions?

Owing to the multifarious nature of this general research question, it could not (and was not meant to) be answered comprehensively. It was not possible to address all the related questions comprehensively. Therefore, I decided to rather focus on certain aspects of particular relevance delimitating various aspects and consequently bring 'the project down to size' (Punch, 2005: 35). The selection of the specific research questions was guided by two principles, namely to keep this study manageable within time and resource constraints following the overall premise of Punch (2005) (that is to rather conduct a smaller project thoroughly than a larger one superficially). and to select those that seem most relevant to firms.

Guided by these principles, and in accordance with the conceptual framework presented in Figure 2.1, two specific research questions were developed during the first phase of this study. With the support of industry experts and academics the scope of the general research question was gradually narrowed. Similar to the process suggested by Punch (2005) various specific research questions were developed, disentangled by splitting them into their components, and ordered so that their focus was progressively developed. After the relevance of these questions was validated within a workshop,[6] I decided to focus particularly on technology auctions as a novel transaction model and two specific aspects of the framework presented in Figure 2.1 (that is the transaction governance structure and properties of the traded asset). Therefore, two specific research questions resulted from this decision process.[7]

In order to develop an understanding of the choice of determinants for transaction models from the relational branch of transaction cost theory, the choice framework for governance structures was employed. For the purpose of applying this framework for technology auctions, it became necessary to first understand the transaction governance structure of technology auctions, for which the first specific research question reads:

1. How is the transaction governance structure of technology auctions characterized?

This first research question is mainly descriptive in nature with the unit of analysis being the transaction in line with transaction cost theory research. This specific research question is addressed in Section 5.3 where the results of my qualitative empirical observations are synthesized (supported by relevant literature from technology transactions and) into a description of the transaction governance structure of technology auctions.

Having addressed the first research question in a sequential and cumulative manner, I explored and tested determinants for the choice of patented technologies that can be transacted in an economically profitable way by means of technology auctions. This led to the study's second specific research question:

2. Which patented technologies should be preferably auctioned?

With the unit of analysis being the traded asset (that is patented technologies), this question appears explorative from its wording (Yin, 2003). However, in this study, the nature of the research question should be considered confirmative.[8] To address it, hypotheses were derived from the results of the first specific research question which were then tested by means of inferential statistical techniques (different types of multivariate regressions). The second specific research question is based on one critical assumption that is worth noting. When operationalizing the measures, I assume that each technology involves patents that are unique by definition, but that also have some properties in common (for example, patents can be filed on the same day and thus have the same age; they can be filed in the same IPC class and subsequently relate to similar technology domains). The technology properties were tested following a quantitative empirical approach which is presented in Chapter 10.

Jointly, both specific research questions cover both aspects of the choice framework for transaction governance structures, which is presented in

Chapter 7. These aspects relate to (i) the governance structure of the transaction and (ii) the properties of the traded asset. Both research questions reflect what Popper (1969) (cited according to Gerring (2007: 39)) points out that 'social science research involves a quest for new theories as well as a testing of existing theories and thus comprises "conjectures and refutations"'.

2.2 RELEVANCE AND CONTRIBUTIONS

In the following section, arguments are provided for the relevance to study TMI, transactions of patented technologies involving TMI, and, in particular, public technology auctions. Moreover, this study's contributions to different stakeholders are outlined. The primary stakeholders of this study include IIP managers of TBF, TMI, policy makers involved in innovation and IP policy and the academic community.

Relevance

Research on the management of technology transactions on industry or national level is well established.[9] However, why is it relevant to study the role of TMI within and their impact on transactions?

In Chapter 1, some arguments were put forward to address this question. According to those studies quoted above, the institutional structure of the MfTI is still far from optimal and many transaction obstacles are present that inhibit efficient market clearing (see Chapter 3). Thus, the efficiency of the external utilization of technologies and IP assets should still be considered far from optimal and hardly supportive of innovation and growth.[10] Put differently, for many firms, it is of increasing importance to utilize their R&D resources in the most optimal way. Owing to the increasing internationalization, increasing competition, shrinking market lead times, and an increase of technically complex products, firms need to assemble various technologies and related IP assets from different sources outside the firm boundaries.[11] This trend is reflected by the open innovation movement, in which firms increasingly collaborate within networks and exchange (acquire and exploit) technologies and IP assets. Subsequently, the importance of well-functioning MfTI increases and firms are required to have a good understanding of it in order to manage such transactions. Anecdotal evidence and a few statistics indicate that TMI have recently started to emerge, developing novel transaction models that impact the ways in which firms govern transactions. These models attempt to facilitate the exchange of technologies to overcome the transaction challenge but are characterized by different transaction governance structures that have not yet been studied systematically and remain insufficiently understood.

The relevance of conducting studies investigating TMI is further supported by recognized scholars in this field. According to Gambardella et al. (2007: 1181) the 'formation of independent concerns, like Yet2.com ... might improve the managerial efficiency of the technology trade activity... [however,] a more thorough analysis of these issues is left for future research.' In a similar manner, Lichtenthaler (2005: 248) concludes that 'the use of intermediaries for different tasks along the commercialization process' is left for future research. Furthermore, international organizations, including the OECD and EPO call for further research in this field. According to EPO et al. (2006: X), 'the IP marketplace is nowadays in a probe and learn period where the number of intermediaries is rising.' These intermediaries bring us one step closer to 'the development of a market for IP transfers ...[and]... contribute to the maturation of the IP market' (EPO et al. (2006: X)). However, why, among the various TMI models, does this study focus on technology auctions?

Various observations indicate that, among the different models developed by TMI, public technology auctions have a particular potential to positively influence market developments. Numerous industry experts have indicated that they consider technology auctions a promising transaction model, which they do not expect to exit the market soon. Further arguments support the relevance of studying public technology auctions. Patents have certain properties that inhibit efficient trade (see Chapter 6). Since patents are unique and intangible by definition, their market value is commonly difficult to determine. However, other asset classes also face valuation difficulties when being transacted. Previous successful applications of auctions in Business-to-Business transactions of asset classes that share some distinct properties with patents indicate that auctions could possibly be used successfully to transact patents. For instance, in the past auctions have been successfully used for licenses of intangible assets (for example, the 3G UMTS radio spectrum auctions across the world[12]) and have been used for many years to sell unique tangible assets that are difficult to value (for example, used machinery[13] in different technical conditions, art, and wine). Auctions can be considered a well-established trade mechanism that is well understood in theory.

Furthermore, taking a first look at the sales data of technology auctions in this study's dataset indicates that a considerable number of transactions have taken place in a relatively short period of time.[14] While it usually takes several weeks or months to prepare and negotiate transaction contracts via bargaining negotiations, data from the public technology auctions indicate a much shorter duration to conclude transactions, which could mean a significant increase in transaction efficiency.

To my knowledge, aside from a few practitioners' papers (for example, Millien and Laurie, 2007), academic scholars have hardly yet investigated the emerging TMI, the novel transaction models they develop and their systematic impact on transactions of patented technologies – particularly not from a transaction cost perspective. Moreover, no systematic study was found that specifically investigates the governance structures of technology auctions[15] or whether and how they facilitate technology transactions. To be even more specific, no study was found that investigates for which patented technologies auctions represent the preferred transaction governance structure.[16]

Contributions

Enhancing the understanding of TMI in general, their role within the innovation process, the governance structures of the novel transaction models they develop, and particularly technology auctions would benefit different groups of stakeholders.

IIP managers in charge of improving management capabilities or (open) innovation processes at TBF[17] – who (prepare to) engage in the transactions of patented technologies and consider seeking the support of TMI (particularly technology auction firms) can benefit from this study.[18] This study contributes to a better understanding of transactions and governance structures and how TMI impact transactions. These insights support IIP managers by providing them with a conceptual basis when designing and choosing governance structures for technology transactions. In particular, IIP managers can benefit from the analysis of auctioned technologies. The results of this empirical study provided in Chapter 10 support them in determining the future pricing of their IP assets. In the past, the absence of price transparency and comparable market prices has often been highlighted as an obstacle for efficient IP transactions on the MfTI (see Chapter 3).[19] The publication of thoroughly analyzed sales prices from technology auctions represent prices that buyers were willing to pay under a standardized and thus comparable framework. Making this data available publicly facilitates market transparency as called for by different scholars (for example, Chesbrough, 2006).[20]

This study also benefits *TMI* in general and auction firms in particular. TMI can benefit from the analysis of governance structures of technology and IP transactions (discussed in Chapter 9), which could help them to further align and improve their transaction models to fit the needs of the TBF. Moreover, TMI benefit from the technology trade statistics and analyses presented in Section 6.2 in that they gain a better understanding of the MfTI and transaction volumes. To improve the design of their auction models, technology auction firms in particular can benefit from the relevant theoretical literature on auction design presented in Chapter 5 and the qualitative empirical results presented in Section 5.3.

This study also contributes to the *academic community*. First, this study contributes by providing an application of the governance structure choice concept, developed in transaction cost theory, to technology transactions in an open innovation context as a basis for the quantitative study. To my knowledge, this study is among the first, if not the first, to empirically test transaction cost theory predictions for the choice of auctions as a governance structure for technology transactions. With the regression analyses carried out in Section 6.3, this study also indirectly contributes to the ongoing discussion on and empirical validation of indicators for patent valuation. Furthermore, this study contributes a definition for TMI in Chapter 4, which was thoroughly developed from related concepts in the literature. The study additionally provides a theoretical argumentation that sheds light on the relationship between TMI, transactions, and open innovation and the impact of TMI on the innovation process.[21]

This study offers contributions (although limited) to *policy makers* concerned with innovation and IP policy.[22] Governments and international governmental institutions have recently become aware of the importance and possibilities of efficient MfTI and particularly the emerging role of TMI as argued in the introduction. However there is still a lack of solid information on them.[23] Various parts of this study contribute to the enhancement of knowledge in this field (for example, the literature-based development of a definition for TMI proposed in Chapter 4). Furthermore, while scholars have claimed that more transparency of technology prices is needed to facilitate further developments of the MfTI, factual market prices paid for patented technologies in particular are very rare and, if available, are often mere estimates based on proxies instead of representing real market prices.[24] For instance, Gambardella et al. (2006: 4) conclude that 'new data collections are … crucial for setting up systematic indicators of the value of patents in additional years and countries'. Chesbrough (2006: 146) also highlights that there is 'no systematic reporting of previous prices paid for external technologies and their associated IP. This makes it hard for sellers to know what price to expect to receive or what price would be reasonable'. In Section 6.2, I reveal factual market prices paid for patented technologies at technology auctions. This systematic data analysis may contribute to an advanced understanding of monetary patent values as well as their determinants (see also Section 6.3). The various components can support policy makers when designing incentive schemes that aim to facilitate an enhanced development of the MfTI.

2.3 RESEARCH STRATEGY

Focusing on TMI, this study needs to be considered an early-stage study in its empirical field which is characterized by severe confidentiality issues. The choice of a suitable research strategy was very constrained.[25] However, it was of considerable importance for the study to find a research strategy that supports the generation of valid results in the best possible manner. In order to address both research questions in the most comprehensive manner, I developed a sequential multimethod study.[26] This strategy follows the third rationale for the use of multiple methods outlined by Yin (2003: 150f), that is where case studies 'elucidate some underlying process' complemented by means of a quantitative analysis.

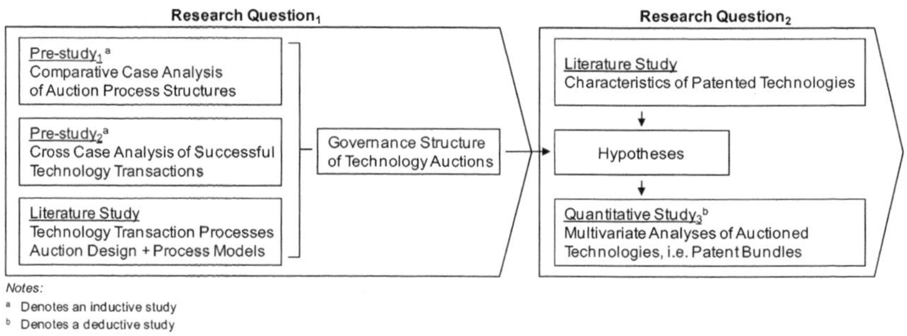

Figure 2.2 - Research strategy

This study's research strategy comprises two complementary, qualitative pre-studies addressing the first specific research question and a quantitative study, which addresses the second research question.[27] While both pre-studies follow an inductive approach, the approach to address the second specific research question is of a rather deductive nature. In the following section, I further motivate my choice of research strategy. Chapter 12 provides methodological details of the three studies. Figure 2.2 illustrates the complementary and sequential multimethod research strategy.

The *first research question* addresses the transaction governance structure of technology auctions. The aims of the first and second pre-studies are similar. Therefore the choice of the research strategy follows a similar argument. The choice is in line with the third of the five major applications for case studies mentioned by Yin (2003: 15).[28] Accordingly, case studies are an appropriate approach to 'illustrate a certain topic within an evaluation... in a descriptive mode.' To phrase it differently, Silverman (1993: 30) recommended case studies when it is 'fundamentally about understanding the routine' which is in line with this study's first specific research question in order to examine transaction processes of technology auctions.

Punch (2005: 147) adds that 'case studies... have a value contribution to make... in situations where my knowledge is shallow, fragmentary, incomplete or non-existent.' Although the research question is mainly descriptive (with its aim to describe the governance structure), it also has an explorative element. In order to describe the governance structure, I had to explore the process structure of technology auctions. Thus, due to this explorative element of the research question, Gerring (2004: 349) argument further supports the use of case studies, since they 'enjoy a natural advantage in research of an exploratory nature.' The limited number of available data observations further supports the use of case studies.[29] Thus, the second rationale for case studies provided by Yin (2003: 40f) applies to this study. He recommends case studies when observations are 'so rare that any single case is worth documenting and analyzing'.[30] According to Rihoux (2006), case studies are often preferred in social science research, where many relevant and interesting objects are naturally limited in number. Yet another argument to document such rare cases was given by Yin (2003: 41) as the fourth rationale to conduct case studies, that is 'when an investigator has an opportunity to observe and analyze a phenomenon previously inaccessible to scientific investigation'.[31]

However, those rationales provided by Yin (2003) and similar ones provided by Silverman (1993) only favor the case study approach, but do not clarify the question on the number of case studies, that is whether to pursue a single or multiple case study design. A possibility for instance would have been to conduct a single case study of the most successful technology auction (for example, measured in terms of monetary returns). Nevertheless, Yin (2003) provides valuable arguments for multiple case studies when he advises that replication of cases is meaningful in two situations: First, when cases have similar results (that is sales of technologies), literal replication serves a similar purpose to multiple experiments; second, when cases have contrasting results, theoretical replication achieves what was predicted explicitly at the outset of the investigation. Although Yin (2003: 53) considers multiple cases 'more expensive and time-consuming to conduct', the evidence from literal replication is generally considered more compelling and robust according to Herriott and Firestone (1983). Multiple case studies enable a cross case analysis which increases the reliability of the results according to Punch (2005) and Stake (1994). Similarly, Yin (2003: 53) argues that, when time and resources are available, 'multiple-case designs may be preferred to single-case designs... [and the] ...benefits from having two (or more) cases may be substantial'.

Conducting multiple-case studies, however, raises the question of how many replications should be selected, which essentially depends on the level of certainty one wants to have about the results (Punch, 2005). Similar to the higher criterion for establishing statistical significance, greater certainty lies with a larger number of cases. However, as mentioned above, for this study, the number of possible case studies was limited either way. This issue is further addressed in Section 4.1, where I elaborate on the case selection for the first and second pre-studies. The case studies are presented in Section 5.1 and 5.2 and the first research question is addressed in Section 5.3. Based on the synthesis of the case studies, complemented with insights from the literature, I developed and tested hypotheses to address this study's second specific research question.

The *second research question* aims to identify which patented technologies are suitable for being transacted using the governance structure of technology auctions identified as the result of the first research question. The nature of the second research question is confirmative with its emphasis on finding proof of causal relationships. Backhaus et al. (2006) differentiate statistical techniques whether they serve confirmatory or exploratory purposes. For confirmatory studies, Backhaus et al. recommend the use of inferential statistical techniques, which I used to test the hypotheses developed from the first two pre-studies and the literature study.

Among the wide variety of inferential statistical techniques, I had to choose the one which is best suited to this study. Backhaus et al. and Hair (2006) provide valuable frameworks for making this decision. Accordingly, the choice of the specific statistical technique depends on the measurement scales of the dependent and independent variables (Love, 2007). Among the two dependent variables that were applied in this study's models (see Section 6.1) one was metric-scaled, while the other variable was nominal-scaled (that is dichotomous / bivariate). All independent variables are either metric-scaled, dichotomous or can be coded as dichotomous dummy variables. For the first model, with the dependent variable being metric-scaled, Backhaus et al. (2006: 13) suggest two analytical techniques (regression analysis; time-series analysis). Since it was not the purpose of this study to investigate time-series, this technique was not applied. However, regression analysis was applied as the 'most important and mostly used multivariate technique… to investigate causal relationships'.[32] According to Hosmer and Lemeshow (1989), regression methods are integral components of any data analysis concerned with describing the relationship between an outcome variable (that is dependent variable) and one or more explanatory variables (that is independent variables).

For the first model, with the dependent variable being dichotomous, Backhaus et al. (2006) propose two techniques (that is discriminant analysis, logistical regression[33]). Accordingly, the discriminant analysis can be used to explain differences in groups. However, I aimed to explain the likelihood of certain technologies being affiliated with certain groups. This purpose rather prefers the use of logistical regression analysis. For a second model, with the dependent variable being the metric-scaled sales price, I employed ordinary least square (OLS) regressions following the suggestion by Backhaus et al.[34] The choice of both models was further supported by the fact that various related patent value studies (see Chapter 6) also apply different types of regression analysis. Therefore, using a similar technique makes the comparison with those studies more reliable. Chapter 8 describes how the analyses were conducted while the results are presented in Chapter 10 addressing the second research question.[35]

To conclude, in line with both research questions, the research strategy is split into two parts (see Figure 2.2). While the first part of this study (with its two pre-studies) is inductive, the second part of the study is deductive, testing the hypotheses developed from the results of the first part of this study. Following the typology of research strategies proposed by McGrath (1981),[36] this study and each of the pre-studies can be classified as a 'field study'. When Scandura and Williams (2000) employed this typology, they proposed to further differentiate this particular category according to the data basis that was used (that is primary or secondary data). Thus, the two pre-studies addressing the first specific research question constitute field studies using primary data while the quantitative study should be regarded as a field study using secondary data.

2.4 DELIMITATIONS AND LIMITATIONS

I follow Cline's (2010: 1) definitions for delimitations as well as limitations. Accordingly, delimitations are understood as the 'characteristics that limit the scope (define the boundaries) of the inquiry as determined by the conscious exclusionary and inclusionary decisions' and limitations are understood as the 'characteristics of design or methodology that set parameters on the application or interpretation of the results of the study; that is, the constraints on generalizability and utility of findings'.

Delimitations

This study investigates few but carefully selected aspects to best contribute to the advancement of this field. I delimitate numerous aspects that would have been of interest and even of relevance, but that could not be covered sufficiently in order to keep the scope of the study to a feasible extent within the given amount of time and resources.

In Chapter 1, I argued that firms are confronted with the make-or-buy[37] decision when they are in need of a technology. Similarly, firms have to decide whether to restrict a technology to internal use or allow it for external technology exploitation (ETE) (that is face a keep-or-sell decision). Despite the fact that these decisions are not at all trivial, they had to be eliminated from this study. Another factor which influences these decisions is valuation. Valuation is often a crucial determinant for make-or-buy and keep-or-sell decisions. For instance, a technology which has a low internal value for a firm might still possess a certain market value. Consequently, it should be considered for the external exploitation. The valuation is also of importance when conducting price negotiations with other parties in external acquisition / exploitation projects. However, although the valuation of technologies still needs further research due to the lack of elaborated and standardized valuation approaches, other scholars have already addressed this phenomenon. Therefore I eliminated this topic from the study. IP valuation of intellectual assets has become a topic in itself, which occupies various scholars.[38] Related to these questions is the commonly applied Resource-based view (RBV) of the firm to manage technologies as assets. I refrained from applying the RBV in this study, although it is referred to where necessary.

In this study, it is assumed that firms have decided in favor of the external acquisition / exploitation mode. When planning to externally exploit / source technologies on / from the MfTI, these firms face the decision whether to govern the transaction primarily with internal resources or to seek support from a TMI. This question is closely related to the theory of the firm (that is vertical integration and outsourcing), but is only marginally discussed in this study. This study is rather concerned with another crucial decision that technology-based firms face when they do not want to govern a transaction with internal resources (for example, because an SME has very limited internal ETE capabilities), but want to rely on the support of TMI. Following this decision, these firms have to choose between the currently increasing number of TMI and the transaction models they have developed. Despite the fact that many of these transactions models need further research, resource constraints have restricted this study to only one of these novel models, namely technology auctions. This study's results – with regard to technology auctions – are then compared on a rather theoretical level to bargaining negotiations as a reference transaction model.

TMI current emergence or their future roles in the MfTI as novel actors in the innovation system are not yet sufficiently understood. However, I was unable to collect sufficient data to cover these aspects comprehensively in this study. I had to refrain from studying TMI history, motivation and business models, and subsequently, I do not provide an explanation for their emergence, how they impact the efficiency of transactions, which obstacles on the MfTI they diminish or eliminate, as well as how and to what extent they contribute to creating innovation more quickly and efficiently.[39] Moreover, I do not address the overarching question of whether there is a need for a private sector or even governmental intermediation in the MfTI. As noted above, some observations point towards this need, however there is currently only limited solid empirical evidence that addresses this question. I consider this question relevant particularly for policy reasons, but had to delimitate it and leave it for future research.[40]

With regard to the use of the auction literature in this study, I applied mainly the auction design literature, but delimitated the auction theory literature to a large extent. Mainly due to missing data on the bidding behavior of prospective buyers, no attempt was made to develop a formal mathematical model, simulation studies or conduct an empirical test (for example, for the winner's curse effect in technology auctions), nor did I attempt to conduct an optimal design analysis. Various other topics that might be of interest to auction theorists were not investigated in this study. These include the risk that participants may explicitly or tacitly collude to avoid bidding up prices and the related topic of rings that bidders might form if a proper auction design does not prohibit it. For instance, ascending auctions can facilitate collusion by offering a mechanism for punishing rivals. Other properties that are eliminated from this study include closing conditions and activity rules. With regard to the auction design, this study focuses on multiple technology auctions that take place offline and within public settings.[41] Thus, I did not investigate auctions that take place in private (closed) settings.[42] Neither did I capture online auctions in this study which might occur on online marketplaces for technology trade (although this is believed to happen rarely). Consequently, many questions are left for future research, since 'the problems of this world are dominated by "practical difficulties" that our knowledge is far away from perfection' (Hayek, 1991: 30).

Limitations

Owing to the study's eliminations, relevant limitations of the results from the pre-studies and the quantitative study should be emphasized. In general, the data collection for both pre-studies was constrained by the early-stage nature of the TMI development and thus the limited availability of possible case studies. Obviously, a higher number of case studies would have raised the reliability of the results from both pre-studies.

For the *first pre-study*, data were collected and combined from different sources to raise the reliability of the results. However, the approach for combining the data could only be systematized to a limited extent. Although the comparative process analysis was carefully conducted based on objective and factual data, the result could have been strengthened if the auction firms had validated them. Hence, the case studies should not be considered in-depth but rather of an explorative nature. However, for the purpose of this study, the approach appeared sufficient as the case studies did not aim to induce theories as proposed by, for example, Eisenhardt (1989).

The results of the *second pre-study* appear limited to the extent that the case studies were selected along similar criteria to account for a possible selection bias. A more heterogeneous sampling approach might have revealed a broader range of insights. However, I rather risked revealing contradicting results instead of being able to benefit from the confirmatory principle when sampling multiple case studies to raise the results' reliability. Furthermore, data were mainly collected by means of a questionnaire, although this was thoroughly based on the existing literature and pre-tested. To reduce the effect of informant bias, the data were collected through a key informant approach by interviewing managers that had been responsible for the specific transactions (Kumar et al., 1993). However, additional interviews would certainly have strengthened the results. The results from an analysis of four case studies are not meant to provide generic evidence, but rather represent specific case-based evidence describing how patent transactions are conducted via technology auctions and how sellers perceive the success of this particular transaction model.

The validity of the results from the second pre-study could have been further raised if the data collected from the sellers were complemented with data collected from the buyers. Results from an approach that corrects the somewhat asymmetric empirical base would reveal whether the sellers' and buyers' perceptions match and would consequently provide a rather holistic picture of the current situation. However, it appeared almost impossible to access contact data from buyers. In December 2007, I addressed a request to the auction firms to either grant me access to the names of buyers or to distribute a questionnaire to them on my behalf and subsequently return the anonymous data to me. However, the request was dismissed due to firm confidentiality interests that these firms had guaranteed their clients. Thus, I dismissed the possibility of taking buyers' opinions on technology auctions into account. Owing to the lack of insights into the buyers' perspective, one might criticize this study's shortage of a major element to draw a comprehensive picture. However, I would like to remind the reader of the technology and IP acquisition literature available by providing insights into the buyer's perspective.

Furthermore, it should be noted that the quantitative scheme used to compare the individual case study results in Section 9.2 was not meant to give the limited results a more solid character, but seemed favorable as the questions were addressed by Likert-scaled questions. As common descriptive measures would be meaningless due to the small number of cases, I developed a conservative triple measure to derive some meaningful results.

In order to derive hypotheses from both pre-studies that were to be tested in the quantitative study, I discuss the results from both pre-studies in comparison to idiosyncratically designed bargaining negotiations. Obviously these negotiations represent a rather theoretical form of technology transactions. Many transactions are not fully idiosyncratic. For instance, licensing departments of large firms might develop and reuse standardized contracts that are only marginally adjusted on a case-by-case basis. However, data constraints prohibit insights into other governance structures of technology transactions, thus limiting this study's discussion on TGS from technology auctions in contrast to transactions conducted via bargaining negotiations.

The interpretation of the results from the *quantitative study* is also subject to limitations. First, the selection of testable hypotheses was limited through the use of secondary data. In contrast to a questionnaire study, it was not possible to collect the data needed; therefore I had to rely on available data. It would have been meaningless to develop hypotheses that could not have been tested due to missing empirical data. If further resources had been available, I would have been able to access additional databases, which would have complemented the data which were available from public sources, primarily the patent office databases. Consequently, the process turned out to be an iterative backward and forward cycling between the available empirical data and the hypotheses that could be developed to be testable.

A further limitation relates to the operationalization of the five technology properties. While the number of testable variables was limited by the available data, certain data that were available from patent documents (for example, the exact number of forward citations or opposition data) could not be used, since I could only derive data from publicly available databases. Furthermore, although I attempted to use measures for the operationalization that had been developed and tested in previous studies, they could only be used to a limited extent. This was mainly due to the fact that previously developed measures were either only available for individual patents (for example, to operationalize technological impact by forwarding citations of individual patents) or could not be operationalized with patent data (for example, for technological complexity as discussed in Rycroft and Kash (1999)). Although conceptual approaches proposed in previous studies could have been used for this study's measures, I had to adjust these measures to fit the operationalization with patents or the available data and the combination of various patents into bundles. Hence, the reliability of these measures may be limited.

Furthermore, it is important to note that I operationalized the technology properties by using only one-dimensional indicators in order to link this study and the hypotheses to prior literature. While for certain variables this approach was also used in previous studies (for example, the technological impact of patents is commonly operationalized by forward citations), for some variables, a rather multidimensional operationalization would have increased reliability. For instance, the reliability of the measure for technological quality could have been improved if additional indicators than the size of the technology development team were considered. Thus, one has to be careful when discussing and interpreting the results, particularly when comparing the results of this study to those of future studies that use other measures for the operationalization and hence conclude different results. I have operationalized the theoretical concept to the best of my knowledge, however mainly one-dimensional indicators were used. One should therefore be careful not to overinterpret the results because of the distance between the theoretical concepts and empirical observations.

This study's results would have also gained further reliability, if I had included independent or control variables in the empirical models in addition to those operationalized by patent data and auction catalogues. For instance, data on sellers' and buyers' motives to exploit or acquire certain technologies could have been collected from an additional survey but had to be excluded from this study, since these data were not available from the auction firms. Additional data on the bidding behavior for the individual technologies would have also enriched this study's results. For instance, the auction literature (see Chapter 5) indicates that sales prices commonly depend on the number of active bidders. This effect could not be controlled for in this study as there were no such data available from the auction firms. Certainly, the multivariate regression models used in this study could also be enriched by controlling for moderator or mediator effects. However, the sample size somehow limited the use of variables in the models.

Although I believe that the results from the quantitative regression analyses can be considered robust for technology auctions with regard to the sample size and number of variables included in these models, one should refrain from generalizing the results to all technologies and patents. Instead, an analysis of transacted technologies might not be representative of the full population but rather biased. Even though this bias is made somewhat transparent in the descriptive data section in Chapter 10, the generalizability of the results to related fields (for example, to patent value studies) remains limited. This is also due to the fact that I investigated technologies as bundles of individual patents while these studies commonly investigate individual patents. Consequently, the expected and realized sales prices of technologies in this study can be compared only to a limited extent to the values of individual patents. Furthermore, the comparability is also limited, since I could not use standardized industry classifications (for example, NACE) although different coherence tables have been provided in earlier literature (for example, OECD (2009)). A consolidation exercise of IPC classes from various individual patents into one measure for a patent bundle revealed various problems.[43] Consequently, this study's results cannot be easily compared to official industry classifications. To account for this effect, I have included a concordance table in Annex II.

Furthermore, this study's results are constrained by the fact that it was conducted for early auctions. Within the six auctions in the study's dataset, three were first auctions conducted in the US, Germany and UK. It remains to be seen how these auctions develop. A later-stage study – after the learning effects have reached an equilibrium stage with robust auction designs – might reveal diverging results. The study's results are also limited by the fact that the dataset includes primarily US patents. As patent jurisdictions differ, the results might be different for technology auctions of technologies protected by non-US patents. However, a control variable was included in this study's models to account for this effect.

Having reflected on the various limitations of this study, the reader might still notice a small degree of subjectivity, which should be taken into account when interpreting the study's results. As Schumpeter (1949: 348) pointed out in his presidential speech to the American Economic Association, every economic study remains to be influenced by the author's 'scientific ideology … [and the]… scientific worker's choice of problems and approaches'.

NOTES

1 See, for example, Bauman and Klein (2010) as well as Katz and Rosen (1998).

2 Granstrand (2000a) provides more details on various ways to develop or acquire technologies.

3 Obviously actors can even acquire and exploit technologies at the same time, that is act on the supply and demand side. Acquisition and exploitation can be regarded as 'two sides of the same coin.'

4 OECD et al. (2005) provide insights into – and a good starting point for those interested in – the institutional environment.

5 Please note, although certain types of brokers and dealers cannot be regarded as new, those specific types of intermediaries which are subject to this study (see the definition in Chapter

4) have started to emerge recently as rather new institutional types (for example, auction firms for technology transactions).

[6] With 25 participants, the workshop was held jointly with the Hamburg Chamber of Commerce on the topic 'exploiting technologies successfully' on 22.03.2007. Among others, participants included large firms (Blohm und Voss GmbH), SME (for example, EuroRapid Gmbh, Imawis GmbH, DCS Innovative Diagnostik Systeme GmbH, and Wöhlk-Contactlinsen GmbH), the German Federal Ministry of Economics and Industry (BMWI), the TTO of a large federal German research center (DESY), the DPMA, a TMI (IP Bewertungs AG), and patent law firms (Meissner Bolte & Partner, White & Case LLP).

[7] Punch (2005: 46) specifies five criteria for 'good' research questions. It is believed that this study's general as well as specific research questions fulfill these criteria, that is they are clear, specific, answerable, interconnected, and substantively relevant.

[8] According to Gerring (2004: 349), 'the world of social science may be usefully divided according to the predominant strategy of research undertaken, exploratory or confirmatory/disconfirmatory.' Brewer and Hunter (1989) propose a typology of six types of units which can be studied in social science research (individuals; attributes of individuals; actions and interactions; residues and artefacts of behavior; settings, incidents and events; and collectivities). Accordingly, the chosen units of analysis can be translated to actions and interactions (that is transactions) in the first research question and attributes of individuals (that is patented technologies) in the second research question.

[9] Various studies have been conducted that investigate technology trade on a national level (for example, Gambardella, 2002; Granstrand, 2004).

[10] The relevance to study the role of TMIs within transactions was furthermore confirmed by more than 20 industrial participants from large firms, SME and technology transfer offices (TTO) at a workshop held jointly with the Hamburg Chamber of Commerce with the topic 'marketing technologies successfully' on 25 March 2007. The German Engineering Federation's (VDMA) and the electrical engineering and electronics industry's (ZVEI) support of a small research project in cause of this study is indicative of the study's relevance (Tietze et al., 2007).

[11] The importance of technology licensing has long been recognized by the managerial and the industrial economic literature. Early studies on licensing, especially in the industrial economic tradition, have already emphasized the importance of licensing for the diffusion of technology in order to avoid the duplication of research and product market competition (for example, Shephard (1987); Rockett (1990); Gallini (1984)). The positive impact of licensing on innovation performance has just recently been proven by Reichstein (2009).

[12] For instance, in 1994, the American Federal Communication Commission (FCC) started radio spectrum auctions to sell technology licenses. Milgrom (2005: 1) notes that the 'early US spectrum auctions had evolved into a world standard ... and by the end of 2001 auction sales totaled more than $100 billion.'

[13] A large, international 'second hand' market exists for unique industrial machinery for which auctions are a common trading model (for example, http://www.maschinenportal24.de, http://www.machinery-market.co.uk/). Machinery is often sold to less developed countries after a new machinery generation is available in a developed country. Another example of the use of auctions in B2B transactions is procurement auctions, which OEMs use in the automotive industry to negotiate delivery contracts with suppliers.

[14] This study's database includes 156 technologies sold for a total of about €30 million at six auctions. This means that 26 technologies were transacted at every auction, each with a duration (including planning phase) of between five and 12 weeks (see Section 9.1).

[15] Birkenmeier (2003) is one of the few scholars who has briefly touched on the auction concept when he mentioned auctions among the market-based approaches for the utility oriented market-based determination of rewards.

[16] However, the functioning of technology auctions is not taken for granted. Furthermore, carrying out this study I question whether auctions can be used for technology transactions.

By investigating them, this study aims to detect whether this novel transaction model can survive as well as if and how auctions can be further improved.

[17] Although their institutions might have different governance structures, similar aspects of the study should also be of relevance to managers of university TTO or other governmental institution as well as for and non-profit research institutions and independent inventors. In Chapter 10, I demonstrate that various universities and independent inventors have – to some extent – participated quite successfully in the auctions, that is they have sold patents for above average prices.

[18] This study is not conducted for large firms or SME in particular. Nevertheless, large firms in general have more advanced IIP management systems in place than SME. Therefore, they might benefit more from rather specific transaction knowledge as presented in this study. However, SME frequently own unused technologies and IP assets that they could exploit externally to better appropriate their returns from R&D (Institut der deutschen Wirtschaft Köln, 2006).

[19] Studying patent is usually difficult. Since research is still currently conducted to develop and establish commonly accepted valuation approaches, most previous studies have rather used proxies (for example, citations). For an overview, see OECD (2009).

[20] Obviously, one has to keep in mind that auctioned technologies might not be representative of the whole population of technologies due to selection bias (see for example, Table 10.10). Thus, the results of this study should not be over-generalized.

[21] In Tietze and Herstatt (2010), we contribute a theory-driven explanation for the recent emergence of TMIs based essentially on Stigler (1951), Williamson (1979) and North (1996).

[22] Investigating innovation policy Bessant and Rush (1995: 100) recommends that a 'possible element in such policies is the use of consultants as intermediaries to assist and advise firms to compensate for the lack of managerial capabilities.'

[23] For the first time in 2006, OECD (2006) included a full chapter on technology markets in their bi-annual Science and Technology Outlook entitled 'Patent Licensing Markets and Innovation.' OECD (2006: 13) refers to 'private-sector intermediaries… [playing a] …leading role in developing technology licensing markets…' Furthermore, concerning recent developments, the OECD, together with the EPO and German Federal Ministry of Economics and Technology, mentions that 'following the expansion of technology markets, the role of technology intermediaries appears to be growing in importance and will likely continue to do so' (OECD et al., 2005: 9).

[24] The most prominent example is the PatVal studies (for example, Giuri et al. (2007)).

[25] Owing to the industry's limited experience with TMIs access to valid and reliable data was restricted. After two pre-studies showed insufficient data quality, I dismissed the possibility to conduct a large scale study collecting primary data. A first questionnaire study was designed and addressed to about 70 TMIs, but resulted in a response rate of below 10 per cent. A second questionnaire study was addressed to sellers and buyers at the IPA auction in Germany in May 2007, but fewer than 10 questionnaires were returned. A request to the TMIs organizing technology auctions to distribute the questionnaire to their participants was dismissed due to the strict confidentiality interests, particularly the technology auctions firms had guaranteed the buyers.

[26] According to Scandura and Williams (2000: 1248) the 'impact of management studies depends upon the appropriateness and rigor of the research methods chosen. Design choices about instrumentation, data analysis and construct validation and more may affect the types of conclusions that are drawn.'

[27] To gain initial insights into the field, I conducted a first explorative pre-study reported in Tietze (2010) in which I analyze selected TMIs and their business models, applying Osterwalder (2004) 'nine business model building blocks' and Herman and Malone (2003) 'business models archetypes' in order to propose a typology for TMIs.

[28] According to Yin (2003: 23), case studies are defined as an 'empirical enquiry that investigates a contemporary phenomenon within its real-life context when the boundaries between

the phenomenon and context are not clearly evident; and in which multiple sources of evidence are used.' According to Punch (2005: 144) a 'case study aims to understand the case in depth and in its natural setting, recognizing its complexity and its context... it has a holistic focus, aiming to preserve and understand the wholeness and unity of the case.'

[29] When the data were collected in autumn 2007, very few public technology auctions had taken place. To my knowledge, only two firms (OT and IPA) had conducted public technology auctions.

[30] A similar rationale is provided by Gerring (2004: 352) who argues that case studies are preferred 'when the strategy of research is exploratory, rather than confirmatory and ... when useful variance is available for only a single unit or a small number of units.'

[31] The case studies represent 'contemporary descriptions of recent events' according to Eisenhardt and Graebner (2007: 25). Particularly within this early stage of this research, it appeared necessary to first gain insights into the technology auction process and transactions before conducting a large scale study. Furthermore, with this approach, I react to critics of pure quantitative approaches as, for example, raised by Gerring (2004), who argues that an in-depth understanding of the studied phenomenon is required but often missing when large scale quantitative approaches are favored.

[32] Note: Own translation.

[33] As an alternative, the probit regression is often recommended. However, in various patent econometric studies, logistic regression models are more widely applied. Therefore, this study does not further discuss probit models.

[34] I conducted various tests which revealed that his data were suitable to conduct an OLS regression. If this had not been the case, other regression types would have been used (for example, GLS - Generalized least squares). The tests and results are presented in Chapter 10. Manning and Mullahy (2001) provide a valuable discussion on the choice of the estimator.

[35] I would like to acknowledge the support of Dr. C. Barkmann who provided valuable comments on the choice of the statistical technique.

[36] This typology has proven its validity in an extensive literature review of methodological approaches used in management studies by Scandura and Williams (2000).

[37] If done systematically, the 'Keep-or-Sell' decision (see for example, Brodbeck (1999)) follows a portfolio valuation of various technologies (see for example, Tschirky and Koruna (1998)).

[38] For a comparison between valuation approaches, see Pitkethly (1999). For details on cost-based approaches see, for example, Mittag (1985: 258) or Schultz (1980: 123ff). For detailed discussions on valuation based on experience see, for example, Megantz (2002: 63f) or Poley and Deutscher Wirtschaftsdienst (1981: 35).

[39] Few of these aspects are covered to some extent in Tietze and Herstatt (2010), Tietze (2010), and Tietze and Barreto (2007).

[40] Furthermore, it is worth mentioning that TMI definition delimitates non-private entities, for example, university TTO that have different motives and incentive schemes; although, I believe that results could still apply to them to a certain extent.

[41] The adjectives 'public' and 'offline' are neglected in this study. However, when referring to 'technology auction', I essentially mean public, offline technology auctions. See Chapter 5 for descriptions of different auction types.

[42] Although often not explicitly labeled as such, one might argue that technology auctions have been taking place for a long time, although primarily in closed settings. A well-known example is the auction of the patent portfolio of the internet company CommerceOne, Inc., who auctioned 39 patents independently from its physical assets after it had to file for Chapter 11 bankruptcy. The patents were sold for US$15.1 million, while the physical assets sold for US$4.1 million. It turned out that Novel, Inc. had bought the patents to ensure that it could uphold its promise to its users of XML technology, to whom it had granted an open source license. According to Bill Smith, the then-vice president for investor relations, 'these patents were purchased consistent with my published policy statement to use my patent portfolio to

defend against those who might assert patent claims against open-source products marketed and supported by Novell' (Markoff, 2005).

[43] The 'international patent classification (IPC), established by the Strasbourg Agreement 1971, provides for a hierarchical system of language independent symbols for the classification of patents and utility models according to the different areas of technology to which they pertain' (WIPO, 2009).

PART II

Technology Transactions and Auctions

3. A Firm Perspective on Technology Transactions

3.1 INTRODUCTION

Since it is becoming increasingly important to open up innovation processes beyond the own firm boundaries, firms are becoming more and more aware of the importance to strategically manage their IP portfolios along open and distributed innovation processes. For instance, IP management includes the selection of internally developed inventions as well as determining whether they should be patented, and, if so, in which countries at what time. IP management also includes the continuous management of firms' IP portfolios (determining which patents should be kept or be dropped), and poses questions that involve parties outside the own firm boundaries. Those tasks relate to the management of the firm's technology base and include the make-or-buy and keep-or-sell decisions. Hence, firms have to decide whether a technology should be developed internally or whether it can be acquired or in-licensed more economically as well as whether an own technology should be kept internally or be externally exploited (for example, sold or out-licensed).

Figure 3.1 illustrates the development of an innovation and its diffusion through a distributed innovation process with different transactions taking place along the product development lifecycle phases.

In the management literature, transactions from the seller's perspective (which is the focus of this study) are closely related to the 'external technology exploitation' (ETE) concept that was developed in the 1970s by, among others, Anderson (1979), Lien (1979), Marcy (1979), as well as Ford and Ryan (1977). In particular the contributions by Ford (1988), Ford and Ryan (1981, 1977) coined the concept 'technology marketing' as a holistic approach to external knowledge exploitation formulating three purposes. They argue that technologies can be used in the manufacturing or design process of a physical product, for the marketing of a service based on a technology (for example, the sale of a design or testing service), and sold as a 'whole' (for example, the sale of a particular bit of electronic knowledge to another party for its own use).

Before the 1990s, however, the concept had not developed very much, nor did it receive considerable attention from academic scholars apart from a few, for example, Mittag (1985). Based on an initial idea published in Granstrand (1982), Granstrand et al. (1992) present the technology-based concept including a typology of technology exploitation and acquisition strategies in the context of their analysis of the multi-technology corporations. According to Granstrand (2000b), since the late 1990s, the topic has received increasing attention in accordance with the growing importance of IP due to the shift towards IP capitalism and the 'knowledge society' and, according to Chesbrough (2003a), the open innovation movement in the early 21st century. The notion of 'profiting from innovation' was coined by Teece (2006).

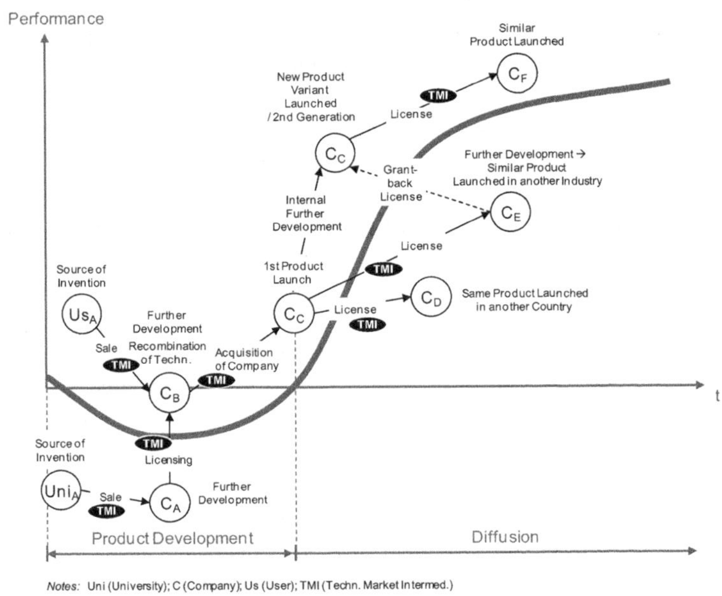

Figure 3.1 - Transactions along the innovation process involving TMI

Although this study investigates technology transactions in general, I primarily apply the seller's perspective as motivated in Chapter 2. In the following section, I develop a definition for external technology exploitation that is applied throughout this study. In order to derive a suitable definition, I reviewed the literature and consequently identified a comprehensive review conducted by Lichtenthaler (2005) to which I predominantly refer in the following section.[1] Focusing specifically on technology exploitation, Lichtenthaler reviews prior literature from early studies from the 1970s until

2004. Aside from various conceptual publications, Lichtenthaler identifies six empirical studies with a 'noteworthy large enough sample' to provide reliable results. I complement Lichtenthaler's review by taking more recent contributions into account. These include three empirical studies by Hentschel (2007), Lichtenthaler (2006a) who uses his 2005 literature review as a pre-study, and Escher (2005), although the latter draws on a comparatively small and qualitative empirical basis. Furthermore, this study takes Granstrand and Sjölander (1990) into account, since it covers an exceptionally large share of Swedish, Japanese and US R&D spendings using their selective (top-tail) sampling of large R&D spenders, despite the small number of firms included. With their study, they are also considered pioneers in this research field having developed the technology base concept with a typology of exploitation (and acquisition) strategies. Thus, their study conducted among Swedish and Japanese firms should receive particular attention despite their comparatively small sample. Table 3.1 provides an overview of the ten empirical studies, including the respective sample sizes considered by Lichtenthaler (2005) in his review and my attempt to develop a suitable ETE definition for this study.

Table 3.1 - Selected empirical ETE studies since the mid-1980s

Authors	Sample Size	Scope of the Study
Ford (1985) *	N=152	US Companies Probably in the Beginning of the 1980s
Mittag (1985) *	N=276	German Companies in 1981 (only 98 Companies with Licensing-out Activities)
Vickery (1988) *	N=119	Companies Worldwide in 1985/86
Granstrand et al. (1992)	N=42	Japanese, Swedish and US Large Corporations
Brodbeck (1999) *	N=281	German and Swiss Companies in 1996
Elton, Shah et al. (2002) *	N>40	(Sample Unknown)
Birkenmeier (2003) *	N=281	German and Swiss Firms
Escher (2005)	N=29	Interviews plus one Workshop
Lichtenthaler (2006)	N=155	Survey among 500 Largest Firms from Germany, plus 100 Largest Swiss and Austrian Firms each
Hentschel (2007)	N=228	Questionnaires and 18 Interviews

Note: Studies marked with an asterisk were reviewed by Lichtenthaler (2005)

From the literature review, drawing mainly on Boyens (1998)[2], Ford (1985) and Ford and Ryan (1977), Lichtenthaler (2005: 233) derives the following definition:

> External knowledge commercialization (exploitation) describes an organization's deliberate commercializing of knowledge assets to another independent organization involving a contractual obligation for compensation in monetary or non-monetary terms.

Accordingly, ETE involves *inter-organizational technology transfer*, that is the transfer of technological knowledge from one legally and economically independent organization to another. According to the notion of Mansfield and Romeo (1980), inter-firm technology transfer is regarded as a subform of technology trade.[3] While the transfer of technologies between different business and functional units of a single organization may represent a transaction from the perspective of a business unit or functional unit, this type of transaction does not constitute an external commercialization of technological knowledge from a corporate perspective.

Furthermore, the external exploitation of technologies is regarded as a *deliberate action* of a firm. Accordingly, it refers to the intended transfer of technologies and thus does not take into account the unplanned loss and leakage of information, that is the 'leakage exploitation strategy' considered by Granstrand and Sjölander (1990) is excluded from the definition.

According to the definition by Lichtenthaler (2005), the actual economic good that is transferred is characterized as *disembodied technological knowledge*[4] similar to the notion of Granstrand (2000a), who identified six properties of technology within the framework of the technology-based theory of the firm. The main focus of the transferred knowledge, however, is on its asset nature as a resource instead of its learning properties.

The abovementioned ETE definition further includes a certain *contractual obligation*. While non-formal ways of externally exploiting technologies, such as informal knowledge trading, are important in practice according to, for example, Schrader (1991) or von Hippel (1987), they are commonly initiated by individuals and, according to Boyens (1998), they often do not follow the explicit intention of a firm. As such, these nonformal types of ETE can hardly be integrated into a firm's technology strategy and are therefore also excluded from the definition, which is consistent with earlier works by, for example, Ford (1985).

A contractual obligation already indicates that a firm will usually receive some form of *compensation* for the technology transfer. In the commercialization of products or services, monetary compensation is by far the most common form of recompense, whereas the exploitation of disembodied knowledge offers various other possibilities, such as bi-directional technology transfer (for example, cross licensing and the mutual exchange of technologies through patent pools), which are frequently used in practice according to Koruna (2004), Grindley and Teece (1997), and Brockhoff et al. (1991).

In addition to Lichtenthaler's (2005) definition of ETE, I reviewed Hentschel (2007), Lichtenthaler (2006a) and Escher (2005). Hentschel (2007) focuses on specific forms of ETE but does not provide a specific definition for ETE. Lichtenthaler (2006a) applies a slightly different, but essentially similar definition to the one he had proposed based on his literature review mentioned above.[5] Escher (2005) does not use Lichtenthaler's (2005) 'external technology commercialization' concept, but, following Ford and Ryan's (1977) traditional terminology, labels ETE 'technology deployment'. Escher (2005: 24) defines technology marketing as 'a process for pursuing normative, strategic and operational enterprise objectives concerning technology acquisition and technology deployment markets'. However, Escher (2005: 36) provides no specific but rather a rough definition for the technology deployment concept, namely that 'technology management and intellectual property management are superordinate concepts containing technology marketing as a subset. Thus, the topic technology marketing itself, according to its definition, can be divided into the two fields: external technology acquisition and external technology deployment. Finally, licensing and technology-based spin-offs are subtopics of external technology deployment'.

To conclude, Ford and Ryan (1977) develop the 'technology marketing' concept, which Escher (2005) uses to develop the technology deployment concept. Granstrand and Sjölander (1990) develop a framework for external technology exploitation (and acquisition) strategies. Lichtenthaler (2005) takes Ford and Ryan's (1977) contribution into account to develop the ETE concept, which shows certain similarities to Granstrand and Sjölander's (1990) framework. I decided to focus mainly on Lichtenthaler's (2005) definition of ETE presented above. This definition fits well with the purpose of this study. It is important to note that, in the remainder of this study, I do not focus equally on all the different ETE strategies, but particularly on the 'technology selling' strategy.[6]

3.2 FIRM STRATEGIES FOR EXTERNAL TECHNOLOGY EXPLOITATION

Having presented the definition of ETE that is applied in this study, it is important to understand among which strategies firms have to choose when engaging in technology transactions. In the early 1990s, Granstrand and Sjölander (1990) developed the 'technology base' concept and proposed a framework with five acquisition (sourcing) strategies and six exploitation (commercialization) strategies (see Figure 3.4) based on an initial idea published in Granstrand (1982). This framework presented below provides a suitable concept for this study.

Exploitation (as well as acquisition) strategies are concerned with the 'generation, combination, transformation, adoption, absorption, regeneration, and recombination of technologies'. Consequently, according to Granstrand and Sjölander (1990: 114), they are accompanied by a transfer of explicit (for example, patents, license agreements and technology descriptions), as well as tacit knowledge (for example, personal experience, knowledge embodied in specialists). Granstrand and Sjölander propose that by applying these strategies, firms can develop and manage their technology base as a corporate resource. The strategies differ in their degree of organizational integration or 'openness,' as their different underlying contractual forms imply, thus ranging from closed to open (see Figure 3.4).[7]

The highest level of organizational integration is represented by in-house technology exploitation, for example, firms themselves directly apply certain technologies for further R&D, in production or marketing. Firms can also choose to exploit technologies by creating another firm and subsequently transferring certain technologies to such a 'newly' founded spin-off. Moreover, technologies are complemented with other technologies from partners or even competitors by forming a joint venture. If neither forming a new firm nor forming a joint venture appears to be feasible or advantageous, a firm might decide to sell a certain technology as well as its related IP assets. If no buyer is available for a technology (also for other reasons), the firm can decide to abandon technologies and no longer follow their application and development. Firms often drop or donate technologies to non-commercial institutions, for example, universities. Finally, if there is no current but potential usage in the future, a firm may decide to store the technology, for example, to apply it in later product generation.[8]

Lichtenthaler (2005) reviews the literature on ETE strategies, including Birkenmeier (2003), Granstrand (2000a), Boyens (1998), Vickery (1988), Ford (1988, 1985). Based on this review, Lichtenthaler (2005) attempts to harmonize the heterogeneity of ETE strategies proposed by different scholars and proposes a classification for what he calls 'technology commercialization' strategies. As illustrated in Figure 3.2, Lichtenthaler (2005) defines six strategies from which he labels four as 'formal ETC,' thus excluding the pure internal usage and the unintended leakage of knowledge as also mentioned by Granstrand and Sjölander (1990).

As illustrated in Figure 3.2, Lichtenthaler (2006a) investigates the sale of patents, the award of licenses, cross-licensing, grant-back agreements, contract research, technological cooperation, and the sale of firm units, for example, newly established businesses. Following Carter (1989) and Schrader (1991), Lichtenthaler (2006a) excludes unplanned external exploitations, such as spillovers or those that do not include any contractual obligations for compensation, for example, information knowledge trading. Following Henke (2000), Lichtenthaler (2006a) further excludes the sale of components because it corresponds more to the sales of products. Comparing Lichtenthaler's classification of the six different strategies with the early typology proposed by Granstrand and Sjölander (1990), the similarity becomes obvious. Thus, I conclude that Lichtenthaler's (2005) review validates the framework by Granstrand and Sjölander (1990) and further specifies it.

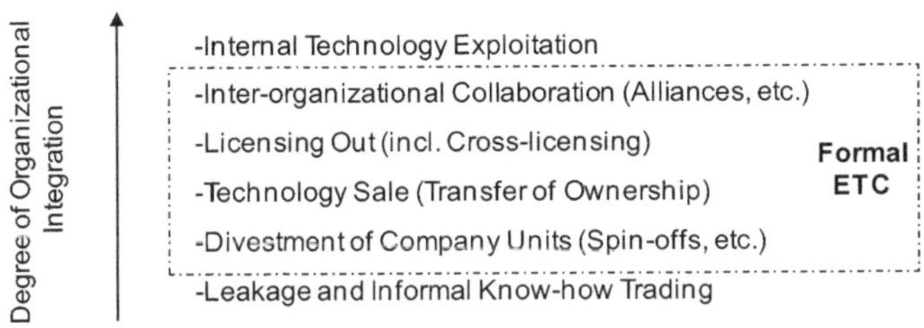

Source: Lichtenthaler (2006a: 19)

Figure 3.2 - Systematization of the forms of technology exploitation

Surprisingly, none of the typologies for ETE strategies account for the role of TMI. It appears that the conceptualizations of ETE strategies rather assume bilateral transactions of the firms exploiting technologies externally – directly with their partners who want to acquire certain technologies. Few conceptual contributions to this issue were found in the literature. Discussing the differences and similarities between product and technology markets, Birkenmeier (2003) briefly touches on this issue. Birkenmeier (2003) proposes to differentiate between direct and indirect communication channels for ETE. According to him, technologies can be distributed with own organizational resources (for example, R&D personnel) or via particularly dedicated departments founded solely for this purpose or by using intermediaries. However, Birkenmeier (2003) defines communication as the tasks that need to be carried out after a contract has been signed. He provides a suitable illustration (see Figure 3.3) of different models for technology trade, distinguishing between direct and indirect 'distribution channels'.[9]

Birkenmeier (2003) argues that the distribution of technologies can generally take place via two modes. He differentiates between direct (including personal, written, and telephone contacts) and indirect distribution channels (including exhibitions, scientific publications, and data pools) to address potential customers. While direct distribution via permanent or temporary organizational institutions bears certain advantages (for example, control of confidential information to avoid leakage), it has certain disadvantages too (for example, it requires direct contact between persons from prospective buyers and the risk of concentrating only on known firms within the closer context of the own firm). Thus, according to Birkenmeier (2003), indirect channels via technology transfer firms or market intermediaries should be favored if the firm lacks adequate knowledge regarding potential customers and faces difficulties to reduce this information deficit, even via a systematic search process, or if the firm faces financial obstacles.

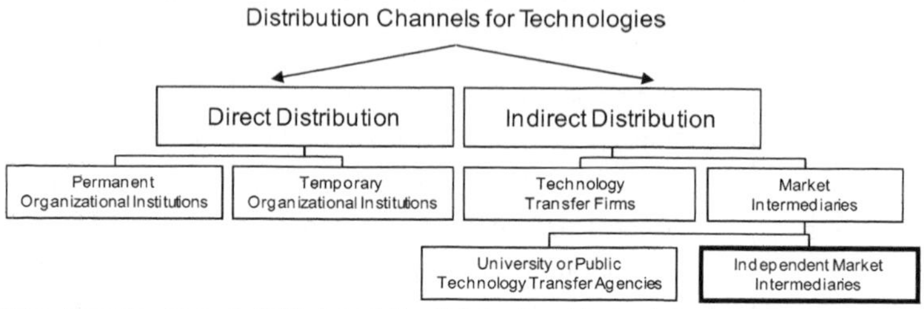

Notes: Adapted from Birkenmeier (2003) who labeled the distribution channels "communication channels"

Figure 3.3 - Distribution channels for technologies

To summarize, different typologies for external exploitation strategies have been proposed and categorized along a continuum of organizational integration. The degree of organizational integration appears closely related to the concept of vertical integration, which is an integral part of a firm's theory. The two extremes of the continuum (high = internal only; low = market transaction) can be further related to the two extreme forms of governance structures for transactions (hierarchy = internally; market = transaction involving other actors) as presented in Chapter 7. Such a classification also relates to the trichotomy followed by Delmas (1999), who investigates the choice of alternative governance structures (including internal development, alliances, and acquisition by contract).

To conclude, following the differentiation of direct and indirect communication channels proposed by Birkenmeier (2003), the concept of Granstrand et al. (1992) can be extended by integrating 'exploitation channels' for technology selling. Figure 3.4 illustrates that technology transactions can be conducted with all necessary resources internally (for example, patent and licensing departments) or with certain tasks being outsourced to TMI. Accordingly, firms can either decide to govern a technology transaction with internal resources bilaterally with the other party or seek the support of a TMI.

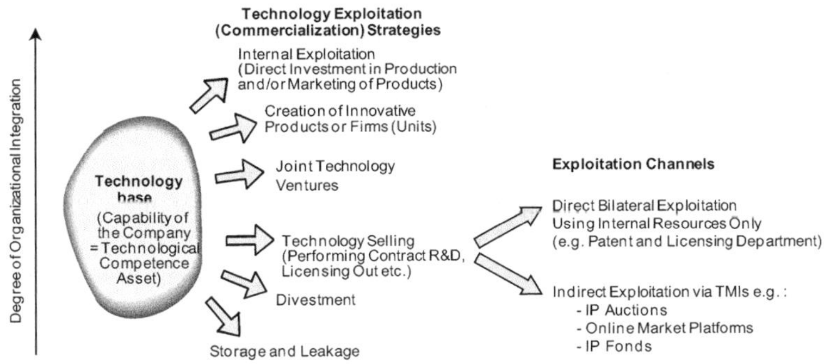

Figure 3.4 - Extended framework incorporating exploitation channels

With an increasing number of TMI (for example, auction firms) developing novel transaction models, firms face an increasingly complex choice of an exploitation strategy as well as an 'exploitation channel', that is whether to govern the transaction with internal resources only or whether to outsource it to TMI, and, if so, to what extent (that is certain specific tasks).

3.3 THE STRUCTURES OF TECHNOLOGY TRANSACTIONS

As highlighted in the framework for this study (Figure 2.1), the structure of the transaction process is a major element of any transaction aside from the traded asset. A process perspective of transactions is common as most prior studies have adopted it explicitly or by implication (Escher (2001), Tschirky et al. (2000), Ford and Ryan (1981)). The transaction process thus needs further elaboration.

Various sequential process models have been developed, for example by Lichtenthaler (2006a), Escher (2000), Belz et al. (2000), and Mittag (1985). In the following section, I focus on the two models proposed by Lichtenthaler (2006a) and Escher (2005) as they build on the previously developed models to some extent. Although I apply the model proposed by Lichtenthaler (2006a) in the remainder of this study, I present two models below in order to highlight their commonalities and differences.

As a part of the 'Integrated Technology Marketing Concept' developed by Birkenmeier (2003), which is based on contributions by Gassmann et al. (2003), Ndonzuau et al. (2002), Megantz (1996), and Mittag (1985), Escher

(2005) proposes a 'technology deployment process' model, for which deployment is used as a concept similar to exploitation. According to Escher (2005: 99), this process is 'on many levels ... the counterpart of the technology acquisition process'. Similar to external technology acquisition, a certain number of pre-processing tasks need to be managed and decisions need to be made before the process can be initiated (for example, a team with a distinct leader has to be organized, the object of the investigation must be determined, and strategic objectives have to be determined). According to Escher (2005), the initiation of the deployment process is triggered by different internal or external, intended or unplanned events (for example, external sourcing requests made by third parties, innovation planning, a change of the corporate strategy or infringements on the firm's patents).

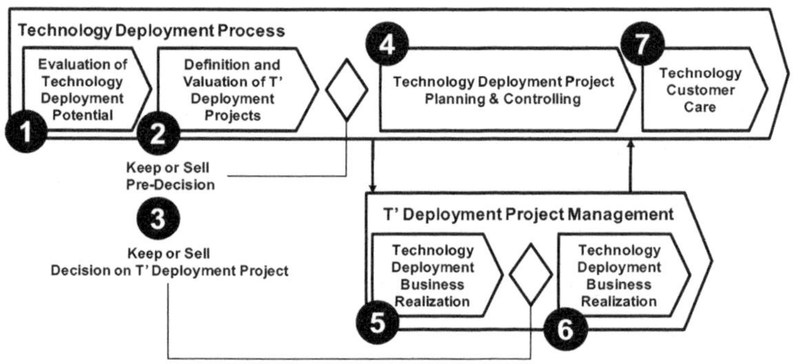

Source: Escher (2005: 100)

Figure 3.5 - External technology deployment process model

The technology deployment process proposed by Escher (2005) comprises seven phases as illustrated in Figure 3.5. The first phase is concerned with the evaluation of technologies that the firm intends to deploy externally. Thus, according to Birkenmeier (2003), the firm needs to examine the technology in terms of its overall business and technology strategy, for example by using the opportunity / threat analysis in four different areas (technology properties, customer properties, the firm's internal settings, and its environment). Additionally, the firm needs to identify the technology's most promising application fields, based on market attractiveness, customer suitability, and concurrent technologies. Whereas this phase of the process determines the deployment potential of the technology and the effectiveness objectives, the second phase is concerned with the determination of specific deployment projects. It focuses on investigating all issues that are relevant for the subsequent keep-or-sell decision. This process phase involves tasks to define concrete application fields regarding content and geography, technology users, preferred deployment methods, and price expectations. The result of this investigation is a prioritized project list that contains information on the deployment objectives, specifying which technologies will be deployed in which markets and to which users and indicates target prices.

During the following phase, the keep-or-sell decision has to be made on whether to proceed with a certain project or whether it should be terminated. A positive decision to proceed with the project will be likely, if strategic and financial remunerations outweigh the risks and resource requirements of the deployment project. Additionally, it needs to be ensured that the priority requisites are respected in the realization of the project. Throughout the subsequent planning and controlling phase, the issues that must be considered are similar to those for technology acquisition projects. Suitable measures have to be defined for the quantification of the project performance and a team has to be composed. However, in contrast to the team composition for technology acquisition projects, the deployment team is more likely to include external specialized personnel. Following this step, a business plan will be developed and potential partners will be identified, ranging from technology buyers to venture capitalists, technology consultants, and technology brokers. During the course of the negotiations, the team will be guided by the objectives and priorities that were formulated in earlier phases. Having successfully concluded the negotiations with at least one partner, the business plan needs to be implemented according to the objectives of the deployment project and the signed contract. In this phase, additional knowledge transfer, by means of, for example, teaching programs might become necessary. Additionally, the fulfillment of the contract conditions needs to be monitored. Finally, after the technology has been transferred, carrying about the customers becomes important in certain cases, for example, to establish a long-term relationship or collaboration.

Based on own interviews and a 'first version' of the technology deployment process published by Escher (2001) as well on additional works by Reid et al. (2001) and Chiesa et al. (2003), Lichtenthaler (2006a) proposes a process for external technology commercialization (see Figure 3.6).

Although, the process model is illustrated as a sequential process comprising five phases – similar to Escher's (2005) process model – Lichtenthaler (2005: 243) notes that the 'process phases will usually not follow each other sequentially but iteratively with feedback loops… [that they] may have to reiterate.' Furthermore, Lichtenthaler (2006a) notes that the process is appropriate both for a single project and for ETE on a firm level.

The ETE process starts with a *planning phase* as the interface between the general corporate planning and the ETC activities. Of particular importance is the overall ETE coordination of individual ETE projects with other

ETE projects and the overall ETE strategy on the one hand and with the firm's internal technology exploitation strategies on the other. The planning phase comprises four tasks: strategic technology planning, target setting, resource allocation, and customer pre-selection.

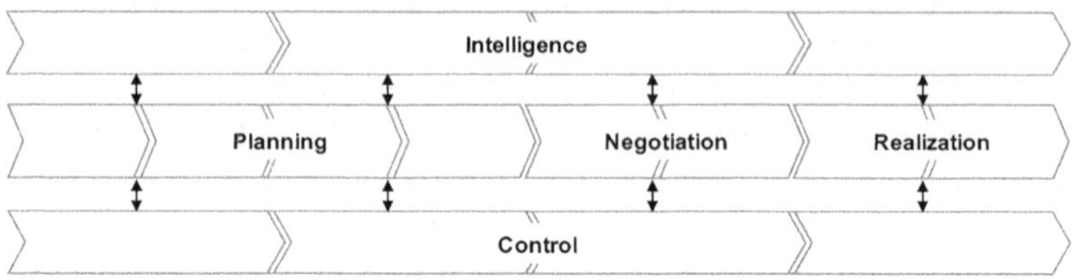

Source: Lichtenthaler (2006a: 115)

Figure 3.6 - External technology commercialization process model

If a firm decides to set up negotiations, a perspective has to be taken that includes the potential ETE users although the focus is still on the management of an individual firm. Lichtenthaler (2006a) differentiates this phase into two phases: the pre-negotiation phase and the detailed *negotiation phase*. The objective of the pre-negotiation phase is to develop an understanding of the firm's compatibility with the potential technology user. During the detailed negotiations, the ETE agreement is set up, which will normally take some type of contractual form. The contract specifies the terms of the transactions, particularly regarding the payment of the seller. The payment can take various forms, ranging from simple lump-sum payments to detailed and complex royalty schemes (for example, with increasing / decreasing royalty rates and granted back clauses for subsequent improvements of the technology).[10] These negotiations are usually an iterative process that can include third parties such as antitrust commissions. As the most important elements in negotiations, Lichtenthaler identifies the project's objective, the project's deadline, each party's expected contribution, the organizational structure, the coordination mechanism, potential interface problems as criteria to assess the project's performance.

When a firm has signed an agreement with a technology user, the end of the negotiations phase is reached and the *realization phase* starts either immediately, or with a certain time lag, which is usually the case. The actual transaction, which is divided into design tasks and implementation tasks, takes place during this phase. During the design phase, the planning of an ETE transaction at an operational level takes place based on the arrangements agreed upon during the negotiation phase. Furthermore, clear objectives are defined, to which the parties have to commit themselves. Additionally, during a detailed process mapping phase, the implementation process is specified and possible interface problems have to be identified. The transfer of the relevant technological knowledge takes place during the implementation phase.

The *intelligence and control phases* run parallel to any ETE project. The intelligence phase comprises the identification of internal technologies that may be externally exploited as well as the scanning and monitoring of the firms' technological environment on the MfTI. Referring to Brockhoff et al. (1991) and Lange (1994), Lichtenthaler (2006a) notes that this phase includes an information search as well as an identification of information needs and the subsequent information evaluation and communication. Owing to the imperfections in these markets, this task is usually considered difficult. Throughout the control phase, Lichtenthaler suggests distinguishing between the identification of information needs, the information generation, the information evaluation, and communication.

To conclude, Escher's (2005) external technology deployment model and Lichtenthaler's (2006a) external technology commercialization model were presented. Both process models provide approaches for managing the external exploitation of technologies. Although the processes are somewhat similar, they also bear certain differences. Most obviously, Lichtenthaler defines two phases that accompany the whole process and specifies explicit subphases. However, in his process description, Escher (2005) includes certain key decisions that have to be prepared and made. However, similar to the ETE strategies, neither of the two models has accounted for the integration of TMI, but rather indicates bilateral, direct transactions between technology owners and prospective buyers.[11] This is in line with the traditional perception of transactions, assuming that technologies are highly specific assets that need an idiosyncratic TGS with individually designed contracts taking place infrequently.

3.4 OBSTACLES PREVENTING EFFICIENT TRANSACTIONS

Aside from the currently observed growth of the MfTI and the various approaches proposed for the management of transactions, evidence proves that the transactions and the MfTI are still hindered by many obstacles (barriers, inhibitors, etc.). On firm level, the obstacles limit the incentives to engage in transactions due to high transaction costs. Consequently, low profitability expectations result in a transaction challenge on firm level.[12] On a market level, these obstacles prevent MfTI from clearing efficiently eventually leading to or being a symptom of market failure. In order to understand whether TMI and the novel transaction models they develop can contribute to a better functioning of the MfTI, I discuss some major obstacles in the following section. First, I present a collection from the scattered evidence, from which I identify three major obstacles that I discuss further.

As pioneers in this field, Ford and Ryan (1977) provide an attempt to explain why ETE opportunities are not realized by many firms and mention various obstacles. According to Ford and Ryan (1977: 370) unrealized opportunities 'may be due to the supposed difficulties of handling the marketing of an intangible product compared with the tangibility of the normal manufactured product. It may also be caused by the difficulties of recognizing a potentially marketable technology among those possessed … by the firm'. Aside of these conceptual, managerial difficulties, Caves et al. (1983) analyze international technology transfers to foreign countries[13] in terms of licensing and provide an argumentation that MfTI, like other markets for intangible knowledge, are susceptible to market failures resulting from five prevalent obstacles. First, they observe that, typically, only very few firms are willing to license a technology they possess, while, on the other hand, the demand of firms that feel a specific need for a certain technology is limited (Contractor, 1981). These few available 'pairs' lead to small-numbers bargaining conditions on the market. Thus, a 'thin' market is in contrast to what Roth (2008: 286) refers to as an essential condition for functioning markets, that is 'to work well, marketplaces need to provide thickness – that is, they need to attract a sufficient proportion of potential market participants to come together ready to transact with one another.' Second, Caves et al. (1983) argue that the different parties involved in a transaction have asymmetrical access to knowledge of the technology, which leads to opportunistic behavior. Third, technologies that are transacted usually still require certain developments in order to be fully utilized so that they work properly at any new location for whatever reason (for example, missing tacit knowledge). Therefore, the technology's economic performance usually remains uncertain at the time of the transaction. Fourth, aside from these specific obstacles, Caves et al. (1983) argue that the actors involved in any transaction usually act risk averse. Since a transfer of a technology usually involves uncertainty of whether the technology will perform as promised, a transaction may threaten the participants due to necessary financial investments. Finally, the preparation and contact costs involved in the transaction can be substantial. Referring to Teece (1977), Caves et al. (1983) find that these costs are between two per cent and 59 per cent (average 19 per cent) of the recipient's total transfer fees, thus transaction costs reduce the attractiveness to engage in technology transactions. In addition, Teece (1986) notes that there are 'difficulties in pricing an intangible asset', due to the unique nature of the transacted good.

When the issue of technology trade started to receive increasing attention by scholars at the end of the 1990s and in the early 21st century, scholars identified and discussed various difficulties involved in technology trade. Throughout their study of technology trade, Arora et al. (2001) identify various reasons why MfTI are inefficient. Major difficulties include the problems of valuing intangible assets without the presence of a market and absence of standard valuation approaches, the context dependency of each and every technology, the stickiness of information, and the opportunistic behavior of the market actors. These difficulties lead to high transaction costs, thereby lowering firms' profit opportunities and their willingness to engage in transactions.

In addition, Lichtenthaler (2004) mentions the only use here (OUH) syndrome, which is present in several firms to a certain degree, due to conflicting internal departments' interests in the resource allocation process. On the other hand, firms that acquire technologies often face the not invented here (NIH) syndrome,[14] which reduces their incentives to embed an external technology into own products or processes. Studying the MfTI primarily in Japan, Chesbrough (2006: 146) finds that there is 'no information standard for technology licensing and associated IP trade'. Without a standard that provides the terms and conditions to trading technologies and IP assets, it is difficult to compile statistics on technology trade. Without these data, it is hard for firms to know what technology is available on the MfTI and what their price ranges are. Additionally, it is very challenging to know how to value available technologies. Many other scholars have also reported on this obstacle. For instance, Troy and Werle (2008: 3) note that 'a commonly accepted method of determining a patent's value is missing'. Furthermore, Chesbrough (2006) lacks a systematic reporting of the previous prices paid for external technologies.[15] To some extent, several other studies have also mentioned obstacles for ETE and technology transactions (for example, Escher (2005), Lichtenthaler (2004), Cesaroni et al. (2004), Chesbrough (2003a), Gambardella (2002), Davis and Harrison (2001), Teece (2000), and Teece (1998a)).

Summarizing the available evidence, I found that three categories reflect most of the obstacles mentioned above.[16] These include various types of uncertainty[17], information asymmetry and high transaction costs. These three categories are intertwined and can hardly be distinguished from one another.

Troy and Werle (2008) are among the few scholars who have discussed uncertainty in detail. Troy and Werle (2008: 3) reveal that 'knowledge markets are far from functioning smoothly… [and are]… confronted with several obstacles, which can be characterized as different facets of uncertainty.' According to Troy and Werle, the different types of uncertainty are included in various stages of the ETE process, including the creation of innovative knowledge, its transformation into a fictitious knowledge commodity (patent), its uniqueness, the strategy of transaction partners, the estimation of the future market potential of final products (based on the patent), and the general problem of incomplete and asymmetric information. Troy and Werle identify two conceptual types of uncertainty. Referring to Dosi and Egidi (1991), they argue that, if fundamental or substantive uncertainty is present, the actors lack all the information necessary to make decisions with predictable outcomes and that this information is unavailable, even in principle, at the time when the decision is to be made. Referring to Schelling (1997) and Akerlof (1970), Troy and Werle (2008) further argue that the other type of uncertainty is of a strategic nature and relates to situations of asymmetric information where certain actors hold information which is unavailable to others that can elicit moral hazard and other forms of defective action.[18] Referring to Beckert (1996), Troy and Werle (2008) also argue that from a functioning market perspective on the exchange of goods, uncertainty has detrimental effects unless 'expectational structures' and 'social devices' emerge which help actors cope with uncertainty in decision-making processes. For the MfTI to overcome this uncertainty, Troy and Werle (2008: 13) argue that 'the efficient functioning of a market requires transparency, homogeneity of goods and complete information on the side of the actors involved in transactions.' In such a market, economic risks would either be completely absent or easy to calculate and absorb. Under these circumstances, actors could anticipate the outcome of their decisions and rationally allocate resources to achieve the most preferred outcome.

In terms of the second category, information asymmetries appear to play a prominent role in preventing efficient market clearing. Information asymmetries are commonly studied in the context of principal-agent problems. In industrial organization and contract theory, information asymmetry transactions are often studied within the context of decisions in which one party has more or better information than the other. Information asymmetry then leads to an imbalance of power in transactions which increases the risk that transactions go awry. Examples of this problem are adverse selection and moral hazard.[19] In adverse selection models, influentially defined by Akerlof (1970), the actor lacks information while negotiating an agreed understanding of or contract to the transaction, whereas in moral hazard, the ignorant actor lacks information about the performance of the agreed-upon transaction or lacks the ability to retaliate for a breach of the agreement.

Akerlof defines two primary solutions to this problem ('the market for lemons'): signaling and screening. Spence (1973) proposes the idea of signaling, that is that, in a situation with information asymmetry, it is possible for people to believably transfer information to the other party and subsequently resolve the asymmetry (signaling). Stigler (1961) pioneered the theory of screening where the under-informed actor can induce the other actor to reveal its information. They can provide a menu of choices in such a way that the choice depends on the other party's private information. However, a general rule applies when information asymmetry is present. According to Howard (2005: 25) 'the greater the information asymmetry between buyers and sellers, the greater the demand for an independent and trusted intermediary'.

In terms of the third category, in the context of transactions on the MfTI – at least to my knowledge – there is only limited evidence that attempts to quantify the concept of transaction costs. However, many scholars have argued that these costs are currently still too high. Hence, for many firms, the expected profits from a transaction are too low to constitute an incentive to proactively pursue transactions, even in spite of the enormous potential of the MfTI. Williamson (1981) identifies three major sources of transaction costs. First, the presence of uncertainty has long been recognized as a disturbing factor in the determination of equilibrium market relations. Williamson combines this environmental condition with the subjective condition of agents' bounded rationality (Simon, 1959). In particular, due to limited computational abilities, as well as information overload, uncertainty is never fully resolvable. Second, a source of transaction costs presents the problem of asset specificity. When a heterogeneous asset is involved in an exchange relationship, the two actors become tied to one another. The condition of perfect and competitive re-contracting, which was prevalent before the exchange had taken place, has since changed. Third, transactions costs rise when opportunistic behavior is present. Opportunism is related to those strategic considerations that dominate self-interest seeking behavior with 'guile'. Strategic manipulation of information and self-disbelieved threats and promises regarding future conduct, may be the source of advantages to some of the actors. Relying only on promises may expose a contract execution and renewal to hazards (Williamson, 1975).

Having reviewed selected research on obstacles inherent in transactions on the MfTI (that result in a transaction challenge for firms) and brief discussions on three major obstacle categories (uncertainty, information asymmetry, and transaction costs), I agree with Lichtenthaler (2007: 242) who points out that 'many imperfections inherent in the markets for intellectual property resulted in the absence of a well-defined demand and supply'.

However, the prevalent obstacles might actually represent a reason for the emergence of TMI. DeGennaro (2005: 107) argues in line with Stigler (1951) that 'market imperfections... generate profit opportunities for entrepreneurs. Institutions or individuals that can lower costs that trace to imperfections have a competitive advantage and can earn economic rents – at least until competing firms adapt'. Accordingly, one can argue that TMI have recently started to recognize the various transaction obstacles leading to high transaction costs and attempt to develop new transaction models in order to reduce transaction costs, thereby supporting firms to overcome the transaction challenge. Reduced transaction costs could ultimately lead to higher profit expectations of firms and thus constitute an incentive to become active on the MfTI when transacting technologies.

NOTES

[1] On this topic other reviews exist wherefore I keep this section short and rather refer the reader to the more detailed elaborations by for example Hentschel (2007) and Escher (2005: 35ff.).

[2] Boyens (1998: XXX) applies a similar definition according to which 'external exploitation of technologies is the planned transfer of technological knowledge from one legally and economically independent company to another and it involves a contractual obligation for compensation, either monetary or in terms of knowledge transfer' (own translation).

[3] Inter-organizational refers to inter-firm transfer as one of five categories of technology transfer (international, regional, cross-industry, inter-firm, and intrafirm) proposed by Khalil (2000: 343f.).

[4] A discussion on the tacit and explicit nature of knowledge is provided by Polanyi (1966).

[5] Lichtenthaler (2006a: 15) defines external technology commercialization (ETC) or external technology exploitation as 'an organization's deliberate actions of commercializing disembodied technological knowledge to another legally and economically independent organization involving a contractual obligation for compensation in monetary or non-monetary terms'.

[6] As previously mentioned, I focus primarily on the organization of technology transactions. Therefore, he refers primarily to the technology management literature, although related topics are also covered in the licensing literature.

[7] With reference to transaction cost theory, the degree of organizational integration relates to the concept of vertical integration. Highly integrated strategies are subject to internal (hierarchical) governance structures, while strategies involving other actors are rather subject to market oriented governance structures, particularly if third party firms are involved to which the actors outsource certain tasks (disintegration).

[8] The previous passage is adopted from Tietze (2004) and relates to Granstrand (2000a).

[9] Czarnitzki et al. (2001) applies a similar notion and differentiated intermediaries that support the ETE process directly or induce an indirect ETE process.

[10] The choice of payment schemes depends on various factors and has been extensively discussed and analyzed for different market constellations by, for example, Bessy et al. (2008),

Lemarié (2005), Sen (2005), Poddar and Sinha (2002), Kamien and Tauman (1986), and Katz and Shapiro (1986).

[11] One exception is that Escher (2005) briefly mentions that brokers (as a type of intermediary) can be identified as partners in the process.

[12] Despite the importance and widely discussed market imperfection, I could not find an applicable typology. To my knowledge, DeGennaro (2005: 117) is among the few who have attempted to systematically 'describe the incredibly broad array of market imperfections'.

[13] See also Mansfield (1994) for a discussion of technology transfer to less developed countries and foreign direct investments with an extension of the discussion provided in Mansfield (2000).

[14] See for instance Lichtenthaler (2006a), as well as Katz and Allen (1982).

[15] I strongly follow this argument, which is also advocated by Hayek (1975), who argues that only the price information on a market represents the necessary information to lead to efficient resource allocation.

[16] Missing a solid empirical typology, I neither aimed to provide a complete overview of these obstacles nor did I claim that these three categories are comprehensive or exhaustive in covering all the obstacles that exist in IP transactions. Referring to DeGennaro (2005: 110) 'there are millions of imperfections and no structure can hope to be complete'. However, I believe that the three proposed categories cover the major obstacles.

[17] Arrow (1962) points out that transactions concerning information cause particularly complex problems. With regard to the trade of technology, he considers these problems complicated due to a high level of uncertainty (about the economic value of invention and the successfulness of the transfer, among other concerns). For elaborations on the concept of uncertainty, which is more generally found in the economic literature, see for example Laffont (1989), Borch and Henrik (1968), and Alchian (1950).

[18] The problem of moral hazard originates from the insurance industry, which means that an economic actor behaves differently if it is insulated from risk than when it is exposed to the risk. See further Arrow (1963), Arrow (1968) and, in connection with licensing, Shapiro (1986).

[19] An example of adverse selection is when people who are at high risk are more likely to buy insurance, because the insurance firm cannot effectively discriminate against them, usually due to lack of information about the particular individual's risk but also sometimes by force of law or other constraints. An example of moral hazard is when people are more likely to behave recklessly after becoming insured, either because the insurer cannot observe this behavior or cannot effectively retaliate against it, for example by failing to renew the insurance.

4. Technology Market Intermediaries

4.1 THE ORIGIN OF THE INTERMEDIARY CONCEPT

According to van Lente et al. (2003: 251) referring to Smits (2002), the 'growing importance of intermediaries in innovation processes relates to a key trend of the last decades to involve more actors and factors in R&D activities'. As auction firms represent a particular type of technology market intermediary (TMI), this study requires a deeper understanding of this actor type. The intermediary concept originates from financial economics. The New Palgrave Dictionary of Money & Finance defines intermediaries according to Tobin (1992: 77f):

> Enterprises in the business of buying and selling financial assets... are not just middlemen like transactioners and brokers whose main business is to execute transactions for clients...[They] do much more than participate in organized markets...[by] adding 'markets' that would not exist without them...[and they] do take risks.

To gain a better understanding of the role of intermediaries, Sauermann (2000) proposes a helpful typology. This typology is grounded in the following argumentation that the share of variable transaction costs are usually lower in highly organized markets than less organized markets. Therefore, a higher organizational degree causes additional (mainly fixed) costs. The optimal ratio of fixed and variable costs is determined mainly by the properties of the traded asset (for example, its degree of standardization and divisibility) as well as of the market actors (for example, the number of actors, their degree of professionalism, and transaction frequency). Closely related to this issue is the configuration of a management system to govern these markets efficiently. Until such a system for regulating credits and debits on accounts is established, markets cannot enable decisions without prohibitive transaction costs via high information efficiency. Sauermann (2000) concludes that, in such situations, intermediaries can serve to develop organized markets and transaction systems (for example, stock exchanges).

In his typology, Sauermann considers four types of intermediaries common in financial markets. First, he argued that intermediaries can act to integrate geographically disconnected markets. On the one hand, efficiency gains result from a lower degree of fixed organizational costs and, on the other hand, from a higher degree of informational transparency, economies of scale, and diversification. Second, according to Sauermann, the monitoring function is carried out by, for example, rating and news agencies. Economic transactions are usually associated with risk and uncertainty. However, investors need to assess and monitor the risk if they want to insure themselves against negative incentives resulting from information asymmetry (monitoring costs). Referring to Eichberger and Harper (1997), Sauermann (2000) argues that these costs can be extremely high, especially for small investors due to the risk of partial market failure. Thus, intermediaries can specialize in the continuous monitoring of risks and opportunities. They might be able to carry out these activities at significantly lower costs by realizing economies of scale and learning their effects. As a third function, intermediaries may carry out transformational functions (for example, as a broker or investment bank). Banks, for example, may serve their clients by shifting funds that are accounted for in the financial statement with certain properties into funds with other properties. For instance, Fabozzi et al. (1994) argue that debts might be activated so that they can be accounted for as own liabilities, particularly related to batch size, risk, and terms. As a fourth function, intermediaries can integrate the abovementioned activities (for example, investment banks and insurance firms) and offer a full service bundle to reduce the total costs. Intermediaries might integrate in order to avoid reduplication and incentives to produce higher quality through internalizing certain defects to create a stronger bargaining position against a capital seeker.

Levine et al. (2000: 32) test and confirm the impact of intermediaries on market developments. They constitute that financial intermediaries 'ameliorate the economic consequences of informational asymmetries, with beneficial implications for resource allocation and economic activity'. Their results support the growth-enhancing view of financial intermediaries represented by, for instance, Hamilton (1781) and Bagehot (1873). According to Hamilton (1781: 36) banks are the 'happiest engines that ever were invented for spurring economic growth' and within the context of innovation, Schumpeter (1912) already argued that intermediaries (for example, banks) spur on economic growth.

4.2 TECHNOLOGY MARKET INTERMEDIARIES AND INNOVATION

In the context of the innovation process, different types of intermediaries have been described by various scholars. However, most authors do not refer to the intermediary concept, but rather use different labels, for example, 'patent brokers' (Benassi and Di Minin, 2009), 'brokers' (Hargadon and Sutton (1997), Provan and Human (1999)), 'bridgers' (Bessant and Rush (1995), McEvily and Zaheer (1999)), 'superstructure organizations' (Lynn et al., 1996), and 'third parties' (Mantel and Rosegger, 1987). Only a few scholars refer to these firms as 'innovation intermediaries' (Lopez and Vanhaverbeke, 2010), 'information intermediaries specifically associated with the exchange of information' (Popp, 2000) or 'intermediary firms' (Stankiewicz, 1995). I do not review all of these studies, but rather review two often cited studies by Bessant and Rush (1995), and Howells (2006) as well as a recent contribution by Benassi and Di Minin (2009).[1]

Bessant and Rush (1995) provide insights into the role of consultants as a particular type of intermediary with a variety of sub-types. According to Bessant and Rush (1995: 98), consultants support the 'multi-dimensional character of technology transfer' processes, which they define 'as non-linear and characterized by multiple interactions, systems integration, and complex networks'. Thus, according to Bessant and Rush (1995: 98), technology transfer is not an 'instantaneous event but a time-based process involving several stages… [and]… involving multiple actors'. These transactions may not always take place on a 'one-to-one [basis] but may also be one-to-many or many-to-many'. Furthermore, the transactions 'may not proceed directly but may often operate through various forms of intermediary'.

Based on the elaboration of the properties of the technology transfer process, Bessant and Rush propose that clients have certain needs in the technology transfer process, during which consultants offer their support as well as bridging activities. Accordingly, consultants support their clients for a variety of generic purposes. To build up certain capabilities, consultants may advise and inform their clients to enable the development of key management capabilities in identifying needs, exploring and selecting innovations, planning, implementation, as well as project management. For 'institution building', these schemes also offer an opportunity for developing strategic capabilities across the supply side, for instance, mobilizing a critical mass of technological knowledge and skills in support of particular technologies. Consultancy services may further help to avoid failures. Providing targeted advice and direct technical and managerial support offers opportunities to reduce the incidence of costly investment failures by means of transferring better innovation management practices, for instance by selecting appropriate applications of new technologies and in project management. Consultants can provide less expensive innovation support through information and an advisory service than their clients can provide internally. Using consultants as intermediaries opens up the possibility of reaching user firms more directly than traditional financial support mechanisms which tend to lack focus and often fail to reach many potential users within a target group. Using consultants can enable a more decentralized mode of operation involving less monitoring and control. Once the broad objectives of a governmental program are set out, it could be largely self-managing, with overall monitoring and quality assurance provided by a small and specialist group of consultants contracted solely for a specific project monitoring role.

Furthermore, Bessant and Rush (1995) provide an overview of consultants' specific functions. Consultants support clients who deem it necessary to articulate specific needs for technologies and throughout the selection of appropriate options. Furthermore, consultants support the clients in the identification of needs, the selection of, as well as the training and development of skills and human resources. Consultants provide financial support to make a business case serving as financial sources (for example, as venture capital funds). Consultants may further support their clients in the identification and development of the business and innovation strategy. Using examples of best practices, consultants may further provide education and serve as a link to external knowledge systems, for example, the identification of knowledge sources in terms of new knowledge for emerging technologies. Consultants may finally serve as specialist resources and provide project management throughout the implementation of external sources, for example, new technologies. Having discussed this, Bessant and Rush present an 'indicative typology' based on five dimensions. Consultants can offer services ranging from expert to process and from sector-specific to general. Consultancy firms can be small 'one man shows' or large, multidisciplinary firms. They can apply specific technologies (for example, total quality) or rather be generalists. Their background can be rather traditional or linked to fairly new phenomena (for example, information technology).

From a review and synthesis of prior literature Howells (2006) develops a typology and framework of the different roles and functions of the intermediation process within innovation and operationalizes the typology within the context of the UK. Howells conducts a set of case studies in the UK that involves semi-structured interviews with managers from the 22 organizations (plus their eight subsidiary firms) based on specific project collaborations, together with their overall strategies and work practices. Throughout the case studies, Howells (2006: 720) applies the following working definition for an innovation intermediary:

> An organization or body that acts as an agent or broker in any aspect of the innovation process between two or more parties. Such intermediary activities include: helping to provide information about potential collaborators; brokering a transaction between two or more parties; acting as a mediator or 'go-between', [between] bodies or organizations that are already collaborating; and helping find advice, funding, and support for the innovation outcomes of such collaborations.

After conducting case studies based on the above working definition and the understanding of innovation intermediaries he had gained from the literature review, Howells (2006: 720) was surprised to find 'considerably more functions than originally conceived'. He identified ten functions, including (i) foresight and diagnostics; (ii) scanning and information processing; (iii) knowledge processing and combination / recombination; (iv) gatekeeping and brokering; (v) testing and validation; (vi) accreditation; (vii) validation and regulation; (viii) protecting the results; (ix) commercialization; and (x) the evaluation of outcomes.

Most recently, Benassi and Di Minin (2009) apply a case-study approach using a snow-ball sampling technique throughout interviews with industry experts. Having interviewed 13 intermediaries, the authors identified seven types of patent brokers along two dimensions (degree of value added and own risk). These include enforcers, aggregators, deal makers, technology promoters, shields, evaluators and consultants, of which the last four are the most prominent. According to Benassi and Di Minin, brokers engage in licensing transactions through (i) IP asset evaluation; (ii) market identification and selection; and (iii) negotiation.[2]

4.3 INTERMEDIARIES AND TECHNOLOGY TRANSACTIONS

To better understand the intermediary concept in the specific context of technology and technology transactions, I discuss the limited and fragmented literature dealing with the roles of intermediaries, specifically in this context from Lien (1979), Czarnitzki et al. (2001), Birkenmeier (2003), Krattiger (2004), and – from a practitioner's point of view – Millien and Laurie (2007).[3]

Lien (1979) defines four functions of the 'middlemen' in the technology transfer process. Middlemen may determine specific opportunities in terms of specific needs, that is to be guided primarily by 'market pull' rather than 'technology push'. In contrast to traditional technology 'shopping', where the buyer chooses technologies among the ones available on the 'shelf', intermediated transactions involve a detailed description of the client's needs. In addition to working as a salesperson when the client has a technology offer, middlemen perform the procurement task when a need is identified. As a second function, Lien proposes that middlemen help to identify appropriate sources of technical breakthroughs, scientific information, and other technological developments that will meet their clients' needs. Once the client's needs are identified, the middleman makes use of his expertise and networking resources to fulfill such needs, which entails an active rather than passive approach. Furthermore, according to Lien, middlemen build bridges between the sources of technologies and those who are able to make use of it. When the two parties of the transaction are identified, the middleman links them through proper presentation and an explanation of how beneficial such transactions will be for both ends. Finally, besides building bridges, middlemen encourage appropriate linking mechanisms and provide other services, skills, and inputs to accelerate the transaction. Thus, in addition to the tasks of an agent or broker, middlemen work as catalysts, providing specialized expertise of intellectual property in the form of supporting services.

Although their study focuses primarily on university technology transfer, Czarnitzki et al. (2001) provide an interesting typology. They differentiate between direct and indirect transactions. Certain intermediaries support transactions directly. These services include consulting and the researching of certain information, as well as offering training services to firms to build up their own competences. Additionally, Czarnitzki et al. identify intermediaries that conduct own R&D and thus add value to technologies. According to Czarnitzki et al., these intermediaries participate in direct transfers. TMI that support transactions indirectly, offer services related to the bridging of the supply and demand side, for example, by providing commercial exploitation of R&D results, services for patent analysis, and technology scouting.

Based on their analysis of problems in the transaction process, Czarnitzki et al. are among the very few scholars who propose 'points of departure' for intermediaries, that is suggestions for how intermediaries can help to develop the MfTI (although primarily focusing on university technology transfer). Although their model is neither exhaustive, nor very detailed or free from overlaps, it is however one of the very few and should be mentioned here. To solve information asymmetries, intermediaries can provide platforms for technology owners to market their technologies, for example, on the internet or in exhibitions. Intermediaries can further consult prospective buyers regarding technologies offered on the market and monitor important trends and the demand for certain technologies. To overcome problems related to high costs for firms interested in acquiring a technology, intermediaries can act as a bridge between supply and demand, can carry out certain searches, prepare reports (for example, due diligence[4]), and offer possibilities for directly contacting firms (for example, seminars, workshops, fairs). To reduce high transaction costs, intermediaries may offer consulting regarding contract design and project management. To reduce uncertainty regarding externalities, intermediaries may facilitate the development of trust between the various actors, carry out 'specific' transaction related tasks, and offer financial support when spinning off firms. Regarding the reduced transfer possibility, intermediaries may offer training courses, create incentives, offer consultancy in innovation management, and support the development of R&D laboratories with strong incentives for transactions.

In addition, Birkenmeier (2003) identified four main functions, which intermediaries support.[5] Accordingly, intermediaries can provide information services regarding technological applications, market data, industries, firms and competitors in terms of existing technological knowledge as well as certain funding sources. Second, intermediaries consult firms regarding their innovation and technology management. Third, intermediaries may support firms in terms of their patent applications, licensing contracts, entrepreneurship, and human resource development. Finally, intermediaries support firms in their project management.

Aside from the very few systematic attempts to develop a typology, there are a few other publications that rather provide lists of various intermediary types. In the following section, I mention a few to illustrate the variety of existing intermediaries, which also illustrates the early stage of the development in the literature.

Krattiger (2004) provides a list of specific intermediaries without discussing their functions and only providing some properties of each intermediary group. According to Krattiger, intermediaries serve as royalty collection agencies, as various types of clearing houses (information, technology, and open-source innovation), as brokers and other types of facilitators, and provide IP management services (law firms and consultants). Furthermore, intermediaries act as commercialization agents, merchant banks or develop patent pools. From a practitioner point of view, Millien and Laurie (2007) provide another collection of various types of intermediaries. These include patent licensing and enforcement firms, institutional patent aggregators / IP acquisition funds, IP / technology development firms, licensing agents, litigation finance / investment firms, patent brokers, IP-based mergers and acquisition advisory, technology auction houses, online IP / technology exchanges / clearinghouses, IP-backed financiers, royalty stream securitization firms, patent rating software and services, university technology transfer intermediaries, as well as some recently 'emerging business models' that include transaction exchanges / trading platforms, defensive patent pools, technology / IP spinout financing, and patent-based public stock indexes.[6]

To conclude, within the specific context of technology and IP transactions, various types of intermediaries have been identified and discussed, but a comprehensive, systematic and widely accepted typology is still absent. With the increasing emergence of TMI, firms will have to understand the properties of the novel transaction models that involve TMI. Firms and IIP managers will have to understand how these models impact the TGS of technology transactions and how they can best utilize the different types of TMI to manage transactions more economically for the creation of innovation. With the ongoing emergence of new types of TMI as well as new transaction models, firms will face the increasingly complex decision of choosing a suitable TMI and its transaction model as an 'exploitation channel' for each technology exploitation project. Technology auctions are among the novel models developed by TMI which I focus on in the remainder of this study.

4.4 DEFINING TECHNOLOGY MARKET INTERMEDIARIES

Based on the previous chapters in which I elaborated on the origin of the intermediary concept and reported results from prior studies of intermediary roles for the creation of innovation, I propose a definition for technology market intermediaries (TMI) which was applied in this study.

TMI are understood as firms that offer knowledge-based services, sometimes labeled 'knowledge intensive business services' (KIBS).[7] Although TMI might purchase technologies from independent inventors, I assume that most of their daily business focuses on business to business related transactions. I suggest adjusting the general definition of intermediaries to specifically focus on intermediaries acting on the MfTI. I thus propose the following TMI definition:

> TMI are for profit, risk taking firms attempting to advance the MfTI through the development of models that facilitate transactions of intangible assets, supporting innovation processes through interaction with primarily technology-based firms, predominantly without adding value or holding property of the asset.

The scope of this definition should be considered as rather narrow, in order to capture the new phenomenon of intermediaries. Hence, this definition also relates to the concept of brokers and dealers that have been present on the MfTI for a long time (for example, labeled 'specialized engineering firms' by Arora et al. (2001)). However, the emphasis is on novel transaction governance structures (for example, auctions) or the use of information and communication (ICT) technologies (that is the internet), for instance the online technology market place Yet2.com. Consequently, this definition excludes traditional patent law firms (PLF), as their concept is rather loosely related to the broker concept as discussed by, for instance, Hargadon and Sutton (1997), that is firms that 'execute transactions for clients.' Although PLF can be considered KIBSs and engage in transactions offering mainly case-by-case services (for example, designing idiosyncratic contracts), TMI are concerned with the development of novel transaction models. Nevertheless, it is important to note that major PLF have recently started to enlarge their portfolio of activities by including novel services and, for instance, have started to staff management positions. Such firms are included in the definition, while I exclude traditional PLF offering only legal services (for example, patent application and litigation support).[8]

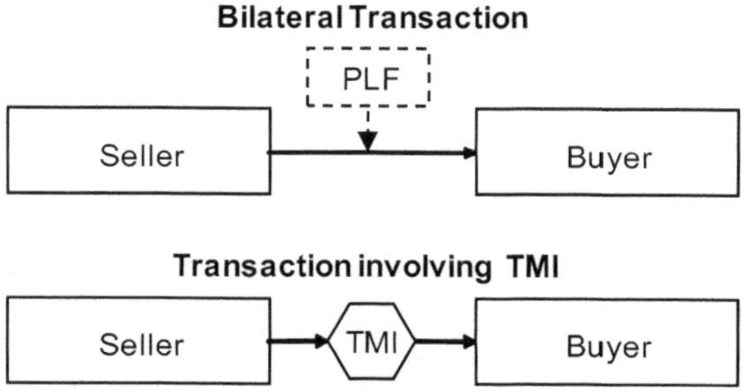

Figure 4.1 - The transaction nature without / with TMI involvement

Figure 4.1 illustrates the difference between bilateral transactions, in which sellers and buyers interact directly, but are supported by PLF. In contrast, the lower illustration in Figure 4.1 shows a transaction in which a TMI is involved in – and maybe even taking ownership of – IP assets.

Accordingly, when TMI become involved in the transaction, one can expect the transaction governance structure to change.

My rather narrow definition of TMI further excludes government support vehicles and technology transfer offices (TTO) of universities due to their different governance structure and incentive schemes. However, my definition and the implications of this study could be of relevance to them. My definition includes the concept of 'non-producing' (or practicing) entities – sometimes referred to as 'trolls' – as mentioned for instance by Reitzig et al. (2010), Hall (2009), and in recent practitioner literature.

NOTES

[1] For a more detailed review and a proposal of an own typology, see Lopez and Vanhaverbeke (2010), Mortara (2010), as well as Tietze (2010).

[2] Applying a broader definition, Lopez and Vanhaverbeke (2010) also propose a typology with four categories, including 'innovation consultants', 'innovation incubators', 'innovation mediators', and 'innovation traders'.

[3] The intermediary concept is further discussed in the specific context of IP transactions by, for example, Pollard (2006), who discusses the role of technology transfer intermediaries in the processes of innovation in networks involving universities, research and development centers, and business firms from an international perspective; Fu and Perkins (1995), who identify tasks that involve the help of 'external experts' (including evaluation, training, financial negotiations, legal / technology / strategy negotiations, tailoring technology, approaching, and identifying prospects); and Mittag (1985), who discusses various forms of specialized intermediaries. Moreover, Granstrand (2004a) discusses the specific situation when large multinational firms spin-off their own licensing departments to act as independent brokers and dealers.

[4] Due diligence is a routine practice with mergers and acquisitions. It covers the process of contacting, gathering information and auditing the object of purchase with a view to conclude an agreement (Parr, 2006). As with mergers and acquisitions, in the case of patent transactions, Rivette and Kline (2000: 169) suggest 'rigorous and informed due diligence of IP assets'.

[5] Please note that Birkenmeier (2003) does not provide a detailed account of how he developed this typology; neither does he provide any detailed explanations for these functions.

[6] For an extensive discussion of patent-based financing models, see de Vries (2011).

[7] Also see, for example, Aslesen and Isaksen (2007), Flikkema et al. (2007), Koch and Stahlecker (2006), Strambach (2001), Den Hertog and Bilderbeek (2000), Hipp (1999), and Miles et al. (1995).

[8] For instance, Zacco A/S, a northern-European leading IP firm, has recently started to adapt their business model. Zacco regards itself not anymore 'as a traditional patent firm, but as a modern Intellectual Property Consultancy ...[and]... offers a full range of portfolio management services, including validation of granted European patents' to its clients.

5. Auctions for Technology Transactions

5.1 INTRODUCTION TO AUCTION THEORY

Auctions are among the oldest negotiation mechanisms for price formation; they appear in various fields and in multiple contexts to trade a wide range of assets. Traditionally, among the most prominent assets traded in auctions are art, antiques, and wine.[1] However, in recent years, auctions have been applied for an increasing range of tangible (for example, second hand industrial machinery exported to third world countries, real estate properties, etc.[2]) as well as intangible assets, including, for example, various types of services (for example, craftsmen services for house repairs, travel services, software developing services, and problem solving services) or IP assets (for example, patents and 3G licenses).[3] Recently, auctions have been developed by TMI to facilitate technology transactions which firms might use for the more efficient creation of innovations. In the context of ETE, Birkenmeier (2003) was among the few scholars to label auctions a 'market-based approach' to determine technologies' values in addition to other approaches (for example, existing market prices, and exact utility-based approaches).

Although auctions are among the oldest and most widely used trading mechanisms, their theoretical foundations have just recently been established. According to Berz (2007), Milgrom (2005), Klemperer (2004), and Vickrey (1961) auction theory was largely derived from game theory. The widespread use of auctions as a trade or a negotiation mechanism for price formation, has drawn scholars' attention to this field and led to the foundation of a new field in economics some fifty years ago. According to Berz (2007) and Klemperer (2004), the theoretical foundation of auctions, that is auction theory, has only been present since the second half of the 20th century. With his widely recognized publication, Friedman (1956) is often considered one of the seminal contributors to auction theory. Since then, auction theory has been established as a special case of bargaining theory, which in turn appears to be a sub-discipline of game theory (Berz, 2007).

Auctions are commonly differentiated according to the direction of the trade, that is whether sellers offer assets (sales auctions) or whether buyers want to purchase assets (procurement auctions). However, according to Klemperer (2004: 15), there are 'no formal distinctions between normal auctions, in which the auctioneer is the seller and the bidders are buyers who have values for the object(s) sold and procurement auctions, where the auction firm is a buyer and the bidders are sellers who have costs of supplying the object(s) bought.' The same argumentation often applies when just mirroring the attributes. However, with regard to this study's focus on the seller's perspective, that is firms exploiting technologies, I mainly refer to sales auctions in the following section.[4] Furthermore, this study focuses on public auctions, thus private auctions in closed environments are excluded from the study's discussion (see, for instance, the CommerceOne case mentioned above). Private auctions have been used for technology transactions even before the public technology auction model was developed; however, data have been notoriously difficult to access due to confidentiality matters. Furthermore, auctions can be differentiated according to whether they take place offline or online. However, my empirical analysis focuses on offline auctions. Moreover, as multiple lots are usually offered within an auction, the following question arises: Should lots be offered sequentially or simultaneously (that is in parallel)? This empirical analysis focuses solely on sequential auctions.

5.1.1 Fundamentals of Bargaining Theory

Aiming to clear the market and reach an agreement between sellers and buyers (that is agree on a sales price that lies within both price windows of the seller and buyer), technology transactions can be interpreted as bargaining situations. In particular, the ways in which technology transactions are traditionally conducted – which are commonly designed on a case-by-case basis (for example, with idiosyncratic contracts) – must be regarded as bargaining negotiations that take place primarily bilaterally between sellers and buyers although they are sometimes supported by a PLF.

As auction theory is part of bargaining theory, this chapter provides a brief introduction to the main concepts of bargaining theory as the mechanism behind any auction. Among the large body of literature on bargaining theory (for example, Muthoo (2002), Napel (2002), Singh and Bhatia (2000), Coles and Wright (1994), Canning (1989), Sutton (1986), as well as Stahl and Economic Research Institute (1972)), I refer primarily to Muthoo (2002), who provides a valuable introduction to the fundamental principles of bargaining theory.

I follow Muthoo's (2000: 146) definition of a bargaining situation. Accordingly, any 'exchange situation, in which a pair of individuals (or organizations) can engage in mutually beneficial trade but have conflicting interests over the terms of the trade can be regarded as a bargaining situation'. Put differently, a bargaining situation is a situation in which two or more actors (for example, an individual person or an organization, such as a firm, a political party or a country) have a common interest to cooperate, but have conflicting interests over exactly how to cooperate. Thus, the main issue that confronts the actors in a bargaining situation is the need to reach an agreement over exactly how to cooperate (that is to agree on a price within the price windows of both parties).[5]

One can assume that each actor, subscribing to the rules of the bargaining ex-ante, wants to reach an agreement, which is as favorable as possible to him or her. The bargaining process is typically time consuming and involves actors making offers and counter-offers. A main focus of any bargaining theory is the efficiency and distribution properties of the outcome of bargaining. The efficiency property relates to the possibility of the actors failing to reach an agreement in the agreed time window. The distribution property relates to the issue of exactly how gains from cooperation are divided between the actors (resource allocation). To phrase it differently, the actors have a common interest to reach an agreement over the partition of a surplus, but have conflicting interests over the exact partition (in that each actor would like to obtain as large a share of the surplus as possible).

Muthoo (2000) proposes five principles of bargaining theory that should be regarded as key concepts which impact any bargaining outcome (and its efficiency and distribution properties). As such, these concepts determine the sources of an actor's bargaining power. These principles refer to the impatience of the actors, risk of breakdown, options (outside and inside), commitment tactics, and information asymmetries.

During most bargaining situations, each actor values time, that is prefers to reach an agreement on any particular price today rather than tomorrow. An actor's value of time, however, depends on a variety of factors (for example, personal circumstances, such as income and wealth and market interest rates). Thus, it is very likely that each actor values time differently. Hence, although actors usually have an incentive to compromise and reach an agreement without delay, the higher the price at which the trade occurs, the less *impatient* a prospective buyer appears to be relative to the others. Thus, an actor's bargaining power can be assumed to be greater the more patient he or she is relative to the other negotiators. From this principle, one can draw the conclusion that an actor might try to decrease own 'haggling cost' and / or increase the other negotiators' haggling costs to enhance his or her own bargaining power. Patience during the process of negotiations confers bargaining power, while risk aversion affects it adversely.

While bargaining, the actors may perceive that negotiations *break down* into disagreement due to some exogenous and uncontrollable factors, for example, when an 'intervention' by a third party diminishes the gains from the cooperation that exists between the actors. For instance, while two firms bargain over how to divide the returns from a new technology, an outside firm may discover a superior technology that makes their technology obsolete. Muthoo concludes that the higher an actor's bargaining power is, the higher his or her profit (or payoff) will be following the occurrence of the exogenous and uncontrollable factor that triggers a breakdown in the negotiations. Similarly, the higher an actor's payoff is, the lower the other actor's bargaining power will be in that eventuality. According to Muthoo the amount of profit is related to the degree of the actor's risk aversion. The more risk averse the actor is, the more eager he or she will be to minimize the risk of a breakdown. This is exploited by the less risk-averse actor who will demand a larger share of the net surplus. If both parties are equally risk averse and equally impatient, they are likely to split the net surplus equally between themselves.

Third, actors' bargaining power depends on their other purchase options. If an actor has relatively small (low) *outside options*, this concept has very little impact on the bargaining price (more precisely, they are too insignificant to be beneficial). However, if an actor faces an outside option that promises to be beneficial, this concept has an impact, that is the price has to equal it. Thus, an actor's outside option might increase the bargaining power if the outside option is sufficiently attractive. The logic behind this conclusion is linked to the issue of credibility (or otherwise) of threats in the bargaining process. This outside option principle indicates that an outside option does not necessarily increase the bargaining power and, thus, negotiators should not allow themselves to be influenced by empty threats (or promises) – in the sense that threats and promises would not be carried out when and if the time came to do so. An actor's outside option only enhances bargaining power if it is sufficiently attractive and therefore credible. If both actors' outside options are sufficiently attractive, then it is likely that there will be no gains from cooperation (and that the parties may thus prefer to exercise their respective outside options). Provided that all actors' outside options are not attractive enough, actors' bargaining power will increase in accordance with how large (high) their *inside options* are, that is the payoff that the actor obtains during the bargaining process (that is, what is the reason for the actor's negotiations or temporary disagreement?). When both actors' outside options are unattractive, the actor with the more attractive inside options will have greater bargaining power. However, when one actor's outside option is sufficiently attractive, both actors' inside options have no impact on the bargaining outcome; the actor with the more attractive outside option gets the more favorable transaction and, if both actors' outside options are sufficiently attractive, then it is mutually beneficial for the actors to exercise them.[6]

Muthoo (2000) further reports that, in many bargaining situations, the actors often take actions prior to and / or during the negotiation process that partially commit them to certain strategically chosen bargaining positions (or demands). Although such *commitment tactics* are usually revocable, revoking such commitments (that is backing down from a demand) can be costly. If – and only if – the costs of backing down from a demand are sufficiently large, commitment tactics enhance actors' bargaining power. Commitment tactics are most effective when a negotiator is bargaining on behalf of a constituency. However, strong commitments with substantial backing down costs risk that the bargaining process will end up in a stalemate.

In bargaining situations, the outcome is further commonly affected by the presence of *information asymmetries* between the actors. A main consequence of information asymmetry is that an agreement may not be reached – when, in fact, it would be mutually beneficial in the absence of complete information – if one or more actors in the negotiation possess information on relevant variables (or factors), which the other actors do not. This in turn leads to inefficient bargaining outcomes, that is, outcomes with disagreements and / or costly delayed agreements, unless the information is credibly communicated to the uninformed actor(s). Knowledge is veritable power in negotiations and enhances the bargaining strength of those who are better informed.

According to Muthoo (2000), the above principles apply to various bargaining situations, while various others are just as important in determining bargaining outcomes. These include the procedures involved in and format of negotiations (particularly the initiative of making offers, rather than responding to them), actors' reputation (their strength as well as their resolve and staying power, etc.), populism, public interest litigation, and judicial activism that may paralyze negotiations unless bargaining is undertaken in complete confidentiality and the actors are well aware of the multifaceted social, economic, political, and other aspects of a bargaining situation and the eventual need for accountability and transparency. With regard to technology transactions – as previously mentioned – information asymmetries are regarded as an obstacle on the MfTI (see Chapter 3). It remains to be seen what technology auctions can contribute in order to achieve improved bargaining outcomes and resource allocations on the MfTI.

5.1.2 Determinants for Auction Choice

Although related to bargaining negotiations, this study is specifically concerned with auctions, which are subsequently discussed in this chapter. According to Kjerstad (2005: 1241), auctions can be considered markets with 'explicit trading rules that specify precisely how market clearing determines prices' that need to be designed properly to assure optimal resource allocation (Roth (2008), Milgrom (2005), Klemperer (2004)). In contrast to bargaining negotiations, auctions induce an element of competition, which is an essential theoretical criterion. Bulow and Klemperer (1996) state that, according to the theory of optimal auctions, the seller is treated as a monopolist who can choose any mechanism, such as establishing a minimum sale (or reserve) price, to maximize their expected profit. As in monopoly theory, optimal auction theory assigns all bargaining power to the seller, subject to the constraint that the seller has no access to the buyers' private information on their asset's value. According to Kjerstad (2005), auctions often serve as a response to asymmetric information problems. Milgrom and Weber (1982) prove that, in general, sellers can raise the expected price by adapting a policy of providing expert appraisals of the quality of the objects on sale, regardless of the auction type. In the absence of market prices, a supplier can also use auctions to extract the minimum reserve price (MRP) of prospective buyers. However, an auction requires strong commitment from all actors that a contract will be signed, in case the auction is successfully completed following the auction rules specified ex-ante. Additionally, although an auction can even take place with just one bidder, usually two or more bidders should participate to provide a real competitive situation. Moreover, the quality of the bidders matters. Therefore, bidders might be pre-selected or the number of allowed bidders could be restricted by the auction firm (Berz, 2007).[7]

Firms planning to exploit a technology externally need to understand in which situations (that is for which technologies) they should prefer auctions as 'exploitation channels' over bargaining negotiations. In the following section, selected results are reported that examine this issue, that is the choice between auctions and bargaining negotiations.

Goldberg (1977) was among the first scholars to investigate under which circumstances auctions or bargaining should be preferred. He generally argues that, for non-standard complex transactions bargaining, negotiations should be preferred, since auctions can prevent the exchange of important pre-contract information. Manelli and Vincent (1995) develop a framework in which the buyer of an asset cares about both quality and costs. In this two-dimensional framework, Manelli and Vincent show that, when quality concerns become strong enough, negotiations are better than an auction. Bulow and Klemperer (1996) use a standard auction model to show that, in most cases, a seller should prefer using a simple (no MRP) auction to the best possible bargaining negotiation with one less buyer. Within their often cited paper, Bulow and Klemperer are of the few scholars who mathematically prove their finding. However, Rosenkranz and Schmitz (2007: 4) find that Bulow and Klemperer's (1996) results do not 'hold in the presence of a reference point', that is an MRP, which was not used in their model. In a complementary study that builds on the theoretical work of Manelli and Vincent (1995), Bonaccorsi et al. (2003) conducted an empirical analysis of auctions and bargaining as alternative procurement mechanisms using data on Italian hospitals' procurement of medical devices. Bonaccorsi et al. were able to confirm Manelli and Vincent's (1995) results, namely that quality concerns affect the choice of award mechanisms. From an OLS regression analysis of hospital procurement contracts, Kjerstad (2005: 1241) found evidence that 'compared with negotiations, auctions do not result in significantly lower prices compared to the outcome of negotiating contracts'. Surprisingly, this finding was in contrast to the results of, for example, Milgrom (1989). Kjerstad (2005) argues that both auctions and bargaining are likely to involve some form of ex-ante and ex-post costs for preparation and contract administration, for instance. Therefore, bargaining costs are likely to increase with the number of negotiators and thus need to be compared with the preparation costs associated with auctions. Kjerstad further reports a possible argument in favor of bargaining, that is the potential for determining the most preferred quality before any transaction is closed. Auctions may well have a higher probability of missing quality targets, although bidders might still be able to test products and conduct due diligences prior to the bidding. In a recent study, Leffler et al. (2007) gathered data from private sales of timber tracts in North Carolina, US. Referring to Goldberg's (1977) and Bajari et al.'s (2007) results on the effects of complexity, Leffler et al. (2007) show a positive relationship between their measures of complexity and the use of bargaining negotiations and proved that more available buyers make the use of auctions more likely. Using a discrete choice econometric model by means of a study on building contracts in Northern California, US from 1995 to 2001, Bajari et al. (2008) report three results: First, more complex projects (that is where ex-ante information is valuable and ex-post changes could be anticipated) are more likely to benefit from bargaining negotiations than auctions. Second, the use of auctions is counter-cyclical, consistent with the increased benefits of auctions when more bidders are available, that is in environments that favor more bidders auctions are preferred. Third, negotiation outcomes appear to be biased towards larger, more experienced contractors. Furthermore, Bajari et al. question the standard assumption of well-defined products, which is central to the mechanism design and auction literature. When ex-ante information is valuable and when ex-post changes are anticipated, the use of auctions, which often requires fixed-price contracts, may be inefficient.

Concluding on the above review of selected results, I find that, across the various studies, the results appear consistent to a large extent although the results from the review appear limited. In contrast to bargaining negotiations I can only draw conclusions on results regarding three determinants for the choice of auctions. Manelli and Vincent (1995) as well as Goldberg (1977) prove that auctions should be preferred for products where quality is of little importance. Bajari et al. (2008), Leffler et al. (2007), and Bulow and Klemperer (1996) prove that auctions should be preferred when the number of prospective buyers is relatively high. Bajari et al. (2008), Leffler et al. (2007), Kjerstad (2005), and Goldberg (1977) find that auctions are less preferred when the project is highly complex, that is ex-ante information is important and ex-post changes can be expected. With respect to the various existing auction types (and also controversial results from some studies), the results should still be regarded as insufficient. This opinion is also shared by the scholars who conducted the studies reviewed above. According to Bajari et al. (2008: 2), the question of whether auctions lead to revenue maximization and, if so, under which circumstances still 'received considerably less attention' and Kjerstad (2005: 1240) notes that 'surprisingly, little empirical work has been done to compare the relative merits of auctions versus negotiations'.

Thus, to develop my hypotheses, I could draw only on limited available results from the auction literature. Hence, although I do not apply the auction theory perspective, the results and their discussions of technology auctions – in contrast to bargaining negotiations – should contribute to this topic and support future research.

5.2 AUCTION PROCESS DESIGN

As outlined in Chapter 2, one of this study's objectives is to generate an understanding of the transaction governance structure (TGS) of technology auctions. To generate an understanding of the TGS, I apply insights from transaction process models discussed in the management literature which are complemented by insights from the auction design literature presented below.[8]

In practice, there is a wide variety of auctions. For instance, auctions can apply different bidding procedures (for example, English ascending) and implement specific design concepts (for example, MRP). Various scholars have reviewed different auction types and propose taxonomies for categorizing them. For instance, Friedman (1993) and Engelbrecht-Wiggans (1980) organize auctions into a hierarchical taxonomy and Wurman et al. (2001: 304) propose a 'parameterization of the auction design space… that captures the essential similarities and differences of many auction mechanisms in a format more descriptive and useful than simple taxonomies'.

I do not review the literature on auction design or the wide range of design concepts, but rather elaborate on the concepts that are relevant to this study. In the following section, a generic auction process model is presented as proposed by Kumar and Feldman (1998). To discuss and assess the performance of any auction, it is important to understand the design of the bidding procedures as a major element in any auction process with various design concepts involved (for example, price development, price logic, and submission of bids). Subsequently, I explain the four most prominent bidding process formats (English, Dutch, first-price sealed-bid, and Vickrey auctions) of which the English bidding procedure was used predominantly in the observed technology auctions. Next, I discuss the nature of auctioned assets' different properties that are commonly examined in the auction literature. This issue is relevant to the hypotheses that are developed to determine the preference for technology auctions as exploitation channel in contrast to bargaining negotiations (see Section 9.3).

5.2.1 A Generic Auction Process Model

In the auction literature, the bidding procedure is often conceptualized as the auction itself. However, according to Milgrom (2005: 251), the 'auction itself … is just one part of the transaction, [and its] success … depends even more on what happens before and after the auction'. Thus, in this chapter, I present a generic auction process model that encompasses different phases before and after the bidding procedure.

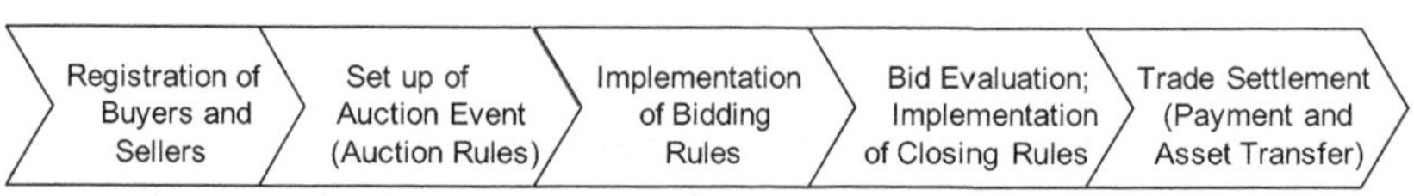

Notes: Own illustration based on Kumar and Feldman (1998)

Figure 5.1 - Schematic view of a generic auction process

As outlined above, auction processes are commonly conceptualized as specific types of bargaining negotiations. As with all negotiations, participants (that is buyers, sellers, auction firms, brokers, etc.) exchange messages to modify a negotiable transaction, from which they eventually reach a final transaction with conditions that can be accepted by the seller and buyer, within a defined set of rules. In contrast to bargaining negotiations, auctions often involve multiple actors among either buyers or sellers or both, which induces an element of 'competitive bidding' into the negotiation process (Bajari et al., 2008). There are different conceptual models which describe auction processes. For the purpose of this study, Kumar and Feldman's (1998) conceptual model appears to be suitable, since it characterizes the transaction process along phases similar to the transaction models of Lichtenthaler (2006a) and Escher (2005). As illustrated in Figure 5.1, the auction process model proposed by Kumar and Feldman (1998) comprises five phases.

The *first phase* includes the authentication and initial registration of the auction participants (particularly prospective buyers and sellers). The *second phase* relates to the set-up of the auction event. This phase includes particularly the description of the offered assets, the definition of the rules of the auction and parameters for negotiation (including bid control rules, fees, delivery dates, options to purchase more or less and contractual issues including the terms of payment). With regard to the auction rules, five particular issues are of importance. The first issue is the selection of the auction type, respectively the bidding procedure. As discussed below there are different types of bidding procedures and, within each method, there are several variations, thus selecting the best suitable bidding procedure can be a challenging task. The second issue relates to the scheduling of the auction event. According to Kumar and Feldman, one of the crucial success factors of an auction firm is to bring the maximum number of prospective buyers to the auction event. Alerting prospective buyers comprises the third issue in this phase. Kumar and Feldman recommend that prospective buyers preregister with the auction firm and indicate their interest in any item being scheduled for an auction so they can be alerted when an auction is scheduled, which is of interest to them. A fourth issue relates to security considerations, for example, to ensure that information on prospective buyers as well as details on the items offered remain secret. Finally, the auction firm can restrict access to the auction. The auction house policy and the instructions from the seller dictate whether an auction is accessible to the public at large (open auction framework), to the buyers / sellers registered with the auction services or only to specific invited buyers / sellers, following a selection process (closed auction framework).

The *third phase* is specifically concerned with the design of the bidding procedure being a key element in the auction process. Again Kumar and

Feldman mention five relevant issues. First, it is important to develop a notification mechanism to update bidders on the latest bidding actions. This issue is easily overcome in classical offline auctions in which the bidders are physically present. However, if bidders are not present in the bidding room (for example, absentee bids via telephone or internet), they need to be informed about the developments of the bidding. Notification possibilities include phone or internet connections, which are either permanently held by each bidder for the duration of the auction or absentees can be manually updated upon client request. Second, the auction firm needs to make the terms and conditions available to sellers and buyers. If the bidding process allows the buyers to request changes of the payment or shipping terms, these terms should be treated as part of the bid. The auction chart (commonly presented in an auction catalogue) should display the terms and conditions offered alongside the bids shown. Furthermore, when creating the product description, the seller should specify the range of acceptable terms and conditions and indicate how they are factored into the bid evaluation. Third, the auction firm has to decide on the deposits for the bids. Depending on various issues, a deposit may be required or a bond may have to be posted before a bidder can participate in an auction. Throughout the bidding, certain security measures should be considered. Particularly in sealed bid processes, the auction firm has to ensure that a submitted bid is not disclosed to other bidders in violation of the auction rules. In open outcry auctions, spurious bids injected by the seller or auction firm to prompt the highest bidder to further increase his or her bids, should be prevented. Fourth, the retraction of bids needs to be defined. During the bidding phase, under certain conditions, the seller may be allowed to stop or withdraw the auction or modify the rules. Similarly, under certain conditions, the bidders should be allowed to withdraw or modify their bids. Finally, defining the closing of the auction is an important issue. Normally, the auction would close according to the closing rules specified. At that time, the winning bid can be treated as and if needed translated to, traditional purchase orders.

The *fourth phase* relates to the evaluation of the bids with notification of bidders particularly with regard to the closing rules of the auction. Two issues are of particular relevance. First, the auction firm must communicate the results of the auction to the bidders. Depending on the auction policy, some information will be made available publicly, some common information will be made available to all bidders and some will be communicated only to bidders to whom the information is relevant. Once again, security and privacy rules might be needed. If the auction firm is not the seller itself, the auction results must be communicated to the seller as well. The auction results comprise the winning bidder's or top few bidders' names and addresses, the bid amounts and the shipping and payment terms if the bidding process allowed changes to these terms. Second, the auction firm must prove to the bidders and the seller that the auction was conducted fairly (record retention). The seller needs to maintain it for his own internal bookkeeping requirements and to prove that government rules and regulations, such as fair trade practices, were met. The record retention policy will vary and will also depend on the monetary value of the transaction and commodity being sold. The records usually include the product description, when and where it was posted and for what period. It could include the list of bidders who participated and the log of the bids. The evaluation method should also be retained as part of the records. Finally, the *fifth phase* is concerned with the trade settlement and includes the payment to the seller, the transfer of goods to the buyer, and – if the seller is not the auction firm – payment of commission fees to the auction firm.

5.2.2 Bidding Process Design

Within the third phase of Kumar and Feldman's (1998) auction process model (although already determined within the second phase), the design of the bidding procedure is of central importance to the outcome of an auction. A central element for maximizing the outcome of an auction is the price negotiations in the form of the bidding procedure, which starts with the 'opening bid' and ends with the 'final bid'. However, the final bid may not necessarily lead to a successful transaction if its value is below a declared minimum sales price determined by the seller.

In the literature, there are various taxonomies defining various types of bidding procedures (for example, Wurman et al. (2001), Friedman (1993), and Engelbrecht-Wiggans (1980)). In the following section, I differentiate between four characteristics of bidding procedures.

The *first property* is related to the development of the price during an auction. In ascending auctions, the bids increase with every bid being made, while in descending auctions, each following bid is lower than the previous one.[9] The intervals by which the bids ascend or descend can be either fixed by the auction firm (for example, through a ticker) or can be left open for the bidder to decide. In the latter case, the bidder decides by how much to raise / lower the bid, although often a minimum interval (for example, 10%, €1) is defined in the auction rules.

A *second property* is the price logic for which there are two main categories. Either the winner has to pay the final bid price (first price) or the winner has to pay the second highest bid price (second price). First price auctions have been criticized for being biased. They do not reveal the buyers' indifference prices or the purchased items' true market values because the winning bid needs to be just marginally higher than the second highest bid in order to win. In order to improve the outcome of the auction, Vickrey (1961) suggests second price auctions. In second price auctions, the winner pays the second highest bid rather than the highest. Thus, bidders are incentivized to bid closer to their indifference price, which optimizes the outcome of the auction.[10]

A *third property* is related to the ways in which bids are submitted. Bids are either submitted as open bids, thus bids from all bidders are known to the other bidders, or closed or sealed bids. Unlike sealed bid auctions, 'open outcry' auctions are 'open' or fully transparent as the identity of all bidders is disclosed to each other bidder during the auction. There is also a mixed form, in which case the auction firm announces all the different bids but without disclosing the respective bidder's name.

A *fourth property* to distinguish bidding procedures is the duration of the auction that can be set according to a time limit or by the number of bidding rounds. A particular distinction has to be made between the cases in which bidders are allowed to bid only once and in which they may bid multiple times. Auctions can last one or several rounds, and can last anything from minutes, weeks or even sometimes months. The auction firm can set a maximum duration of the auction or rather determine maximum durations for the single rounds. The auction firm can also decide to set breaks between the rounds. Breaks can be preferred when the bidders need time to rethink their indifference prices depending on additional information gathered during the bidding, for example, knowing the bids of the other bidders or at least the announcement of anonymous bids by the auction firm after closing a round.

Table 5.1 - Four common bidding procedure types

	English auction	Dutch auction	First-price sealed-bid auction	Second-price sealed-bid auction (Vickrey)
Price development	Ascending	Descending	Sealed	Sealed
Price logic	Second price	First price	First price	Second price
Submission of bids	Open	Open	Sealed bid	Sealed bid

Klemperer (2004) claims that four basic types of bidding procedures are widely used and analyzed; the ascending-bid auction (also called the open, oral or English auction), the descending-bid auction (also called the Dutch auction), the first-price sealed-bid auction, and the second-price sealed-bid auction (also called the Vickrey auction).[11] I describe these four auction types according to the first three properties discussed above. The fourth property is dismissed, since all its specifications can be equally fitted to all bidding procedure types. English auctions are applied largely by the technology auctions (see Chapter 9).

In the *English auction*, the price is successively raised until only one bidder remains and that bidder wins the object at the final bid price. Thus, it involves an iterative process of adjusting the price in a direction that is unfavorable to the bidders. Submitted bids are commonly visible to all bidders. This auction can be run by having the seller announce prices or by having the bidders call out (open out-cry) prices themselves, or having bids submitted electronically with the best current bid posted. While the highest bidder pays the final bid, an English auction is termed second-price, since the winning bidder only needs to outbid the next highest bidder by a minimum increment. Thus, in effect, the winner pays an amount equal to, or slightly higher than the second highest bid. Two additional rules are often applied to this setting. One rule does not allow bidders to step into the bidding again, if they have dropped out once. The other rule, which is commonly applied, is that bidders are not allowed to make large 'jump bids'. Antiques and artworks as well as houses are commonly sold using versions of the ascending auction. Different types of ascending auctions are discussed, for instance by Bikhchandani and Riley (1991).

The *descending or Dutch auction* works in the opposite way. The auction firm initially indicates a price for the asset for sale, which is substantially higher than any bidder is willing to pay and then lowers the price continuously over a set time span, commonly using a 'clock'. The first bidder who calls out to accept the current price wins the asset at the bid price, thus the price paid is the first-price. Classical examples of applications of Dutch auctions are flower auctions in the Netherlands, fish auctions in Israel or tobacco auctions in Canada (Katok and Roth (2004), van den Berg et al. (2001)).

The *first-price sealed-bid auction* is a derivativ of the first-price auction. Each bidder independently but simultaneously submits a single bid, without seeing others' bids and the object is sold to the bidder who makes the highest bid. No bidder knows the bid of any other participant. The winner pays the 'first-price' bid. As bids remain anonymous, bidders cannot adjust their own bids accordingly. First-price sealed-bid auctions are commonly used for auctioning mineral rights in government-owned land. In the *second-price sealed-bid auction*, also known as Vickrey (1961) auctions, each bidder independently but simultaneously submits a single bid, without seeing the others' bids. The asset is sold to the bidder who makes the highest bid, however paying the second-highest bidder's bid. In practice, the price paid is often that of the lowest winning bidder. According to Lucking-Reiley

(2000), English auctions that employ proxy-bidding systems (for example, on websites) often resemble Vickrey sealed-bid auctions. This auction type is used less commonly in practice.[12] A rare example of Vickrey auctions is stamp auctions by mail.

In addition to the different types of bidding procedures, specific attention needs to be drawn to an additional property that can be applied to all types of auctions and which has a substantial impact on the auction outcome according to Ockenfels et al. (2006), Dewally and Ederington (2004), and Bajari and Hortacsu (2002). The MRP can be regarded as the minimum value of the asset for which the seller is willing to sell to a bidder – thus the indifference price of the seller or, put differently, the seller's minimum willingness-to-sell. If the bids do not reach the level of the MRP, the asset remains unsold and is said to be 'bought in'. The MRP can be regarded as a security mechanism in case of weak competition to avoid that an asset is sold below a minimum price. The MRP thus represents the seller's floor price. Sellers are commonly offered the option to choose whether they want to set an MRP for which they will be charged an extra fee. If an MRP is set, bidders are usually informed about the fact, although not necessarily about its amount. Whether the amount of the MRP is revealed usually depends on the auction firm or the auction design.[13]

While a revealed MRP serves a similar function as a minimum bid, a secret MRP functions as an invisible, additional bidder that bids up to the level of his or her independence price.[14] Whether the MRP should be kept secret (for example, an announcement could discourage some bidders from participating) or revealed to the bidders has been discussed controversially by Horstmann and LaCasse (1997), Vincent (1995), Ashenfelter (1989), as well as Milgrom and Weber (1982). This decision depends on a variety of factors. Foremost, it depends on the design of the bidding procedure, that is whether first or second price auctions are used. Bajari and Hortacsu (2002: 26) find that 84 per cent of sellers in their dataset were reluctant to use a secret MRP, which they perceive to be attributed to the fact that bidders were 'frequently confused and frustrated'.

When sellers or auction firms want to use an MRP, they are confronted with the challenge of having to understand the effects of the MRP on the auction outcome and to determine its optimal level. However, according to Englmaier and Schmöller (2008), there is little evidence on when sellers use MRPs, how sellers set MRPs and when they prefer disclosed or anonymous MRPs.[15]

According to Englmaier and Schmöller (2008), there is mixed evidence on the effect of an MRP on sales prices. While Kamins et al. (2004) as well as Ariely and Simonson (2003) find positive effects of MRPs on sales prices, Hoppe and Sadrieh (2007) as well as Bajari and Hortacsu (2002) find no effect at all. Bajari and Hortacsu (2002) find that sellers tend to limit the use of MRPs to high value assets. It appears that the MRP increases the complexity for the sellers as well as for the buyers and thus transaction costs, which in turn bears a risk for certain asset types. Various studies have discussed how to determine an optimal MRP, for example, Rosenkranz and Schmitz (2007), Reiley (2006), Myerson (1981), and Riley and Samuelson (1981). However, their results can hardly be generalized. Instead, one can generally conclude that an MRP is associated with a 'reference-point effect'. Thus, a higher MRP can have a positive impact on the bidders' willingness-to-pay. However, raising the MRP above a certain ceiling usually increases the probability that trade will not take place at all (Rosenkranz and Schmitz, 2007). Klemperer (2004: 7) notes that the failure to set a proper MRP 'increases the incentives for predation and may encourage collusion that would not otherwise have been in all bidders' interests'. In general, one can only conclude that the MRP needs to be credible (that is not too high, or just high enough to explore an assets market value by letting the bidders reveal their indifference prices). According to Klemperer (2004: 112), if an MRP is 'not a genuine commitment to not sell an object, if it does not reach its reserve, then it has no meaning and bidders will treat it as such'.[16]

5.3 THE NATURE OF THE ASSET'S VALUE

Auctions are used to trade a wide range of assets, as outlined in the introduction to this chapter. Traditionally, auctioned assets include, for instance, art, antiques, and wines while recently auctions have become applied for an increasing range of assets (for example, second-hand industrial machinery, real estate properties, travel and software developer services, and 3G licenses). Recently TMI have also applied auctions to facilitate technology transactions on the MfTI. Complementary to Chapter 6, which discusses technology properties derived from the management literature, this chapter discusses asset properties that I consider relevant in the auction literature. Primarily based on standard auction literature (for example, Berz (2007), Ashenfelter and Graddy (2005), Milgrom (2005), and Klemperer (2004)), I found that auctioned assets can be differentiated according to four criteria. Three relate to the asset and one particularly to the asset's value with one effect worth noting, the 'winner's curse'.

First, assets can be differentiated according to their *physical properties*, that is whether they are tangible (for example, houses, cars, and machinery) or intangible (for example, services, technologies, or other IP asset types).

The second criterion refers to whether assets can be distinguished as *homogeneous* or *heterogeneous*. Homogeneous assets have uniform properties. Thus any unit of it is interchangeable with any other, thus, such items are qualitatively identical or at least viewed as identical by buyers (for example, agricultural products, metals, and energy goods) (Black, 2002). In contrast, heterogeneous assets differ in their specifications or quality (for example, bear different brand names which affect customers' confidence). Heterogeneous assets might be considered special, if there are only few substitutes.[17] This criterion can apply either to the asset in general or to its current condition. While the first refers, for example, to machinery specifically tailored for a manufacturing system of a particular firm, the latter refers to whether an asset can be considered brand-new (that is sold directly from the OEM) or used (for example, after depreciation, a firm can resell machinery on the secondary market to third world countries). In an extreme case, if there is no substitute, a special asset could be considered unique with regard to the amount of units produced (that is being equal to 1) or with respect to its current condition.[18] If an asset is not unique, various amounts of it might be auctioned. Auctions where various 'lots' contain and trade the same asset are commonly regarded as multi-unit auctions.[19] The tulip auctions in the Netherlands are a prominent example for this type of auction. In contrast, by definition, unique assets can only be traded once and are thus regarded as single-unit auctions.

A third criterion is related to an asset's *relation to other assets*. Assets are usually distinguished according to whether they are complementary to other assets (that is they function or work more efficiently with other assets) or independent (that is they can be fully utilized solely – without the need for other assets) (Berz, 2007: Chapter 3.2).

The fourth criterion refers to the *uncertainty of the value* of an asset. While certain assets have low uncertainty with regard to their market price (for example, homogeneous assets based on a long trade history), other assets lack a reliable market price (for example, heterogeneous assets with a short trade history). Thus, the value concept of auctioned assets frequently further distinguishes between assets that have a 'private value' or assets that have a 'common value'. For *private value assets*, it is assumed that the asset has a specific and different value to each bidder. While each bidder knows the specific value an asset has to him or her, the bidders do not know the value of the asset to the other bidders. Thus, each bidder has an own indifference price. Bidders are assumed to behave competitively; therefore, such auctions are commonly treated as non-cooperative games among bidders (Milgrom and Weber, 1982). According to Milgrom and Weber (1982: 1093), the private value assumption is 'most nearly satisfied in auctions for non-durable consumer goods'. They argue that the 'satisfaction derived from consuming such goods is reasonably regarded as a personal matter, so it is plausible that a bidder may know the value of the asset herself and may allow that others could value the asset differently'. According to Ashenfelter and Graddy (2002) and Goetzmann and Spiegel (1995), private value assets are commonly assumed in art auctions . In contrast, *common value auctions* constitute a market setting in which participants may be particularly susceptible to judgment failures that affect market outcomes (Kagel and Levin, 1986). Common value auctions take place in an environment in which information on the value of the asset for sale is dispersed among bidders, that is when an asset has the same value to all bidders, although none may know the exact value. Unlike the private value models, in common value auctions, any bidder's indifference price depends on the values of the other bidders. Any information that a bidder receives on the value estimation of the other bidders will be considered in the determination of the own indifference price. According to Milgrom and Weber (1982), the common value model is often applied by the government when selling mineral or oil rights. However, in practice, neither pure private nor pure common value auctions take place. There is usually a common value element to most auctions. For instance, bidders for a painting may want to purchase it for their own pleasure (a private value element), or they may also bid for investment and eventual resale purposes (a common value element).[20]

Furthermore, in common value auctions, the *winner's curse effect* may appear.[21] According to Kagel and Levin (1986) as well as Lorenz and Dougherts (1983), the winner's curse, a disequilibrium phenomenon, can be defined as a judgmental error of the highest bidder's systematic overestimation of the asset's value.[22] Assuming that all bidders make unbiased estimates, the bidder with the largest estimate will win. If, on average, bidders are estimating correctly, the highest bid will tend to have been placed by someone who overestimated the asset's value. Thus, the winner will have overestimated (on average) the value of the asset. In particular, weaker bidders (for example, small firms when bidding against large firms) have to be especially cautious, since they must recognize that they are only likely to win when they have overestimated the value by even more than usual (Klemperer, 1998). In contrast, an 'advantaged bidder' can be less cautious, since beating very cautious opponents need not imply one has overestimated the value. Hence, because the winner's curse affects weak firms more than strong ones and it is self-reinforcing, the advantaged bidder mostly wins in such auctions. Furthermore, these bidders generally pay a low price if they win, because their rivals bid extremely cautiously. Of course, the winner's curse effect exacerbates the problem that weaker bidders may not bother to participate in an ascending auction. A strong bidder also has an incentive to create a reputation for aggressiveness, which reinforces his or her advantage. Thus, an ascending auction often effectively blocks the entry of weaker bidders and, at the same time, encourages stronger bidders to bid jointly or to collude; after all, they know that no-one else can enter the auction to seal the collusive rents they create. Lorenz and Dougherts (1983) argue that the winner's curse effect can be corrected given sufficient time and the right kind of information feedback. The winner's curse is ubiquitous and is applied fairly uniformly across individuals at early stages of the learning process (Kagel and Levin, 1986). Auction firms might attempt to counteract the winner's curse by providing sufficient information to the bidders, for example, in data rooms or by means of expert opinions.

Applying these identified criteria from the auction literature to classify patented technologies, they can be classified as follows: With respect to their physical properties, patented technologies represent intangible assets (as outlined in Chapter 6). They must be considered heterogeneous, since the patents that protect their different components are unique by definition. Patented technologies often have a need for complementary assets (either other technical components in complex technologies or discrete technologies might require marketing investments for the promotion of the innovation). Technologies bear a private value element, since they commonly have a value for the acquiring firm, even if it is only to block competitors. However, technologies often bear a common value element, for instance, if a technology is established as an industry standard and could be widely used by other firms.

NOTES

1 For instance, Sotheby's and Christie's are among the world's most famous art auction houses.

2 See, for instance, Azasu (2006) as well as Shenkar and Arikan (2006).

3 Scholars' interest in auctions was largely triggered by the emergence of the internet, mainly because the internet offers the seller access to a wider group of buyers than bargaining negotiations can. Auctions have thus gone through a renaissance recently. Various types of online trading platforms have emerged that address business-to-business relations (for example, set up by automotive OEMs for procuring items from their suppliers) as well as private customers (for example, eBay). According to Lucking-Reiley (2000: 38), the volume of auction re-

lated 'transactions on [the internet, in] the order of $1 billion ... in 1998, is impressive – especially for a market with such a brief history'.

[4] For a reading on procurement auctions, I recommend Bajari et al. (2008).

[5] According to Arévalo (2004: 193), Nash (1950) defines a classical bargaining problem in game theory 'as being a set of joint allocations of utility, some of which will correspond to what the actors would obtain if they reach an agreement and another which represents what they would get if they failed to do so'.

[6] A related concept is the 'Best Alternative to a Negotiated Agreement' (BATNA) which is defined as an alternative action that will be taken should a proposed agreement with another market actor result in an unsatisfactory agreement or when an agreement fails to materialize (Fisher et al., 1991). See further, for example, Tekin-Koru (2009), Peppet (2003), and Brett and Okumura (1998).

[7] Since there are several surveys on auction theory, I do not provide a detailed literature review. These include Kagel and Levin (forthcoming), Kagel (1997), Rothkopf and Harstad (1994), Milgrom (1989), McAfee and McMillan (1987a), as well as Engelbrecht-Wiggans (1980). For a chronological bibliography of selected, prominent contributions to auction theory and design, see Mossetto and Vecco (2002: 21ff). A general bibliography of the literature on competitive bidding has further been collected by Rothkopf and Stark (1979) as well as a collection of articles in Engelbrecht-Wiggans et al. (1983), and the bibliography by Stark and Rothkopf (1979). In addition, Cassady (1980) surveyed the history and practice of auctions.

[8] As previously mentioned, this study focuses on offline auctions, although there are online websites for technology auctions. For studies of online auctions in general, see, for example, Parente et al. (2004), Dewally and Ederington (2004), Bajari and Hortacsu (2002), as well as Geng et al. (2001), and, for a comparison between online and 'live' auctions, see Kazumori and McMillan (2005).

[9] The word 'auction' is derived from the Latin 'auctio,' which means 'to ascend'.

[10] For a detailed reasoning of this mechanism, see, for example, Klemperer (2004).

[11] Descriptions and exemplary applications of different types can be found in various auction survey papers, including Milgrom (2005), Friedman (1993), McAfee and McMillan (1987b), Milgrom and Weber (1982), and Engelbrecht-Wiggans (1980).

[12] Rothkopf et al. (1990) tested seven reasons why Vickrey auctions are so rarely used. They found two main reasons (bidder fear of bid taker cheating and bidder resistance to truth revealing strategies) and rejected five reasons (multiple objects for sale, bidder risk aversion, bidder asymmetry, non-independent values, and inertia). However, among economists, Vickrey auctions are often studied because of their attractive theoretical properties.

[13] In practice, the MRP is often kept secret, as, for example, by auction houses, such as Christie's and Sotheby's (Ashenfelter (1989)).

[14] The MRP should not be mistaken for a 'phantom' bid, which can be used by the seller to raise the bid price, although this only applies to second-price auctions. According to Berz (2007: 120ff.), an MRP has no influence in first-price auctions.

[15] In contrast, there is considerable (although conflicting) evidence of the effect of MRPs on bidding behavior. Since bidding behavior is not subject to this study, I do not report on these. For a similar reason, he does report results from studies that investigated the relationship between the MRP and the opening bid (for example, Bajari and Hortacsu (2002)).

[16] A stronger bidder in an ascending auction has a choice between either colluding to end the auction quickly at a low price or forcing the price up to drive out weaker bidders. The lower the MRP is at which the auction can be concluded, the more attractive the first option is (Klemperer, 2004).

[17] In microeconomic theory, assets that can be substituted for one another at a constant rate are considered 'perfect substitutes.' With regard to substitutes, the economic concept of indifference curves is of importance. For perfect substitutes, the indifference curve is a straight line, that is perfect substitutes have a constant marginal rate of substitution (Katz and Rosen, 1998).

[18] For auctioning specialized assets, Milgrom (2005) recommends that the seller attracts the 'most likely purchasers' and defines targeted marketing for such bidders as 'especially valuable'.

[19] For a discussion on multi-unit auctions, see Ausubel and Cramton (1995 (revised 22 Jul 2002)), Kagel and Levin (2001), Maskin and Riley (1989), as well as Engelbrecht-Wiggans and Kahn (1998).

[20] Hence, if assets are not only bought for enjoyment, but there is an investment component to the buyers' motives; that investment component leads one to classify the value of the asset as having a common-value component (Ashenfelter and Graddy, 2002).

[21] Although, traditionally, auction theory predicted the winner's curse effect only in common value auctions, there are indications that it might be present in private value auctions as well. For instance, investigating whether the 'private value component of art induces buyers to overpay for paintings', Goetzmann and Spiegel (1995: 549) findings indicate that 'the private value return may be significant and positive'.

[22] The winner's curse was first proven by Capen et al. (1971) who claim that this effect is responsible for oil firms' low profit earnings on offshore tracts in the 1960s. As examples of cases where the winner's curse effect occur, Kagel and Levin (1986) and Roll (1986) report corporate takeover battles, outer continental shelf oil leases, professional baseball's free agency market (Cassing and Douglas, 1980), and in auctions for book publications (Dessauer, 1977). The winner's curse is sometimes discussed similarly to the adverse selection or Akerlof (1970) 'lemons' example.

6. Technology Properties

6.1 DEFINING TECHNOLOGY

Schmookler (1966: vii) defines technology as 'the social pool of knowledge of the industrial arts. Any piece of technological knowledge available to someone anywhere is included in this pool by definition' and inventions are 'the "bits" that are added to the existing stock of knowledge'.[1] Delmas (1999: 638), referring to Loveridge (1990) and Burgelman and Rosenbloom (1989), defines technology as a 'set of competencies. It is the practical knowledge, know-how, skills, and artifacts that can be used to develop a product / service and / or production / delivery system. Technologies can be embodied in people, materials, cognitive and physical processes, plant, equipment, and tools. Key elements of technology may be implicit, existing only in an embedded form (like trade secrets based on knowhow) and may have a large tacit component'. Drawing on the knowledge or the resource-based view of the firm, Granstrand (2000a: 548) defines technology in the context of the technology-based theory of the firm as a 'body of knowledge about techniques and technical relationships, typically regarding ways to transform material matter to achieve more desirable physical effects. The body of knowledge referred to may be more or less specialized'. Following the definition proposed by Tschirky and Koruna (1998: 227), Hentschel (2007: 5), Escher (2005: 17), and Birkenmeier (2003: 17) state that 'technologies enclose specific and collective knowledge in explicit and implicit forms for product and process-oriented usage based on natural, social and engineering-scientific knowledge'.[2] OECD (2007: 780) defines technology rather generically as 'the state of knowledge concerning ways of converting resources into outputs'.

However, I consider the rather specific definition proposed by Singh (1997) particularly valuable to this study, since it highlights the combinatory effect of various complementary components which a technology consists of and thus refers to the increasing complexity of technical products. Singh (1997: 340f), building on Simon (1969), defines a complex technology 'as an applied system whose components have multiple interactions and constitute a non-decomposable whole... [with its]... elemental units or components, usually organized in hierarchies of subsystems. Though interactions between individual components are often simple and direct, multiple interactions and feedback between components within subsystems, between components across subsystems, and between subsystems at various hierarchical levels create a complicated network of 'non-simple' relationships ('architecture of a system'). ... It cannot be separated into its components without seriously degrading its capabilities or performance'. In a similar manner Aggarwal and Walden (2009: 27) provide a recent definition of technology that is particularly related to this study, since it explicitly refers to patents (which each represent an invention) as components that are combined into a technology similarly to the technologies studied and operationalized in my quantitative study. Accordingly, they define technology as a set of 'inventions... [that] are protected and owned separately by way of IPR. ...Thus, innovators must coordinate the disparate interests of various IPR owners in order to create useful technology. If some of the elements of the bundle are beyond the boundaries of the firm, they need to be transacted (for example, in-licensed). Sometimes the cost paid for individual components in the bundle can overshadow the value of the technology itself'.

When one understands technology as a bundle of components, the following question arises: Which of these components need to be transferred within a technology transaction? Hence, it is important to understand how the different elements of technologies (that is components and patents) relate to each other. To address this question, the concept of Ewald (1989) appears valuable, since he characterizes technology primarily as a bundle of knowledge components. Ewald [3] argues that different informational carriers bear the technical knowledge embedded in a technology and he subsequently introduces the concept of 'information carrier of knowledge'. Ewald (1989) distinguished three categories of carriers that are illustrated in Figure 6.1.

Ewald categorizes researcher, scientists, manager, staff as well as lawyers as personal carriers. Materialistic carriers comprise prototypes, lab equipment, products, systems, and means of production, while informational carriers comprise patents, licenses, construction drawings, norms, software tools, and various forms of documentation. The concept proposed by Ewald illustrates that patents can be regarded as informational carriers that comprise explicit knowledge codified into written text. As discussed in the literature, this aspect is reflected by the difficulty that trading patents does not constitute a technology transfer that enables a buyer to fully utilize a technology. In addition to the transfer of ownership rights only (for example, patents), buyers often require a further transfer of additional codified knowledge through materialistic carriers (for example, prototypes, technical drawings, etc.) as well as of tacit knowledge transferred primarily through personal carriers (for example, training by lead inventors). The extent to which different carriers are required to enable a buyer to utilize an acquired technology depends on various factors, for example, the technological complexity of the item, its uncertainty and the extent to which the tacit knowledge has been codified (Cowan and Foray, 1997). According to von Hippel (1994), codified knowledge may have low transmission costs and thus be slippery and hard to contain, but that is largely irrelevant if what one needs is its 'sticky', tacit counterpart.[4]

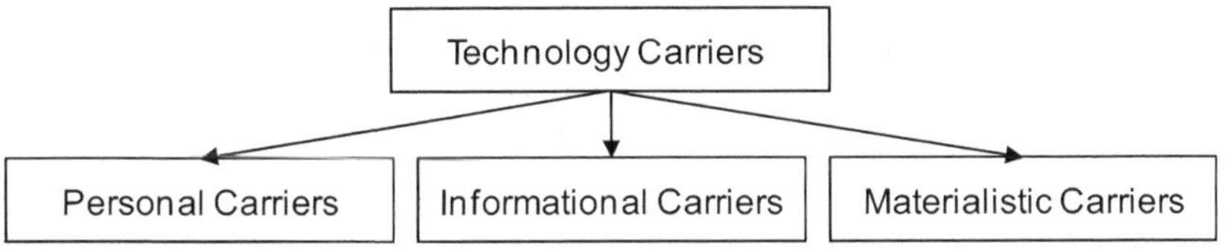

Source: Ewald (1989: 40)

Figure 6.1 - Types of carriers for technological information

Combining the understanding of different knowledge components (explicit and tacit / sticky) with the definition of Aggarwal and Walden (2009) and the concept of Ewald (1989), I define technology in this study:

> A technology is a bundle of inventions that are partly protected by patents, which are regarded as informational carrier comprising explicit knowledge, where each patent represents a sub-system in the bundle of components that jointly enables the utilization of the technology.

Following this understanding, I further elaborate on some specific properties of patents. According to the definition of the EPO as presented by Granstrand (2000a: 71), a patent is a 'legal title granting its holder the exclusive right to make use of an invention in a limited era and time by stopping others from, among other things, making, using or selling it without authorization'. Thus, for its owner, a patent serves as a temporary monopolistic right[5] that protects a technical invention from imitation. Hence, the monopolistic right granted by society functions as an incentive for a firm to appropriate sufficient returns to make R&D investments worthwhile (Levin et al., 1987). Compared to other types of Intellectual Property Right (IPR) (for example, copyrights, trademarks, designs, etc.), patents are often considered the key type of IPR to protect technical inventions as they offer the strongest legal certainty. Once a patent is granted, it is regarded as the intellectual property of the applicant who is free to transfer the ownership or certain uses of it through licenses.[6]

Although a patent provides the patent holder with protection against the commercial exploitation of the invention by others, it is not a legal title granting its holder the exclusive right to exploit the invention. A patent is rather a legal title granting its holder the exclusive right to stop others from using or making the invention. If a third party wants to use an invention protected by a patent, that party needs to acquire ownership of the patent or a license to use it, which is obtained from the owner of the patent.

In almost all jurisdictions, in order to be patentable, a technical invention has to meet three patentability conditions, that is it has to be novel, non-obvious and industrially applicable.[7] Novelty refers to the fact that the patent matter has to be unknown and unpublished prior to the application date. Particularly due to the novelty requirement, any patent is unique by definition. Therefore, according to Williamson (1985, 1979), patents are characterized as having high 'asset specificity' when interpreting them in the context of transaction cost theory. The non-obvious condition refers to the fact that the invention should not have been conceived by someone who has 'ordinary skills in the art' without undue experimentation (inventive step requirement). Industrial applicability refers to the fact that the invention has to be of a technical nature (for example, not a literary text) and that an invention must be industrially applicable. In addition to those three patenting criteria, patent jurisdictions commonly require that the way in which the patent application is written must reveal to someone who is skilled in the technology how to replicate and practice the invention. Patents can be granted for products (for example, parts, applications or chemical substances) or processes (for example, manufacturing procedures).[8]

Furthermore, an important legal property of the patent system is the fact that it is commonly a national system, that is the patent systems of each country allows only to grant a monopolistic right in that respective country. A set of patents (or applications) in different countries (subsequent filings) following the original patent application (first or priority filing) is commonly known as a 'patent family' (OECD, 2009).[9] Hence, inventors need to apply for a patent in each single country where they claim patent protection in order to gain international patent protection. Although harmonization efforts to smoothen the application process (for example, the EPO and PCT application systems) have been ongoing for many years (Granstrand and Holgersson (forthcoming)), no single world patent yet exists. Differences in the national patent systems and procedures can lead to differences in the scope of the protection applied for and granted in the first and subsequent filings. However, as legal aspects of patents are not of primary concern in this study, I rather elaborate on the technical and economic properties of patented technologies in detail in the following.

6.2 PROPERTIES OF PATENTED TECHNOLOGIES

In the following the emphasis is particularly on patented technologies as they represent the specific asset type transacted via the technology auctions that have been subject to this quantitative study (that is that addresses this study's second specific research question to understand which technologies should be auctioned). I focus particularly on the discussion of the properties that are operationalized in the empirical study based on patent data in Chapter 10. The properties discussed in the following include (i) technologies' technological complexity, (ii) technological uncertainty, (iii) technological impact on subsequent technical progress, (iv) technology quality, and (v) technology value, and the (vi) sources from which they originate, that is where they are developed.

6.2.1 Technological Complexity

In Chapter 1, I have argued that the complexity of products in many industries has been growing recently. For instance, modern cell-phone generations – nowadays labeled 'smart-phones' – include not only voice transmission technologies, but cameras, touchscreens, GPS, and wifi modules (Lindmark, 2002). Within the open innovation framework, the understanding of the complexity concept is important because it is closely linked to firms' necessity to manage transactions across firm boundaries (for example, acquire technologies or components that a firm decided not to develop internally in order to accelerate innovation speed).[10] The complexity of technology is also reflected in Singh's (1997) definition of technology (see above).[11]

According to Simon (1962: 468) a complex system is 'made up of a large number of parts that interact in a non-simple way... [and is] composed of subsystems that, in turn, have their own subsystems and so on... [Its organization]... frequently takes the form of [a] hierarchy'. Accordingly, complexity concerns both the number of the parts involved in the system and the nature of the interconnections between those parts (Simon, 1969). In the context of patents, Merges and Nelson (1990) distinguish between discrete and complex (multi-component) technologies. Discrete technologies are characterized by a relatively strong product-patent link (for example, in pharmaceuticals or chemistry), whereas complex technologies are commonly protected by a large number of patents, often held by various parties. In the literature, this distinction has been widely accepted (Hall (2005), Cohen et al. (2000), Kusunoki et al. (1998)). Kogut and Zander (1992) refer to technological complexity as 'the number of critical and interacting elements embraced by an entity or activity'.

Other definitions (for example, by Langlois (2002), Soh and Roberts (2003), Simonin (1999)) refer to similar notions of technological complexity. Singh (1997: 340) definition reflects the major elements of the previous definitions He defines a complex technology as an 'applied system whose components have multiple interactions and constitute a non-decomposable whole'. Singh's definition appears to be particularly valuable to this study, since it summarizes the major elements of complexity but also explains them in a systematic, detailed way. He refers to three aspects of complex technologies, each of which is necessary but insufficient.

First, the 'systemic' aspect implies that a complex technology comprises elemental units or components, usually organized into hierarchies of subsystems. Owing to the hierarchical structure, the performance of each subsystem depends on the performance of its components, while simultaneously influencing and being dependent on the performance of higher-order systems. Second, a complicated network of 'non-simple relationships' is created though interactions between individual components that are often simple and direct, multiple interactions and feedback between components within subsystems, between components across subsystems and between subsystems at various hierarchical levels (Simon, 1969). These interactions represent the interface between the components and subsystems and are referred to as the 'architecture of a system' (Henderson and Clark, 1990). Third, a technology is non-decomposable if it cannot be separated into its components without genuinely degrading its capabilities or performance. Thus, if 'a set of subsystems is richly joined, each variable is as much affected by variables in other subsystems as by those in its own. When this occurs, the division of the whole into subsystems ceases to have any natural basis' (Ashby, 1960: 213). According to Holland and Miller (1991), it is hardly possible to decompose complex technologies into subsystems while maintaining the technology at or close to optimal performance levels. Modularity is often referred to as a way to manage complexity by grouping elements into a smaller number of subsystems (Simon, 1969).

Therefore, aside from these three aspects of complex technologies, another property becomes relevant when transferring a technology. As mentioned above, the degree of tacit (sticky) knowledge involved in a technology has an impact on its transferability. With an increasing degree of codification, the complexity of a technology is reduced and it can be more easily understood and utilized. According to Foray (2004), through codification, technology is becoming more like a commodity. It can be more precisely described and specified in terms of content and intellectual property and it is thus easier to be transacted. The technology becomes transferable independently of the transfer of other carriers, such as lead inventors who possess tacit knowledge. Consequently, when transferring a low complex technology, tacit knowledge needs to be transferred only to a limited extent. It might be sufficient to transfer only a few documents, maybe even only the property rights (for example, patents), that contain explicit, codified knowledge. However, when transferring a highly complex technology, the amount of tacit knowledge that needs to be transferred in order to enable the acquirer to fully utilize the technology, would be much higher. For instance, additional training by the lead inventors might be necessary. Therefore, transacting complex technologies requires the drawing up of complicated contracts that cover all specifications and interdependencies (Delmas, 1999).

6.2.2 Technological Uncertainty

Uncertainty is among the main determinants for the choice of TGS (see Chapter 7). Consequently, it has been investigated in the quantitative study and is subsequently discussed in the context of technology properties.[12] The literature distinguishes between uncertainty, which commonly follows the dichotomy suggested by Abernathy (1978) as well as 'demand uncertainty' and 'technological uncertainty' as formally specified by Walker and Weber (1984). This study focuses on technological uncertainty that has been defined quite consistently in the literature of governance structure choices.[13]

Walker and Weber (1984) define technological uncertainty as the unanticipated changes in circumstances surrounding technology (for example, new technology generations); that is the probability of a technology being improved which might render the current technology development efforts obsolete. I apply this definition as it is commonly used, for instance, by Arranz and Arroyabe (2008), Holcomb and Hitt (2007), Folta (1998), Robertson and Gatignon (1998), as well as Balakrishnan and Wernerfelt (1986). In a similar manner Bensaou and Anderson (1999) define technological uncertainty as the market's inability to accurately forecast the acceptance of a particular technology. Flygansvær et al. (2002) define technological uncertainty as corresponding to the unpredictability of the environment and as the inability to accurately forecast the technical requirements in a transaction relationship. Technological unpredictability may result from changes in the component design (Walker and Weber, 1984), changes in the end products (Heide and John, 1990) or changes in the underlying production process (Stump and Heide, 1996).

With regard to the choice of TGS, Walker and Weber (1984) postulate that, as technological uncertainty increases, a buy decision (purchase from an outside firm) becomes less likely. However, their empirical results point in an opposite (although non-significant) direction, that is that high technological uncertainty is actually more likely to be associated with a buy decision. Since then, Holcomb and Hitt (2007) have argued that – at higher levels of technological uncertainty – larger information deficits increase the likelihood of opportunism, making it costly to handle exchanges through markets. At higher levels of technological uncertainty, firms find it difficult to accurately predict ex-ante the combination of possible events and outcomes that are likely to emerge from production (March et al., 1958). According to Rindfleisch and Heide (1997), firms will favor an arms-length relation when they are exposed to technological uncertainty. Perry et al. (2004) argue that, in the face of considerable technological uncertainty, a high level of trust in the partner firm makes increased commitment possible. Having reviewed the somewhat controversial literature until 2001, Flygansvær et al. (2002) conclude that – even though the results are to some extent inconclusive – the results primarily indicate that arms-length relationships are the preferred mode of governance when technological uncertainty is high. Having defined technological uncertainty, one might question whether it remains constant or changes over time (Delmas, 1999).

Similar to the notion of Kuhn (1962), in which scientific progress is known to take a staged trajectory, technological development is commonly accepted to happen along lifecycle stages (Brockhoff, 1992). There is consensus in the literature on the interpretation of the S-shape relationship of technology's evolution throughout lifecycle stages (for example, Campbell (1983)). In the early lifecycle stage, fundamental scientific and technological problems are resolved that are associated with high R&D risk and technological uncertainty. For instance, emerging technologies are uncertain in terms of their potential value and success (Dosi, 1982). With only a small number of pioneering firms willing to bear high technological uncertainty at such an early stage, the number of patent applications is typically low, and, while it is slowly increasing, it remains centered around a small number of firms. When the basic technological and market uncertainties have vanished and a dominant design has been established (Abernathy and Utterback, 1978),[14] a broad range of market applications can be developed. A common prediction in the literature is that technological uncertainty will decrease when a dominant design appears. At that stage, the R&D risk and technological uncertainty will decrease while the number of patent applications will increase, which addresses predominantly small technical problems that result in incremental innovations. Within this growth stage, the patent concentration ratio decreases typically because new competitors appear and the absolute number of patent applications per year is on the rise. In the following stage of maturity, the number of patent applications (typically incremental innovations) remains constant. Subsequently, when the potential for new product innovations on the basis of the technology, and therefore the number of annual patent applications, starts decreasing, the decline stage begins. This argumentation has been repeatedly supported by various scholars (for example, Utterback (1996), Henderson and Clark (1990), Tushman and Anderson (1986), Abernathy and Clark (1985), Dosi (1982)).

A main reason for the changing degree of technological uncertainty along a technology's lifecycle is the changing shares of tacit and codified knowledge involved in a technology. Bessy and Brousseau (1998), who argued that the degree of a technology's commonality depends on its lifecycle stage, claim that, in an early stage of their lifecycle, technologies often constitute a large share of tacit knowledge. Thus, in an early lifecycle stage, a particularly large share of knowledge components, which are necessary to utilize a technology, remains with the inventing firm (that is these knowledge components are particularly sticky to the internal inventors). In such a situation, the expert community outside the inventing firm (for example, engineers from competing firms) lacks these tacit knowledge components to utilize the novel technology. Hence, to transfer such a novel technology to another firm so that it can be fully utilized, the technology requires much more to transfer than just the related IPR.[15]

Along the lifecycle (that is with an increasing diffusion of a technology), a larger share of the tacit knowledge is made explicit (for example, codified into technical manuals). Hence, at a later stage of a technology lifecycle (for example, after a dominant design has emerged), one can assume that the major principles of the formerly novel technology have diffused into and are known to the expert community. Ultimately, when a technology is characterized as 'common' (that is it has become part of the scientific commons), little tacit knowledge remains with the original inventor and the IP assets are often the only remaining assets (Bessy and Brousseau, 1998). Therefore, a later technology cycle stage is characterized by a higher level of commonality or a lower level of technological uncertainty. Thus, when transferring a technology at such a late stage, the seller has to transfer complementary tacit knowledge to the buyer, only to a limited extent.

6.2.3 Technological Impact

In this study's introduction, I argue that the economy is becoming increasingly reliant on the production, refinement, and accumulation of ideas (Powell and Snellman, 2004), with all technology developments contributing to the stock of knowledge (Schmookler, 1966). A growing stock of technologies provides increasing opportunities to recombine more technologies into innovations.[16] Modern models of innovation thus recognize that innovation is a cumulative process, which builds upon previous discoveries (Gallini, 2002), generating a continuum of improvements on some pioneer inventions (Scotchmer (1991), Merges and Nelson (1990)).[17] Hence, a technology that contributes to subsequent technical progress has a technological impact on future technology developments and the creation of innovations.[18]

Scotchmer (1991: 29) reasons that 'most innovators stand on the shoulders of giants and never more so than in the current evolution of high technologies, where almost all technical progress builds on a foundation provided by earlier innovators.'[19] Similarly, Aghion et al. (2001), Harris and Vickers (1987) and Budd et al. (1993) interpret innovation as knowledge production, and technological progress as emerging from a dynamic step-by-step process in which the outputs from one step are inputs into the next. Knott and Posen (2005: 629) also mention that 'under the spillover effect, firms passively benefit from the cumulative activity of other firms in the market'. Murmann and Frenken (2006: 944), referring to Sahal (1985), Dosi (1982), and Nelson and Winter (1982) highlight that 'most rapid improvements in technological history have typically taken place along well-defined technological trajectories', which were later labeled 'path-dependent'.[20] According to Murray and O'Mahony (2007: 1008), 'for innovation to occur, knowledge must not just flow: Innovators must have the ability to actually combine or accumulate knowledge'.

Following this principle, Murray and O'Mahony (2007: 1006) define innovation as 'cumulative' when it follows a 'process of building on the ideas of others to create new innovations'.[21] The accumulation of knowledge is thus not inherent to the innovation process but can be either supported or limited by the context in which innovation occurs (Mokyr, 2004). For instance, networks and technical communities play a role in helping innovators to build on one another's ideas (for example, as discussed by Tietz and Herstatt (2007), in open source software developments by Schweisfurth et al. (forthcoming), as well as Dahlander and Magnusson (2005)) and represent an important aspect of knowledge accumulation in organizational theory as studied, for example, by Fleming (2001), Hansen (1999), Knorr-Cetina (1999), Hargadon and Sutton (1997), and Van de Ven (1993).[22]

In a similar manner, Riis and Shi (2009: 1) define innovation as 'cumulative if later innovations are built on previous innovations... [or]... future innovations will necessarily build on the current technology'. The emphasis on cumulative innovation may, at first glance, seem in contrast to the more traditional interpretation of the disruptive nature of innovation (Christensen,

1997) and radical innovations' contributions to progress. However, an innovation that accumulates disparate technologies from various sources may be as disruptive as an innovation that builds on a narrow stream of prior technologies and vice versa.

Dosi and Nelson (2010: 25ff) note 'that in most fields of technology, progress is cumulative, with yesterday's efforts, both the failures and the successes, setting the stage for today's efforts and achievements... [Or,] in a cumulative technology, today's technical advances draw from and improve upon the technology that was available at the start of the period and tomorrow's in turn builds on today's'. Recent evidence (for example, Garud and Karnøe (2003), Katila and Ahuja (2002), and Fleming (2001)) proves that innovations are, more often than not, cumulative to the extent they incorporate prior knowledge from multiple sources. Murray and O'Mahony (2007) point out that the framework of cumulative innovation is particularly well suited, but not limited to knowledge-intensive fields with a distributed 'locus of innovation', such as software, biotechnology, and creative industries (Powell et al., 1996). However, sources of innovation are increasingly recognized as diverse and distributed – also in sectors such as semiconductors, oil and gas exploration, consumer products, and electronics (Jeppesen and Frederiksen (2006), Bessen (2004), Chesbrough (2003c), Linder et al. (2003), and von Hippel (1988)).

Scotchmer (1991: 29) provides further examples of cumulative innovations. She explains for instance that 'most molecular biologists use the basic technique for inserting genes into bacteria, as pioneered by Herbert Boyer and Stanley Cohen in the early 1970s, and many use a technique for causing bacteria to express human proteins that was pioneered at Genentech. In pharmaceuticals, many drugs like insulin, antibiotics and anti-clotting drugs have been progressively improved as later innovators bettered previous technologies.' Two well-documented examples from the British industrial revolution, include the blast furnace in the iron industry (Allen, 1983) and the Cornish pumping engine in the mining industry (Nuvolari, 2004). More recent examples include the invention of lasers and spreadsheets (Scotchmer, 2004). The laser was inspired by its predecessor, the maser, while Microsoft Excel's spreadsheet was built on its predecessor, Lotus 1-2-3, while Lotus itself was an improvement upon the first electronic spreadsheet, VisiCalc (Bessen and Maskin, 2009).

Cumulative innovation can appear on a micro level if firms focus their R&D efforts on certain stages of the innovation process on which firms build that are situated further down the value chain. Such a 'division of labor' is common in many industries (Allain et al. (2008), Arora et al. (2001)). For instance, small biotechnology firms might have a comparative advantage in achieving early stage discoveries in certain fields, while large pharmaceutical firms are considered more efficient in conducting later stage clinical testing (Arora and Merges, 2004). In such situations, firms have to innovate openly, that is transact technologies across firm boundaries towards those firms situated at the next stage of the innovation process. Thus, technologies that influence developments further down the innovation process (for example, become used in manufacturing processes or embedded into products) have a technological impact. However, the extent of a technology's impact varies and might be temporary (that is diminish as the technology becomes obsolete or substituted by a novel technology) although to a varying extent. Not all technologies contribute equally to future technological progress. While some technologies are specific to a technical problem in a particular firm of a single industry, other technologies are generic, multi-purpose technologies (for example, a 'general purpose technology' with a wide impact even on future technological developments outside the own firm domain (Hall and Trajtenberg, 2006)). For instance, the TFT display technology development has emerged from small scale applications along a technology trajectory across firm and industry boundaries and is today applied in various domains, ranging from medical equipment, TV screens to airplanes (Tietze et al., 2009). Such technology becomes an indispensable part of the technology stock and essentially contributes to the accumulated technological progress. A technology that proves itself helpful for subsequent technological developments is considered to have a technological impact on future progress.

6.2.4 Technology Quality

Technology quality is a multi-faceted concept. In this study, higher quality technology is defined as superior to other lower quality technologies developed as a solution to a similar problem. The quality of a multi-component technology is also related to the quality of the patents that protect the underlying inventions. If patents are high quality, they would be deemed valid if actually tested in court (Lemley and Shapiro (2005), Parchomovsky and Wagner (2005)). Hence, low quality patents are patents that would not have been issued had the examiner considered the entire universe of relevant prior art.

An extensive study by Wuchty et al. (2007) shows that a patent's quality is strongly linked to the development efforts and particularly the development team. Technologies can originate from the development efforts by either a single inventor or collaboratively by inventor teams. Jones (2009), referring to Malone (1995), provides a valuable example. The invention of the microprocessor was by necessity the work of a team. A researcher named Ted Hoff, who joined Stan Mazor in the development, came up with the original idea. However, neither of them knew how to translate the architecture into a working chip design. Only very few persons possessed the knowledge and skills to perform the next step. Hoff and Mazor subsequently involved Federico Faggin, a physicist and electrical engineer. Accordingly, the microprocessor should not be considered one person's idea, but a team innovation. Singh and Fleming (forthcoming) explain why inventive teams produce inventions of higher quality. According to Singh and Fleming (forthcoming: 9), 'a collaborative team will consider the invention from a greater variety of viewpoints and potential applications; such broader consideration is more likely to uncover problems. Given the typically greater diversity of experience on a collaborative team, some members are more likely to recall having seen a problem with a similar invention and argue to abandon or modify the approach. In short, collaborative creativity will subject individually conceived ideas to a more rigorous selection process, so that fewer poor ideas are pursued'.

There is strong evidence that teamwork has become more common and that it is more fruitful than working individually. Giuri et al. (2005: 13) report evidence on the distribution of individual and team inventions. Drawing on the PatVal sample, they find that only one third of patents are developed by single inventors. They subsequently conclude that a 'patented innovation is typically the result of teamwork'. However, they note that patent documents do not indicate whether collaborations take place among inventors belonging to the same or different organizations. Neither do patent documents indicate the type of collaboration, which can range from close, formal collaborations towards rather weakly tied, informal joint development activities.[23] However, Giuri et al. (2005: 13) find that in the EU, slightly more than 20 per cent of all patents (Netherlands, 34.5 per cent; Germany, 13.3 per cent) have been developed jointly with other partners. In an extensive study of 19.9 million academic papers over five decades and all registered US patents (2.1 million patents from 1975 to 1999) Wuchty et al. (2007: 2) confirm that inventions by teams increasingly dominate solo inventions across virtually all fields. From their data they observed an 'annualized increase in the propensity for teamwork. The average team size has risen from 1.7 to 2.3 inventors per patent, with the positive trend towards larger teams continuing.'

In addition to the increasing tendency that inventions result from team efforts, different studies have similarly concluded that teams also produce inventions of higher quality. Wuchty et al. (2007) also prove that teams produce more highly cited research than individuals and that this advantage is on the increase. Measuring the quality of team inventions Wuchty et al. (2007: 2) find that 'there is a broad tendency for teams to produce more highly cited work than individual scholars, that is produce exceptionally high quality research output, even where that distinction was once the domain of solo scholars'. Furthermore, Wuchty et al. (2007: 3) make it clear that 'individuals were never more likely than teams to produce the more influential work'. Considering a regional sample of 570 Italian inventors, Schettino et al. (2008: 3) essentially confirm the results of Wuchty et al. (2007) and find that 'the presence of an inventive team is effective in raising both the quantity and quality of inventions [and] exerts a positive and very significant impact on inventors' productivity'. Moreover, they find that 'patent quality is positively affected by different individual properties, such as age, level of education, and gender while, among the organizational properties, only the presence of an inventive team exerts a significant positive impact' (Schettino et al., 2008: 22). Furthermore, they report that, among the different sources of invention, inventive teams tend to be more present in larger firms with wider patent portfolios. Jones (2009) also confirms a rise in teamwork. There is strong evidence – more than any other technological property – that teamwork leads to higher quality of the resulting patents.

6.2.5 Technology Value

Prior to transactions, IIP managers of the seller as well as of the prospective buyers need to assess the value of a technology or its patents as a basis for cost calculations and negotiations. Assuming that all prior R&D investments are sunk costs, for any transaction to be economically successful, its transaction costs (TC) should not exceed its expected returns (see also Figure 9.4). As TC vary for different TGS, the estimated value thus has an influence on the choice of a suitable TGS.

However, to assess a technology's value is by no means trivial, neither is the definition of the value concept. Commonly, one would distinguish between the value of a technology (respectively the value of the underlying inventions) and its patent value. The former refers to the technological impact of the invention, that is, its contribution to the technology stock – thus its potential for innovative products, processes, services, etc. The latter comprises only the value added due to the fact that the invention is patented; that is, the difference between the value of the invention as it is patented and the value it would have had if it had not been patented (sometimes referred to as 'patent premium' according to Arora et al. (2004)). In the following section, I refer to the value of a technology as a bundle of (patented) inventions.[24]

It is difficult to assess a technology's value, due to the multi-dimensional complexity of such an exercise. To conduct a value assessment, firms have to take into account numerous factors, particularly related to the intended purpose of the technology and the underlying patents, that is their current and future as well as operational and strategic utilization or their impact on future technology developments (Hall and Ziedonis, 2001). According to Troy and Werle (2008), a patent's value is influenced by whether the owner of the patent sees it as a means to exclude potential competitors from the market and obtain a monopoly price for the patented product (monopoly value) or as a bundle of technical knowledge components, which is to be sold or licensed (trade value). A patent's value also depends on whether the patent will be used to acquire capital through, for instance, a bank loan (asset value), to facilitate blocking innovative activities of other firms (blocking value) or to build up particular technological strength (Narin and Noma Ross, 1987). Furthermore, one needs to assess the position of a technology within its lifecycle, which is further complicated by the fact that a technology's position and importance might differ depending on what stage of its lifecycle it is in, that is from the date of the invention, patent application (which determines the patent term) to the date when the technology should be transacted. Difficulties also result from patents' unique (idiosyncratic) nature and their technical, commercial and economic uncertainties according to Reitzig (2003) and Granstrand (2000a). Despite extensive research on patent valuation, there are still 'notorious problems in assessing the value of patent rights' (Harhoff et al., 2003: 1345).

As it is beyond the scope of this study to understand and review existing valuation approaches, the author does not follow this discussion any further.[25] Instead, this study looks into the generic conceptual valuation and pricing model for patents proposed by Granstrand (2000a). This model, which can also be applied to technologies, is helpful to illustrate how sellers and buyers reach agreements on prices. Granstrand argues that the market value of a patent (or a group of patents) can be determined as the value of a corresponding exclusive, non-geographically restricted license. He provides a list of factors that buyers and sellers should consider. Accordingly, sellers can use these to estimate a 'floor price level', while buyers can use them to establish a 'ceiling price level'. If the resulting price window is relatively narrow, buyers and sellers can most likely negotiate a price, which both are willing to accept.

With regard to the value assessment of sellers, it might be worth noting that scholars have discussed whether owners overestimate patent values.

Inventors might overestimate a patent value because of their inherent commitment to the development of the invention. For instance, Heller and Eisenberg (1998: 701) argue that, due to 'cognitive biases ... people consistently overestimate the likelihood that very low probability events of high salience will occur'. For instance, in biomedical research, such bias would cause owners of upstream patents to overvalue their discoveries. Given the assumption that no patent owner knows ex ante which invention will be the key for future blockbusters, upstream owners will collectively demand higher values than the aggregate market value of their inputs. Thus downstream buyers will decline the offers and the new drug will never be developed. With reference to the Granstrand (2000a) model, such bias would result in a larger price window and reduce the likelihood of negotiating a price that both parties would be willing to accept.[26]

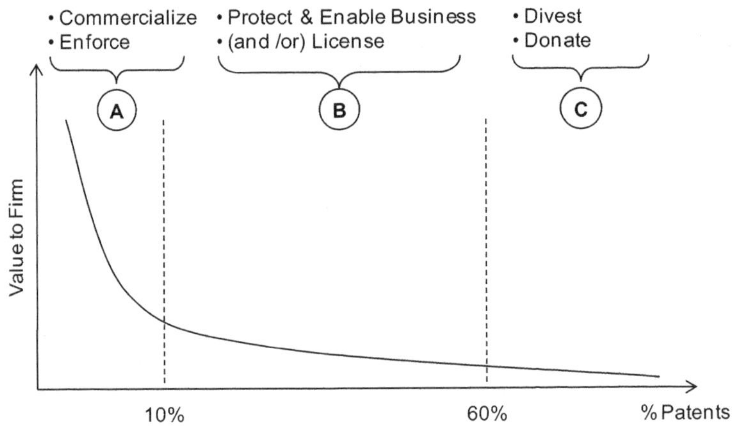

Source: Tao et al. (2005: 54)

Figure 6.2 - Patent portfolio value distribution and recommended actions

However, there is limited evidence on the size of this effect due to the difficulties involved in measuring patent values. Gambardella et al. (2008: 76f) refer to the potential bias of inventors and argue that managers might provide a more realistic value assessment than the inventors themselves, who have a more objective view on the market value and future uses of patents. Comparing inventors' and managers' value assessments, Gambardella et al. are able to show that inventors assess higher values on average than managers do. However, their approach does not allow for determining whether the real patent values have been overestimated by both the inventors and the managers, since – especially in small firms and TTO – 'inventor[s] and manager[s] probably work close to each other... [and thus]... discuss a great deal about the potential value of the patent'.[27]

Aside of the difficulties to assess technology or patent values, an agreement has been reached with regard to the value distribution. Rossman and Sanders (1957), published in Griliches (1990), find a high dispersion of patent values, which was further confirmed by Lanjouw et al. (1998), Grabowski and Vernon (1994), Pakes and Schankerman (1984), as well as Scherer (1965).

Figure 6.2 provides an illustration of the distribution of patent values in firm portfolios. Developed from 70 interviews with industry representatives and from his own industry experience, Tao et al. (2005), in line with previous studies, finds that patent values in firm portfolios are skewed (log-linear distributed).

In contrast to common conceptualizations of value distributions that distinguish between high and low value patents Tao et al. differentiate between three categories.[28] Accordingly, a small portion of patents (~10 per cent) with a high value should be actively enforced against competitors and commercialized at least with own products (Category 'A'). The majority of patents (~50 per cent) with only a moderate value are important to protect own core technologies (surrounding patents) in order to enable business. A significant share of those however should be considered for licensing (Category 'B'). A substantial share of patents in a firm's portfolio (~40 per cent) is commonly of low value. Those patents should be donated (for example, to universities for tax reasons) or divested (for example, lapsed into the public domain) (Category 'C').

In the following section, I present selected evidence of the monetary economic value of patents from econometric patent value studies. However, the evidence on the concrete, quantitative values of technologies and patents is still limited. Griliches (1990) reports an early collection of empirical results on patent values. Among those studies, Rossman and Sanders (1957) estimate the mean value of patents that are in current use at US$577 thousand (median value of US$25 thousand) and of the no gain, loss, and not yet used patents at US$112 thousand (median value close to zero). Harhoff et al. (2003) classify the values of the patented inventions into nine categories, ranging from less than €51 thousand to above €31.3 million. They find that 27 per cent of the patents in their sample have a value of less than €51 thousand. A total of 74.1 per cent of the sample has a value of below €511 thousand and only 5 per cent of the sample has an average value above €2.6 million.

A subsequent study, commonly known as PatVal, was conducted between 2003 and 2004 and comprises a sample of 9 017 patents granted by the EPO between 1993 and 1997 (see Figure 6.3).[29] According to Giuri et al. (2007), patent values were classified into ten classes ranging from less than

€30 thousand to more than €300 million. Approximately 8 per cent of the patents in the PatVal sample are estimated to be worth more than €10 million. About 24 per cent of the patents are estimated to be worth between €1 million and €10 million and a share of about 68 per cent falls into the left sector of the distribution with estimates lower than €1 million. Unfortunately, appropriate measures of location and dispersion are difficult to apply due to the high skewness of the value distribution. However, Gambardella et al. (2006) estimate a median value of €300 thousand per patent. Moreover, from a study of 2 600 randomly selected German manufacturing and service firms, Institut der deutschen Wirtschaft Köln (2006) estimate a rough average value of €147 thousand for unused patents.[30] Furthermore, patent values differ across industries.[31] Accordingly, Giuri et al. (2007) report differences between patent values across five macro technological classes. Inventions that are worth more than €10 million are usually in the chemicals and pharmaceuticals sector (11.7 per cent) compared to all the other sectors and to the overall sample (7.2 per cent). Only 58 per cent of chemical and pharmaceutical patents are worth less than €1 million, while about 70 per cent of electrical engineering, instruments, process engineering, and mechanical engineering patents are worth less than €1 million.

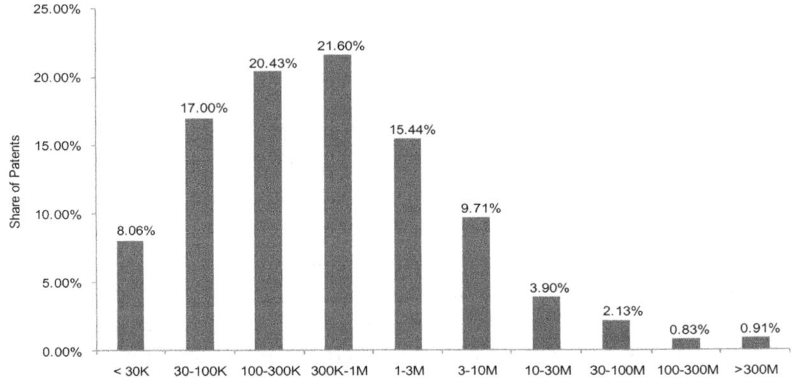

Source: Gambardella et al. (2008); Number of observations: 8 217; Currency: Euro.

Figure 6.3 - The value of European patents across macro-technology classes

Gambardella et al. (2006) report average and median values for patents across 30 technological classes (see Table 6.1). Ranked by median values, the patents in the organic fine chemistry industry rank the highest with a median value of €679 thousand, followed by pharmaceuticals and cosmetics with a median value of €605 thousand. The chemical and petrol, and basic materials chemistry rate third with a median value of €580 thousand. In total, 16 of the 30 industry classes show above-median values. The lowest median patent values, of €170 thousand, can be found in agricultural and food processing, machinery and apparatus.[32]

Table 6.1 - Average and median values of patents by industry

Industry	Median Value [T€]	Average Value [T€]	Industry	Median Value [T€]	Average Value [T€]
Organic fine chemistry	679	6 261	Transport	298	2 782
Pharmaceuticals, cosmetics	605	5 260	Semiconductors	284	2 555
Chemical&petrol, basic materials chem.	580	5 298	Machine tools	276	2 568
Materials, metallurgy	467	4 187	Civil engineering, building, mining	275	2 546
Macromolecular chemistry, polymers	449	3 980	Nuclear engineering	274	2 421
Space technology weapons	414	3 854	Handling, printing	269	2 441
Agriculture, food chemistry	406	3 637	Analysis, measurement, control technology	252	2 358
Materials processing, textiles, paper	367	3 325	Telecommunications	247	2 331
Engines, pumps, turbines	364	3 344	Mechanical Elements	243	2 208
Environmental technology	354	3 250	Audio-visual technology	236	2 296
Medical technology	351	3 279	Information technology	235	2 265
Biotechnology	336	3 134	Electrical devices, engineering, energy	211	1 938
Chemical engineering	332	2 992	Consumer goods and equipment	193	1 814
Thermal processes and apparatus	327	3 012	Agricultural&food proc-machine-apparatus	170	1 544
Surface technology, coating	325	2 924	**Total**	**305**	**3 024**
Optics	308	2 960			

Source: Gambardella et al. (2006).

To conclude, assessing the value of technologies and patents is still a cumbersome exercise for firms and scholars, due to the influence of numerous factors. However, scholars agree that patent values follow highly skewed distributions. Recent but limited evidence reports median values of patents across all technological classes ranging from €170 thousand to €679 thousand with an average of €300 thousand. However, these values were only determined for individual patents that were held full term. Therefore, it appears difficult to extrapolate the value of a technology comprising a bundle of different patented inventions. Does each patent contribute a similar value to the value of a bundle? Or does it appear more reasonable to assume that one or a few patents protect the core of the technology complemented by patents protecting specific parts of the technology?

6.2.6 Technology Source

Technologies can originate from the development efforts of different types of inventors. Throughout the literature three sources of inventions are commonly discussed. These include individuals (independent inventors and users as, for example, differentiated by Hüner et al. (2010)), inventions developed by academic faculties of universities / research institutes and corporate inventions (large firms and SME). Moreover, collaborative inventions appear to be developed jointly by actors from different organizational types.

There is an ongoing discussion in the literature on whether large firms ('bigger is better') or small firms ('smaller is better', that is the entrepreneurial view of Schumpeter) develop 'leapfrogging' (radical or disruptive) innovations (Diamond Jr, 2006). Brodbeck (1999) shows that large firms are more active in externally exploiting technologies than SME, since they focus more on central value adding (manufacturing) activities. However, although SME possess a lower number of patents in absolute terms than large firms, the Institut der deutschen Wirtschaft Köln (2006) finds a similar value distribution for both groups.

Inventions developed at universities / research institutes are invented by their faculty members and are commonly supported by the university's TTO when filing patent applications and exploiting them externally (Markman et al., 2005).[33] While university faculties commonly pursue basic research, their inventions are often not close to being marketable, but are rather perceived as being in an 'embryonic' phase, thus requiring longer lead times and significant complementary investments to be turned into an innovation[34] (Thursby and Thursby (2007), Tijssen (2002), Schilling (1998)).[35] Moreover, the failure rate of innovation projects appears to be closely related to the development stage of technologies that firms acquire from universities. According to Thursby and Thursby (2007), innovation projects based on university inventions have a higher failure rate than inventions licensed from non-university sources. Since licensing is ranked as a service, which does not weigh heavily in tenure and promotion decisions, faculties are reluctant to patent innovations (Markman et al. (2005), Jensen et al. (2003), Agrawal and Henderson (2002), Tijssen (2002)). Furthermore, in contrast to the patent departments, which large firms have been employing for decades, the recently founded TTO are often still inexperienced and lack resources (Taylor and Silberston, 1973). In comparison to corporate inventions, Sapsalis et al. (2006) find that the value distribution of academic patents is very close to that of corporate patents and has similar levels of skewness.

Independent inventors are a heterogeneous group ranging from 'heroes', who contribute substantially to technological progress, to 'hobbyists', who make rather marginal contributions (Dahlin et al., 2004). Rosenberg (1994), in line with Schumpeter, argues that, as inventions have become more industrialized, independent inventors are increasingly being perceived as marginal contributors in the world of innovative activity. However, others argue that users appear to be a particularly important source of inventions in many industries (Fleming (2007), Lettl et al. (2006), Herstatt and von Hippel (1992), von Hippel (1988)). In general, resource constraints make independent inventors less likely to commercialize their inventions (Dahlin et al. (2004), Grown and Bates (1992), Bates (1990), Evans and Jovanovic (1989)).[36]

Table 6.2 - Patent use across different invention sources

	Internal use (%)	Licensing (%)	Cross-licensing (%)	Licensing and use (%)	Blocking competitors (unused) (%)	Sleeping patents (unused) (%)	Total (%)
Large companies	50.0	3.0	3.0	3.2	21.7	19.1	100.0
Medium sized companies	65.6	5.4	1.2	3.6	13.9	10.3	100.0
Small companies	55.8	15.0	3.9	6.9	9.6	8.8	100.0
Private research institutions	16.7	35.4	0.0	6.2	18.8	22.9	100.0
Public research institutions	21.7	23.2	4.3	5.8	10.9	34.1	100.0
Universities	26.2	22.5	5.0	5.0	13.8	27.5	100.0
Other Govt. institutions	41.7	16.7	0.0	8.3	8.3	25.0	100.0
Other	34.0	17.0	4.3	8.5	12.8	23.4	100.0
Total	50.5	6.2	3.1	3.9	18.8	17.5	100.0

Source: Giuri et al. (2007, Table 7); Number of observations: 7 556.

While prior studies have mainly employed small samples (for example Meyer (2006), Ernst et al. (2000), Narin and Breitzman (1995)), Table 6.2 presents the patent use across different sources of invention based on the extensive PatVal sample.[37] Giuri et al. (2007) find that large firms use 50 per cent of their patents internally. They trade less than 10 per cent of them and about 40 per cent is not used. More than half of the unused inventions aim to block competitors. Giuri et al. argue that a large share of unused patents developed by large firms is likely to stem from their lower marginal cost of patenting. Owing to their larger scale, they patent more often. For this reason, larger firms create internal divisions that specialize in patenting or licensing or they have specialized managers or assets dedicated to this task. They subsequently exhibit a higher propensity to patent because of the fixed costs involved. As a result, they also patent minor inventions, which are less likely to be used. In fact, this is consistent with the higher share of unused patents by large firms (40.8 per cent) compared to medium sized firms (24.2 per cent) and small firms (18.4 per cent) shown in Table 6.2. Moreover, medium sized firms have a higher rate of internal use (65.6 per cent) than small firms (55.8 per cent), which have a slightly higher rate of internal use than large firms (50.0 per cent) and a much higher rate of licensing (25.8 per cent vs. 9.2 per cent). Public or private research organizations and universities license a large fraction of their technologies and do not use them internally as previously discussed by, for example, Mowery et al. (2001).

NOTES

[1] An overview of various definitions of technology can be found in Brodbeck (1999) and detailed explanations can be found in, for example, Cardullo (1996), Edosomwan (1989), Rubenstein and Ribenstein (1989), and Mitchell (1988). A discussion on various facets of technology can also be found in Sahal (1982, 1981).

[2] The definition proposed by Tschirky and Koruna (1998) refers specifically to the dichotomy of different knowledge components which technologies consist of. This dichotomy refers to the distinction between tacit and explicit knowledge originally proposed by Polanyi (1966). Even Williamson (1979: 242) refers to Polanyi's (1962) notion, noting that 'even in the modern industries the indefinable knowledge is still an essential part of technology'. According to Delmas (1999: 640) referring to Collins (1974) and Teece (1977) 'when knowledge is tacit, it cannot be effectively transferred in codified form; its exchange must rely on intimate human contact… It is difficult to draft simple contracts governing the sale or licensing… (Mowery, 1983, Pisano, 1990'.

[3] Birkenmeier (2003) also mentions this framework, but does not provide any detailed discussion. Also see the comparison between technical information carriers developed by Granstrand (2010: Table 13.1).

[4] According to Cowan and Foray (1997: 600), there is 'now a consensus that codified and tacit knowledge are complements rather than substitutes. The process of codification does not provide all of the knowledge needed to undertake an action; there will always be some tacit knowledge involved in performing any action. This is why codification cannot be considered as a simple transfer of knowledge from the tacit to the codified domain. It is, rather, the construction of new ensembles of codified and tacit knowledge. In other words, codification is never complete and some forms of tacit knowledge will always continue to play an important role'.

[5] Historically, the patent system was designed as an economic incentive system to facilitate R&D investments eventually leading to innovation (Scotchmer, 2004). According to Arrow (1996: 125), patents 'create artificial scarcities where none exist naturally, though the duration of property rights is limited. These scarcities are intended to create the needed incentives for acquiring information. An alternative path to the innovator is to maintain secrecy and there is some property protection available to trade secrets.' There is currently, an ongoing debate on the real effect of patents on innovation, with a focus on the role of IP in the open innovation paradigm. Arrow (1996: 127) anticipated this problem noting that 'we are just beginning to face the contradictions between the systems of private property and of information acquisition and dissemination'.

[6] See Foray (2004: 131ff.) for a definition of various IPR and delimitation from other types of cultural property rights.

[7] In Europe, following the European Patent Convention – for an invention to be patentable – it must be novel (no prior disclosure or use; Art. 54), inventive (not obvious to one skilled in the art; Art. 56) and industrially applicable (Art. 57). For US patents following the US Patent Act (35 USCS Sects. 1 – 376, ch.10) particularly the third condition is more relaxed. According to the 'statutory requirements for the invention's patentability' in the US, to be patentable an invention has to be novel, non-obvious, and has to have utility. However, according to OECD (2009), the interpretation and scope of the third term is generally the same as that of industrial application. International patent treaties often use 'utility' and 'industrial applicability' synonymously.

[8] Since the emergence of new matters as an integral part of today's GDP in many countries (for example, software, genes, and business models) there is an ongoing and controversial discus-

sion on whether those new matters should be patentable. In the US, software can be patented since 1981 and methods-of-doing-business (MDB) since 1998. In Europe, however, software could not be patented until fall 2003 (software is still only patentable when embedded into 'computer implemented inventions') and the MDB business patents are not yet accepted according to Granstrand (2000a).

[9] Other definitions of the 'patent family' exist from which the most common ones are presented in OECD (2009).

[10] Closely related concepts, such as 'technical complexity' and 'innovation complexity' are used almost synonymously in the literature. See also Ebersberger et al. (2005).

[11] While complexity has been discussed and defined on various levels (including products, technologies, firms, industries and countries), in this study, I focus on the definitions available for technologies.

[12] The concept of risk is closely related to that of uncertainty. However, it is less common in studies related to transaction cost theory. Knight (1921) is often cited as the scholar who first made the distinction between quantifiable uncertainty and unquantifiable uncertainty, while he refers to the former as risk that can be established either a priori (as is the case when rolling a die) or post priori (for example, a car insurance company calculating the accident risk of a model). In both cases, the defining feature is that, although the actual outcome cannot be established in advance, the probability of its occurrence can be. In contrast, according to Knight, there is an 'inherent unknowability' of the probability of occurrence. Knight also notes that the majority of business decisions actually fall into this category. This is because they relate to situations so unique and non-replicable that there is no previously established probability distribution for them.

[13] For an intensive discussion of uncertainty and transactions, see Shin (2003).

[14] According to Nelson (1994), much of the contemporary formulation of what is labeled 'dominant design theory' was provided by Abernathy and Utterback (1978). When they first published the dominant design concept, it was based on a detailed observation of only one industry, namely automobile manufacturing. Since then, the 'basic story line' has been tested and found fitting in a wide range of industries (for example, Tushman and Anderson (1986), and Utterback and Suarez (1993)). However, some scholars, for instance Nelson (1994), are skeptical as to whether the dominant design theory applies to all industries.

[15] In addition, the market uncertainty is higher for an early-stage technology (that is whether it will be established as an industry standard). However, this type of uncertainty is not investigated in detail in this study.

[16] For instance, Mukoyama (2003) studied the effects of imitation and the diffusion of cumulative innovations.

[17] For instance, Rothwell (1994: 26) proposes as fifth generation innovation process so that 'industrial innovation can be depicted as a process of know-how accumulation or learning process, involving elements of internal and external learning'.

[18] As innovation is important for firm growth, the cumulative innovation conceptualization has also been formalized in endogenous growth theory (for example, Romer (1990)), evolutionary growth literature (for example, Nelson (2008)), and neo-Schumpeterian growth models (for example, Denicolò and Zanchettin (2004)).

[19] The notion 'standing on the shoulders of giants' is said to originate from Isaac Newton. Hawking (2003) also titled a bestselling book 'On the Shoulders of Giants' and Google Scholar adopted the notion as its dictum.

[20] See, for instance, Dosi and Nelson (2010: 17) for a discussion on technology trajectories and cumulative innovation. They refer to 'normal technical progress as those advances occurring along a given trajectory – irrespective of how big they are and how fast they occur – while we reserve the name of "radical innovations" to those innovations linked with paradigm changes'.

[21] Murray and O'Mahony (2007) provide a framework for cumulative innovation. The cumulative innovation framework focuses on 'who is able to share, reuse and build on knowledge'

and, most importantly, 'under what conditions.' Disclosure is the first step in a cycle that allows for cumulative innovation. Second, the innovator must be able to access these ideas. Third, both of these conditions depend on rewards; to encourage earlier innovators to both disclose their ideas and provide access so that later generations, usefully integrates these ideas.

[22] Economics also explicitly attends to the factors affecting cumulative innovation but with a strong emphasis on the law, specifically intellectual property rights (Jaffe and Lerner (2004), Scotchmer (1991)). However, according to Murray and O'Mahony (2007: 1007), 'with little exchange between economics and organization theory, we have only a fragmented picture of the accumulation process'.

[23] Hagedoorn (2003) points out that firms often consider partnering sub-optimally due to the legal complexities involved in the management of intellectual properties across firm boundaries and international patent jurisdictions. Hagedoorn also shows that co-patenting was more frequent for chemical and pharmaceutical patenting, where patent protection is stronger and the scope for legal controversies is more limited. Therefore, apart from underestimating the extent to which there is collaboration in R&D, the data on co-patenting may be biased towards specific technologies.

[24] The economic literature further distinguishes between the private and social patent values (OECD, 2009). The private value to the patent owner is defined as the discounted revenue flow created by the patent within its lifetime. The patent's social value is defined as its contribution to society's technology stock. However, both conceptualizations are related. The generated revenues should correspond to the technological contribution. However, part of the social value is not appropriated by the patent owner due to externalities. The knowledge disclosed through the patent, for instance, might be used by other inventors and/or competitors.

[25] For a review on valuation approaches I refer to Kamiyama et al. (2006). They reviewed existing valuation approaches and classified them into three categories. These include qualitative (for example, rating and scoring systems), quantitative (most prominently the cost, market, and income approach), and econometric approaches (for example, large scale regression models). According to Fröhling (2005), the choice of the approach is particularly affected by the size of the patent portfolios. While a small patent portfolio allows for an in-depth annual valuation (monetary) and evaluation (non-monetary valuation), simplified evaluation approaches are preferred for large patent portfolios (for example, more than one thousand patent families and 10 thousand patents and patent applications).

[26] In behavioral economics, the endowment effect (also known as divestiture aversion) is an effect that predicts that people value a good or service more once their property right to it has been established. In other words, people place a higher value on objects they own than objects that they do not. In one experiment, people demanded a higher price for a coffee mug that had been given to them but put a lower price on the one they did not yet own. The endowment effect was described as inconsistent with standard economic theory, which asserts that a person's 'Willingness-to-pay' for a good should be equal to their 'willingness to accept' compensation to be deprived of the good. This hypothesis underlies consumer theory and indifference curves (Plott and Zeiler (2005), Kahneman et al. (1990), and Thaler (1980)).

[27] Although it is beyond the scope of this study to discuss the sellers' value bias, Chapter 10 provides some quantitative insights comparing the sellers' value perceptions to the actually achieved auction prices.

[28] For instance, Gu and Lev (2000) report that 'as a rule of thumb,' ten per cent of the patents in a patent portfolio account for 80 to 90 per cent of the portfolio value, while 90 per cent account for the remaining ten to 20 per cent of the portfolio value.

[29] This and subsequent studies (for example, Harhoff et al. (2003), Gambardella et al. (2007)), in which some of the scholars from the first study had participated again, adopt a similar approach but modify the question slightly (for example, the year of the patent samples changed, thus the year for which the sales price was requested changed) and enlarge the number of questions imposed to the inventors as well as the scope to other countries. Coun-

tries included in Giuri et al. (2007) are France, Germany, Italy, Spain, the Netherlands, and the United Kingdom.

[30] In a survey, the respondents were asked to reveal the minimum value for which they would be willing to sell their unused patents. In German, the study by Institut der deutschen Wirtschaft Köln (2006: 21) formulated the value as: 'derjenige Preis … zu dem ein Unternehmen gerade bereit wäre, seine noch nicht genutzten Patente zu verkaufen.'

[31] Reasons for this relate to, for instance, differences in various incentives to innovate or the effects of patents' and thus firms' use of the patent system across different industries (Levin et al., 1987). Griliches (1990) further notes that industries vary widely in their propensity to patent.

[32] Both studies apply the classification scheme based on 6 digit IPC classes developed by Hinze et al. (1997).

[33] For instance, Zucker and Darby (1996) discuss various types of scientists.

[34] Owing to universities' increasing orientation towards patenting in order to generate licensing royalties, concerns have even been raised that universities are moving towards applied research and away from fundamental research according to Thursby and Thursby (2007). However, the existing evidence draws a mixed picture.

[35] Based on the results of two surveys by TTO managers and industry representatives investigating university inventions that were licensed to the industry, Thursby and Thursby (2007) report that 82 per cent of the technologies that TTO have licensed to firms were in an early stage and corporate managers reported that 74 per cent of the technologies they had licensed from universities in the past were in an early stage.

[36] According to Mariani and Romanelli (2007), the difficulty to obtain information about individual inventors has prevented previous research from performing systematic empirical studies on this matter. However, some studies (for example, Gruber et al. (2009)) investigate determinants for quantitative and qualitative inventor productivity (for example, age, gender, education, and community embeddedness).

[37] Giuri et al. (2007) differentiate between sources of invention by applying the definition of the 'European Commission's convention,' that is small firms (fewer than 100 employees), medium firms (100 – 250 employees), large firms (more than 250 employees), universities, public or private research institutions, and others.

7. Transaction Cost Theory

7.1 INTRODUCTION

The idea that transactions form the basis of economic thinking was introduced by Commons (1931). Transaction cost theory, originally formalized by Williamson (1975), suggests that transaction costs (TC) are crucial for making or buying decisions and hence impact the choice between the firm (vertical integration) and the market (contracting). Put differently, transaction cost theory is an approach for the explanation of institutions, considering the relative merits of conducting transactions within the firm boundaries in contrast to (inter-firm) market transactions (Black, 2002). In transaction cost theory, the unit of the analysis is the unit of activity – the transaction, with its participants. Hence, it is congruent with this study's unit of analysis.

According to Shelanski and Klein (1995) transaction cost theory's relational branch is particularly relevant to this study as it aims to explain how trading partners choose from the set of feasible institutional alternatives. Within the context of open innovation, firms increasingly transfer technologies across their own firm boundaries. Therefore, they need to choose transaction partners. Particularly IIP managers face make-or-buy decisions (that is whether to acquire a technology or to develop it internally) and whether to keep a technology internally or to exploit it externally (that is the keep-or-sell decision). However, this study is not primarily concerned with these decision types. Instead, it contributes to the understanding of the questions that follow when these decisions have been made. Having decided to not develop a technology within the firm's own R&D centers, but to acquire (exploit) a technology, IIP managers have to decide whether to conduct such technology transactions with internal resources or use TMI instead to ensure the most effective and efficient creation of innovations. Furthermore, IIP managers need to decide on the optimal acquisition / exploitation strategy (that is the best suitable governance structure), which is reflected by the transaction models' properties developed by the TMI. These types of questions are particularly addressed by the relational branch of transaction cost theory. In the open innovation literature, various studies have applied transaction cost theory for similar purposes as in this study. For instance, these studies include questions related to technology acquisition (Van de Vrande et al. (2006), Delmas (1999)), make-or-buy decisions (Klein, 2005), licensing contracts (Bessy et al. (2008), Brousseau et al. (2007)), and alliances (Colombo (2003), Oxley (1999)).

However, to my knowledge, very few studies have been conducted closely related to the choice of TMI for technology transactions. In a small scale study, Feller et al. (2009) explore the ways in which firms utilize hierarchical relationships and the market system to supply and acquire innovation capabilities and IP assets from sources external to the firm. Wagner (2006: 3) explores the make-or-buy decision of firms particularly related to IP services or a 'firm's decision between maintaining a sufficiently large IP-department and purchasing a certain share of the services necessary for the management of its IPR on the marketplace'. However, at least to my knowledge, no study has yet investigated when firms should prefer to use TMI instead of own internal resources for transactions or when auctions are preferable over bargaining negotiations. Thus, due to the close fit of the related literature, TC theory and particularly the concept of transaction governance structures are used in this study.

7.2 (DYNAMIC) TRANSACTION COSTS

According to Allen (1999), Demsetz (1964) made the first explicit statement of transaction costs and later defined TC 'as the cost of exchanging ownership titles' (Demsetz, 1968: 35). Since the mid-1970s, TC theory literature has grown continuously. An extensive survey by Macher and Richman (2008), which includes references to the major seminal contributions, estimates the number of empirical papers in transaction cost economics at over 900.[1] According to Macher and Richman (2008: 3), a key insight from TC theory is to 'recognize that, in a world of positive transaction costs, exchange agreements must be governed, and that, contingent on the transactions to be organized, some forms of governance are better than others'.

According to Tirole (2008), referring to Coase (1937) and Williamson (1975), four types of TC can be distinguished of which two occur at the contracting date and two occur later. The first category refers to contingencies which the parties face, but that are not foreseeable at the contracting date, as well as to situations in which there are too many contingencies to be included in a contract, although they might be foreseeable. The second category refers to those transaction costs that occur when monitoring whether the involved parties comply with the contract or occur when enforcing the contract (for example, legal costs). However, nowadays, a number of TC types have become known under particular names, for instance 'search and information costs', 'bargaining costs', and 'policing and enforcement costs'. TC are often further distinguished according to whether they occur inside the firm boundaries (intra-firm) or external to the firm boundaries (extra-firm) (Dahlman, 1979).[2]

Furthermore, it seems noteworthy that the concepts of TC and governance costs are often distinguished.[3] Demsetz (1988) argues that a differentiation should be made 'between the cost of transacting across this market and the cost of in-house managing'. According to Demsetz transaction costs are defined as the cost of transacting and more specifically the costs of coordinating resources through market arrangements. Granstrand (1998) also follows this line of argument and differentiates between governance (management) costs associated with hierarchical transactions inside a firm and reserved the term TC for market transactions. Schlag (1989: 1675) points out that an 'overly expansive view of transaction costs threatens to make the Coase Theorem tautological ... [while] an overly restrictive view of TC can effectively invalidate the theorem'. Williamson, however, remains 'one of the strongest proponents of applying the notion of transaction costs ubiquitously' (Allen, 1999: 900). Having recognized these arguments, I accept that one has to differentiate between the types of costs associated with market and hierarchical transactions. Nevertheless, one can argue that, particularly in large firms, internal technology development that takes place in the own firm's R&D centers constitutes transactions across firms' internal sub-markets (for example, where the R&D center is the supplier to the manufacturing department). In the remainder of this study, I use the term TC. However, where necessary, he distinguishes between the two components, indicating costs related to market transactions as TC_{market} and those associated with internally governed transactions as $TC_{hierarchy}$.

Scholars have additionally argued that transaction costs are dynamic, and hence change over time. According to Langlois (1992: 99), dynamic transaction costs 'are the costs of persuading, negotiating and coordinating with, and teaching others. They arise in the face of change, notably technological and organizational innovation'. Langlois argues that the TC defined by Williamson are not of crucial importance, but rather those labeled as dynamic TC. Accordingly, he points out that 'ultimately, the costs that lead to vertical integration are the (dynamic) transaction costs of persuading, negotiating with and coordinating among, and teaching outside suppliers in the face of economic change or innovation.' In a similar manner, Granstrand

(1998: 471), who distinguishes between governance costs for internal transactions and TC for market transactions, argues that 'managerial learning decreases the governance management costs of the firm faster than any learning associated with recurrent contracting on the market. In a dynamic TC theory, this would explain and predict gradual internalization over time better than would any theory based on comparative static transaction costs as in the usual TCT.' Among others, Jacobides and Winter (2005) and Jacobides (2008) build on Langlois's (1992) arguments and attempt to bridge the gap between TC and capabilities approaches.[4] Jacobides and Winter (2005: 395) propose a 'framework where capability differences interact with changing transaction costs'. Williamson (1999: 1100) has also recognized these scholars' criticisms, noting that 'transaction cost economics is static because it works out of an equilibrium contracting setup'.

7.3 GOVERNANCE STRUCTURE

According to Macher and Richman (2008), the key conceptual assumption of TC economics is to describe firms in organizational terms (that is as governance structures) rather than in neoclassical terms (that is as production functions). Coase (1937) coined the concept of transaction governance structure (TGS).[5] According to Zhang (2006), Williamson (1991, 1981, 1979) extended it and defined TGS as the structure that mediates exchanges of goods or services between technologically separable interfaces or production stages.

Furthermore, Ring and Van de Ven (1989: 173) define TGS as the 'legal forms of governance that apply to different kinds of transactions (ranging from markets to hierarchies) and the structural and procedural safeguards that parties negotiate into a transaction'.[6] In other words, when transacting assets, firms need to choose the most economical governance structure for their transaction. They need to define the inter-organizational relationship that 'occurs when two or more organizations transact resources (money, physical facilities and materials, customer or client referrals, technical staff services) among each other' (Van de Ven, 2000: 25).

According to Williamson (1998: 179), 'each generic mode of organization ... is defined by a distinctive form of contract law.[7] Each mode also differs with respect to incentives and controls and each implements the mix of autonomous and cooperative adaptations differently. The upshot is that each generic mode of organization is defined by an internally consistent syndrome of attributes that gives rise to distinctive strengths and weaknesses'. In a typology of governance structures, Williamson (1979: 247) proposes three broad types of governance structures: non-transaction-specific, semi-specific, and highly specific. The market is the classic non-specific governance structure within which 'faceless buyers and sellers ... meet ... for an instant to exchange standardized goods at equilibrium prices'. In contrast, highly specific structures (idiosyncratic / hierarchical) are tailored to the specific needs of the transaction and the identity of the actors. Semi-specific structures (hybrid) fall in between.

While many studies that apply TC theory approaches use a polar classification to differentiate between market governance and hierarchical governance, the literature has recently seen a 'shift of transaction governance structure from a polar classification toward a continuum,' according to Zhang (2006: 59) referring to Hennart (1993), Heide and Miner (1992) and John and Reve (1982), and also recognized by Williamson (1991). According to Klein (2005: 28), 'a wide variety of contractual and organizational options are available; there are many shades of gray'. According to Macher and Richman (2008), the most common empirical approach nowadays is to conceptualize TGS as one of three discrete types: market transactions, various hybrids or vertically integrated transactions as illustrated in Figure 7.1. Simply put, spot market transactions are at one extreme and vertically integrated transactions at the other extreme; and a variety of hybrid forms fall in between. These include, for instance, joint ventures (Klein et al., 1990), relational contracting (Goldberg, 1980), and bilateral governance (Heide and John, 1990) as well as complex contracting, franchising, reciprocity agreements, equity linkages and the use of hostages, as well as alliances (Oxley, 1999). Oxley proposes a market-hierarchy spectrum ranging from simple unilateral contractual agreements (for example, one-sided licensing agreements, long-term supply contracts and R&D contracts), to bilateral contractual agreements (for example, technology sharing or cross-licensing agreements, joint research agreements) to equity-based alliances (that is joint ventures).

With regard to innovation, an example of hierarchical governance is the development of a technology within an internal R&D center that is exploited only in the firm's own products or services (the left extreme in Figure 7.1). Such transactions refer to the 'internal R&D / exploitation strategy' with the highest degree of organizational integration of the Granstrand et al. (1992) framework (see Section 3.1). Williamson (1979: 241) defines these idiosyncratic transactions as those 'for which the relationship between buyer and supplier is quickly thereafter transformed into one of bilateral monopoly on account of the transaction-specific costs referred to above with profound contracting consequences'. On the other extreme, an acquisition of a technology from an online marketplace with a standardized contract and a lump sum payment could be interpreted as a (spot) market transaction (right extreme in Figure 7.1). Williamson (1979: 241) defines spot market transactions as those 'feasible for standardized transactions (because large-numbers competition is continuously self-policing in these circumstances), where contracting has seriously defective investment incentives where idiosyncratic activities are involved'. Such transactions relate to Granstrand et al.'s (1992) 'technology purchasing / selling strategy' with a low degree of organizational integration.

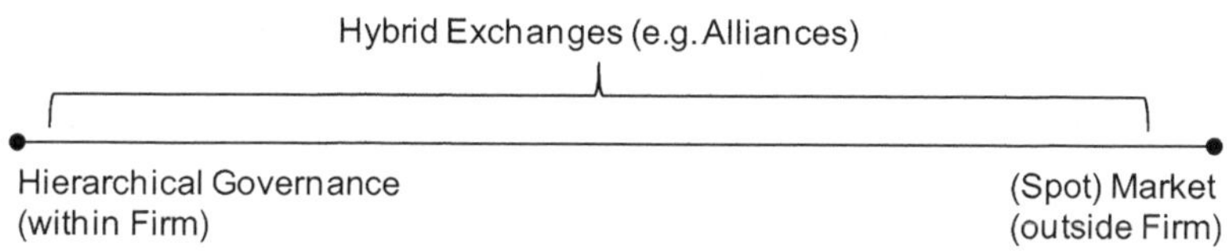

Source: Adapted from Burr (2007: Abb.14) and Pelton et al. (1997: Exhibit 11.2)

Figure 7.1 - Continuum of transaction governance structure types

In line with the above arguments, Granstrand (1979) points out that, in the context of innovation, neither of the extreme forms seem to represent the preferred TGS but rather an intermediate form lying on the continuum between them, particularly when looking at different industries. Granstrand (1979: 371) questions whether there 'exists in some sense optimal intermediate forms of structure between Williamson's polar or undifferentiated types of management and market structures'. An example of a hybrid governance structure would be the joint development of a technology by two firms in an alliance or joint ventures. This example relates to Granstrand et al.'s (1992) 'joint technology venture strategy' with a medium degree of organizational integration.

In addition, TGS are categorized according to the number of actors involved. Aside from transactions involving only a seller and a buyer, Williamson (1985) indicates the possibility of engaging a third party as a 'go-between,' resulting in 'trilateral governance', which will help to enable the exchange between the buyer and the seller (Benassi et al. (2010), Nooteboom (1999)). In the context of technology transactions, Benassi et al. (2010: 16) mention that 'patent brokers are an example of third parties that rely on trilateral governance modes'. If a hybrid or market transaction involves third parties to the extent that they become actively involved, for instance they take temporary ownership of the asset, I consider such transactions trilaterally governed. Particularly when TMI are involved in transactions for the creation of innovations, Feller et al. (2009) expect the importance of TC theory and particularly the transaction governance structure concept to grow in the open innovation context.

7.4 DETERMINANTS OF THE GOVERNANCE STRUCTURE CHOICE

When firms engage in technology transactions on the MfTI, IIP management needs to choose the best suitable TGS. Previous conceptual and empirical studies have presented results that serve as a basis for this study. According to TC theory, the choice of TGS is influenced by the TC to a large extent.

Originally, Williamson (1975) defined five TC determinants, namely transaction frequency, asset specificity, uncertainty, bounded rationality, and opportunistic behavior. According to Yazdanparast et al. (2010) transaction frequency refers to the number of transactions in which parties are involved. Asset specificity refers to the idiosyncratic investments in a partnership that cannot be re-deployed (for example, such as training and special equipment). Uncertainty may be further divided into environmental uncertainty and behavioral uncertainty. Environmental uncertainty refers to circumstances surrounding an exchange that cannot be specified ex-ante, while behavioral uncertainty refers to the difficulty involved in verifying whether compliance with established agreements has occurred. Bounded rationality implies that decision makers have a constraint on their cognitive capabilities (or have a limited information processing ability) and limits to their rationality. Opportunism states that, given the opportunity, decision makers may unscrupulously seek to serve their self-interests and that it is difficult to know a priori who is trustworthy and who is not.

According to Macher and Richman (2008), three of the five determinants suggested by Williamson (1975) are of major importance in various studies. This finding is in line with the argumentation of Williamson (1979), who concludes that the critical dimensions for describing contractual relations are the degree to which investments are idiosyncratic *(asset specificity)* and *uncertain*, and the *frequency* with which transactions recur. Macher and Richman (2008) report results of empirical tests of these three determinants, which are summarized in the following.

First, according to Williamson (1991: 281), *asset specificity* refers to 'the degree to which an asset can be redeployed to alternative uses and by alternative users without sacrificing productive value'. Assets that are unspecialized among users pose few hazards, since buyers in these circumstances can easily turn to alternative sources and suppliers can sell output intended for a specific order to other buyers without difficulty. 'Non-marketability' problems arise when the specific identity of the actors has important cost-bearing consequences. Williamson (1979) refers to transactions of this kind as 'idiosyncratic'. Accordingly, investments can be classified into three categories (nonspecific, mixed, and idiosyncratic). He further proposed six types of asset specificity: (i) site specificity (the advantage of two or more assets jointly located to each other); (ii) physical asset specificity (physical equipment customized to each other); (iii) human asset specificity (training and learning by doing); (iv) brand name capital; (v) dedicated assets (discrete investments in general purpose plant that are made at the behest of a particular customer); and (vi) temporal specificity. According to Macher and Richman (2008), empirical studies have largely confirmed the significant impact of asset specificity as a determinant for the choice of TGS.

Second, Williamson (1979) argues that *transaction frequency* provides a boundary condition separating private ordering (where only the actors themselves are involved in governance) – which is efficient for high-frequency transactions – from trilateral governance in which a third party, such as an arbitrator, is involved in governance that is efficient for low-frequency transactions. The frequency of transactions is commonly labeled as one-time, occasional, and recurrent. Similarly, Nooteboom (1999) argues that – when governance to control transaction costs is necessary but the transactions involved are too small or infrequent to justify the often considerable costs of a 'bilateral' governance scheme (bargaining negotiation) – it can be more efficient to make a simpler overall agreement and engage a third party for arbitration. In this vein, Benassi et al. (2010) argue that trilateral governance should be possible when transactions are occasional and of a mixed and highly specific nature. According to Nooteboom (1999), third parties must have the trust of both protagonists, as well as the competence and intention to judge effectively and fairly. Nooteboom defines six roles for the 'go-between' of which he considers two especially important for innovation, namely that third parties should act as a filter against spillover and an intermediary in building up trust.

Transaction frequency has received far less attention in the empirical literature than asset specificity and uncertainty (Rindfleisch and Heide, 1997). However, the extant empirical studies' results are ambiguous. Some empirical studies have found a significant relationship between transaction frequency and TGS, mainly when measuring transaction frequency using a dichotomous variable (one-time versus recurring) (Klein et al. (1990), John and Weitz (1989)), while others could not prove a positive association between them (Maltz (1994), Anderson (1985), Anderson and Schmittlein (1984)).

The treatment of *uncertainty* in the empirical TC literature is more uneven than that of asset specificity, particularly because of the multitude of uncertainty types. According to Noordeewier et al. (1990: 82), uncertainty generally refers to the 'unanticipated changes in circumstances surrounding an exchange' and typically refers to changes in the environment and future events (Anderson, 1985). A second type of uncertainty has a behavioral foundation. However, this strategic construct has seen far fewer operationalizations in the literature. Harrigan (1986) argues that a plausible explanation for the diverse empirical results around uncertainty is that it needs to be examined in conjunction with asset specificity. When asset specificity is absent, transaction cost economics does not predict that uncertainty leads to more hierarchical forms of governance.

7.5 SELECTING TRANSACTION GOVERNANCE STRUCTURES

In order to select transaction governance structures, Williamson (1979) proposes a framework that is illustrated in Figure 7.2. The framework categorizes TGS according to the three-way classification of contracts by Macneil (1973) and matches asset specificity (investment characteristics) and transaction frequency while it omits discrete transactions that occur only once. This framework is among the core arguments for this study and is of major relevance when matching technology properties (see Chapter 6) with the TGS of technology auctions (see Section 9.3) to decide which technologies should be auctioned.

Oftentimes, specialized TGS are designed for two types of transactions. These are recurring transactions of the mixed and highly specific (idiosyncratic) assets. Idiosyncratic TGS require extensively specialized human and physical assets for production. Consequently there are no obvious economies of scale to be realized through inter-firm trading that the buyers (or sellers) are unable to realize themselves (through vertical integration). However, due to the frequent occurrence of transactions, designing a specialized TGS can pay off. Williamson (1985: 75) defines transactions as *unilateral* when they are 'removed from the market and organized within the firm subject to an authority relation (vertical integration)'. With regard to technology development, this transaction type refers to situations in which firms have a frequent need for highly specific technologies. In such cases, it can pay off to install an own R&D department. According to the Granstrand et al. (1992) framework (see Figure 3.4), such TGS refer to the internal R&D strategy with a high degree of organizational integration. *Bilateral* TGS refer to frequently recurring transactions in which the autonomy of the parties is maintained and the traded asset is less idiosyncratic (mixed). With regard to technology development, this refers to, for instance, situations in which firms set up a licensing department to continuously acquire technologies, where each specific contract is developed internally and negotiated bilaterally. This TGS is similar to the 'technology purchasing / selling strategy defined in Granstrand et al. (1992) framework with a medium degree of organizational integration. Their recurrent licensing activities (and learning effects) permit the costs of the specialized structures for running the licensing department to be recovered.

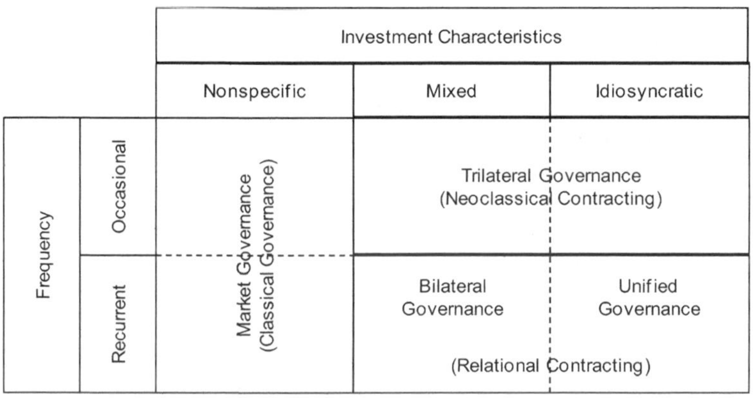

Source: Adapted fromWilliamson (1979, Figure II)

Figure 7.2 - Matching governance structures with commercial transactions

For mixed and highly idiosyncratic transactions that take place less frequently (occasional) Williamson (1979: 249) proposes *trilateral governance* involving mediating third parties. For such transactions, once the actors have entered into a contract, there are 'strong incentives to see the contract through to completion'. Not only have specialized investments been put in place, but the transfer of these assets to a successor supplier pose inordinate difficulties in asset valuation. The principals' interests in sustaining the relationship are especially great for highly idiosyncratic transactions. Market relief is thus unsatisfactory. With regard to technology transactions, for instance, this situation relates to infrequent transactions in which an SME has no internal licensing department, but only maintains the coordination of the transaction and outsources various tasks to a TMI (for example, the identification of a buyer and valuation of the technology). For the SME, due to the low frequency of transactions, running a full internal licensing department would not pay off. In the Granstrand et al. (1992) framework, such a transaction would relate to the technology purchasing / selling strategy with a low degree of organizational integration.

Market governance is the main TGS for nonspecific transactions that occur occasionally and frequently. Market governance is especially efficacious when transactions occur frequently and the traded asset is standardized (nonspecific). In such situations, both actors only need to consult their own experience to engage in the transaction at little expense. For instance, the purchase of a mature technology, which can be easily understood by the acquiring firm's engineers, can be purchased readily on the market (for example, a highly codified chemical substance and its IP assets that are transacted to an acquiring firm and manufactured with existing processes and equipment). Being standardized, alternative purchase and supply arrangements are presumably easy to find. For nonspecific and occasional transactions, buyers (and sellers) are less able to rely only on their own experience and, subsequently, they rather refer to consultants (for example, rating services and PLF). In the Granstrand et al. (1992) framework, this TGS would also relate to the technology purchasing / selling strategy, although with a lower degree of organizational integration, that is a firm would not need a licensing department, but could perform these transactions via its purchase department.

To summarize, the three determinants originally proposed by Williamson (1979) still appear to be the most prevalent after intense empirical scrutiny. Thus, the following argument of Williamson (1979: 239) still seems to be valid. According to him, 'simple governance structures should be used in conjunction with simple contractual relations and complex governance structures reserved for complex relations ... [The use] of a complex structure to govern a simple relation is apt to incur unneeded costs and [the] use of a simple structure for a complex transaction invites strain'. A substantial amount of research has produced consistent results; that is, when asset specificity, uncertainty, and frequency are low, market TGS are more economical; otherwise, hierarchical TGS tend to result in lower transaction costs (Anderson (1985), Heide and Miner (1992), Walker and Weber (1987)). Thus, more integrated TGS are associated with a higher degree of relationship-specific assets, more complex transactions, greater uncertainty (including but not limited to customer, supplier or technological uncertainty) or more frequent exchange.

Furthermore, prior research has proven that asset specificity remains the most important determinant for the choice of TGS. In the innovation context, complex and novel technologies are examples of technologies with high asset specificity, that is, since they are in an early stage of their life-cycle, they cannot be easily adopted by a new owner who rather needs to dedicate specific resources (for example, intensify employee trainings). In contrast, a well-known, discrete technology in a mature lifecycle stage is an example of low asset specificity.

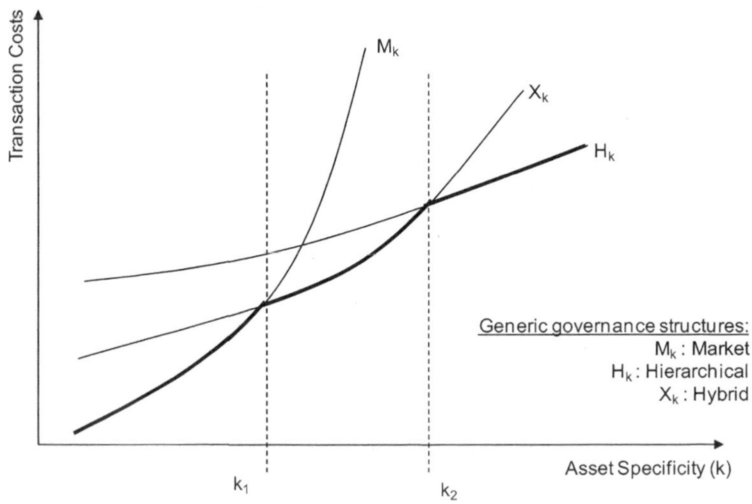

Source: Adapted from Williamson (1979, Figure II) also appearing in Williamson (2002, 1996).

Figure 7.3 - Transaction costs and asset specificity of governance structures

Figure 7.3 illustrates the transaction costs of three generic TGS (both extremes of the continuum illustrated in Figure 7.1 and a hybrid TGS), which depend on the asset specificity as the strongest determinant for the TGS choice. According to Figure 7.3, a market transaction would be less costly than a hybrid transaction for the trade of assets with asset specificity below k_1. For assets with an asset specificity above k_2, however, a hierarchical transaction (that is vertically integrated) would be the most cost-effective. For assets with an asset specificity between k_1 and k_2, a hybrid transaction would be the most cost-effective.

According to Dyer (1997: 535) referring to Williamson (1985), the standard reasoning for the curves is that 'as asset specificity increases, more complex governance structures (that is more complex contracts) are required to eliminate or attenuate costly bargaining over profits from specialized assets'. Thus, transaction costs are presumed to increase with an increase in asset specificity. To date, most research has produced results that are consistent with the logic that asset specificity is greater for transactions within hierarchies than across markets (Masten (1984), Walker and Weber (1984), Monteverde and Teece (1982)).

This is in line with Williamson (1979), who argues that incentives for trading weakens as transactions become more idiosyncratic. The reason for this is that, as the human and physical assets involved in the transaction become more specialized to a single use, and hence less transferable to other uses, the choice of a TGS turns towards that option with superior adaptive properties, that is vertical integration (hierarchy). Scholars have noticed this firm behavior within the closed innovation paradigm where many firms prefer to develop technologies internally (that is vertically integrated) within their own R&D resources.[8] However, in the open innovation paradigm, firms proactively engage in inter-firm transactions of technologies, which are often characterized by a high degree of asset specificity. In such situations, TMI can have positive effects, designing novel (trilateral) TGS. Section 5.3 addresses how the specific TGS of technology auctions is characterized, and thus how it fits into this framework.

Furthermore, referring to Jacobides (2008), Jacobides and Winter (2005), Langlois and Foss (1999), and Langlois (1992), in terms of the choice of TGS, one has to consider that a firm's preference might change over time if TC are dynamic. For instance, when firms continuously establish internal capabilities (for example, learn by participating in transactions) TC for intra-firm transactions (hierarchy) decrease over time while TC of market transactions (inter-firm) might actually remain constant if TMI keep their fees constant (that is although TMI might also learn to optimize their transaction model, hence decreasing internal TC, they have an incentive to keep them high in order to increase their profits). Hence, although market transactions might have appeared economically favorable in a certain period of market development, leading to firms' disintegration (that is outsourcing of transaction governance to TMI), this might be less favorable in a later period. Instead, firms might turn to reintegration due to lower TC of intra-firm transactions if hierarchical TC are reduced below the level of the TC for market transactions.

NOTES

1 For other reviews, see Rindfleisch and Heide (1997) or Shelanski and Klein (1995).

2 Please note that TC must be considered a rather narrow definition of the broader term 'governance costs'.

[3] According to Allen (1999: 903), Demsetz (1964) was the 'first to deal with the breadth of definition used for transaction costs'.

[4] Teece and Pisano (1994) are among the first (if not the first) to initiate discussions on dynamic capabilities.

[5] The term governance derives from the Greek verb κυβερνάω [kubernáo], which means to steer and was already used by Plato in a metaphorical sense (European Commission, 2002).

[6] Other theories (for example, RBV and real options) have also been tested to predict the choice of transaction structures or the make-or-buy decision. However, according to Klein (2005: 28), none of them 'has produced a body of evidence remotely rivaling the transaction cost explanation for vertical integration'. The resource-based (or capabilities) view is frequently applied in the strategic-management literature, but it has not generated a substantial body of empirical work. However, scholars have recently started to advocate an integration of different theories into a comprehensive theoretical framework (Poppo and Zenger, 1998), arguing that the RBV and transaction cost theory should be seen as complementary approaches (Mahoney and Pandian, 1992). For comparative approaches, see, for example, McIvor (2009) and Madhok (2002).

[7] Transaction cost economics further recognizes that different governance modes are supported by different legal regimes, ranging from contract law for market governance to employment law for hierarchies (Williamson (1991), Masten (1988)).

[8] Applying a more complex reasoning that also includes arguments from evolutionary economic theory, such as bounded rationality, Hall (1994) reaches similar conclusions.

PART III

An Empirical Study of Technology Auctions

8. Methodological Approaches

8.1 THE GOVERNANCE STRUCTURE OF TECHNOLOGY AUCTIONS

To derive an understanding of the TGS of technology auctions, based on specifically selected cases[1] I conducted two complementary studies that combine insights from a qualitative auction process structure analysis (1st pre-study) with insights from the sellers' experience of transactions (2nd pre-study). The combination of different data sources for the pre-studies can be considered a form of triangulation. According to Scandura and Williams (2000), a similar aspect (that is the TGS of technology auctions) is investigated from different perspectives, which contributes to an increased reliability of the results.

8.1.1 The Process Structure of Technology Auctions

Section 2.3 explains why I chose to conduct multiple case studies for the first pre-study to address the first specific research question. Primarily following the approach of Yin (2003) to conduct case studies, I compiled descriptive case studies which were analyzed in a comparative manner in order to derive similarities and differences between both processes. In this section, I elaborate mainly on three aspects of this approach, including the selection of the case studies, the data collection, and the analysis.[2]

Case Selection

When the case studies were conducted at the end of 2007, to my knowledge, that is based on the pre-study of TMI reported in Tietze (2008) and various expert interviews, there were only two firms that had conducted public technology auctions. By the end of 2007, the US-based TMI, Ocean Tomo Inc. (OT) had organized five auctions (four in the US and one in the UK) and the German-based daughter firm of IP Bewertungs AG, Intellectual Auctions GmbH (IPA), had held one auction in Germany. Thus, the maximum number of possible case studies was limited to six.[3]

With regard to the choice of cases in multi-case studies' research designs, Yin (2003) proposes different measures to raise the certainty of the case study's results. These measures include replication (literal and theoretical) and choosing cases, which differ to some extent – preferably, they have to be from the extreme ends of the spectrum, that is what he calls 'two-tail' design (for example, good / bad) with some theoretical conditions being deliberately chosen. However, following the proposal of Seawright and Gerring (2008: 294), who conducted a comprehensive review of case selection techniques, attention should be paid 'to purposive modes of sampling'. Being limited to six potential cases, after a first superficial investigation of the auction catalogues of the five OT auctions, I realized that they were similarly designed to a large extent. Even the UK auction was very similar, which seemed to be due to the fact that this auction was organized by the same people from the US branch of the auction firm. It became reasonable to assume that OT went through a learning curve from their first to their fifth auction, since they had made minor adjustments from one auction to the next. Thus, I decided that the results would gain little additional reliability from compiling case studies of each of the five OT auctions, therefore. It was decided to investigate only the most advanced auction process, that is the latest OT auction conducted in October 2007.

To raise the reliability of the study's results, I complemented the results from the latest OT auction with a case study that differed to some extent. Thus, I chose the auction organized by IPA as a contrasting case study. Organized by a firm based in Europe (in contrast to OT being based in the US), this condition allowed for the assumption that the auction design differed enough to justify a second case study. Obviously, the choice of the latest IPA auction was determined by the fact that IPA had conducted only one auction by the end of 2007.[4]

Data Collection

Yin (2003: 85) discusses 'six major sources' for the collection of evidence that are 'highly complementary' (that is documentation, archival records, interviews, direct observation, participant-observation, and physical artifacts). In this study, I used three of these sources to collect data for his case studies, namely documentation, interviews, and direct observations.

As Yin (2003) favors data from documentation, because it can be regarded as stable, unobtrusive, exact, and as having a broad coverage, I collected data from different *documentation* types. The auction catalogues served as upfront and primary sources. Although the catalogues should be regarded as promotional material, the rules and regulations described in the first section of each catalogue can be considered legally binding thus making these catalogues a reliable data source. The catalogues provided detailed information on the legal rules and regulations as well as the institutional framework and chronology of the auction process (for example, the closing date for the hand-in of technologies). Descriptions of rules and regulations in the auction catalogues of the first four OT auctions were further studied to verify the differences and similarities between them and the fifth OT auction. In addition, I collected data for the auction processes from the corporate websites of the auction firms, OT and IPA. Moreover, I used data from various internet sources including comments on the technology auctions by IP experts (lawyers, economists, and IP professionals) who had participated in them.[5]

Silverman (1993: 30) stressed that observation is fundamentally about understanding the routine' (emphasis added). Following this advice, in order to gather firsthand insights into the process of technology auctions, I attended two auctions.[6] During the IPA auction, two observers participated in the auction, contributing to the reliability of the results according to Yin (2003). Thus I was able to realize the benefits mentioned by Yin (2003: 92), that is to cover the events in real time and cover the context of the events.

According to Yin (2003: 92), 'interviews are an essential source of case study evidence.' While attending the technology auctions of IPA in May 2007 and of OT in June 2007, I also interviewed three different participant groups of the technology auctions.[7] Interviews were conducted with organizers, sellers, and prospective buyers of the technologies as well as with suppliers of dedicated services to the technology auctions (for example, valuation service provider).[8] As suggested by Fontana and Frey (1994) I favored unstructured interviews in order to deepen my understanding and to clarify specific questions that remained open after having investigated the auction catalogues, in order to access the setting of the technology auctions and get acquainted with the language of the IP community. According to Punch (2005), this approach is preferred if the interviewer wants to drift into intimate conversations in order to build up trust. Using this approach, I was able to address even critical questions so that respondents could feel comfortable expressing their opinions, for instance, with regard to weaknesses of the technology design.

Case Analysis

Having collected data from three different sources and having clarified open questions and contradictions through interviews, the data analysis (for this but also for the second pre-study) was conducted primarily following the procedure proposed by Yin (2003). However, for certain specific aspects, the contributions of Gibbert et al. (2008), Eisenhardt and Graebner (2007),

McCutcheon and Meredith (1993), Leonard-Barton (1990), Bourgeois III and Eisenhardt (1988), and Bonoma (1985) were taken into account. The analysis was carried out in two stages, that is in line with Price and Erwee (2006: 10) with the aim 'to distinguish the processes and outcomes across cases and to expand the understanding of similarities and differences'.

Throughout the *first stage* of this analysis, I followed the third 'general analytic strategy' proposed by Yin (2003: 109) in order to develop the structure of technology auction processes. I conducted what Eisenhardt (1989) calls 'within-case analyses', which is similar to the fifth case study technique ('chronologies') recommended by Yin (2003) or the sixth analytical technique described by Miles and Huberman (1994). While the auction process generally covers events over time, the data were sorted in chronological order of the auction process to illustrate the causal structure of the process events. According to Miles and Huberman, this technique is particularly suitable for the descriptive purpose of case studies. According to Yin (2003: 125), 'the compiling of chronological events is a frequent technique in case studies and may be considered a special form of time-series analysis.' While drafting the case reports of the technology auction processes, I was aware of avoiding the common pitfall raised by Yin (2003: 154), that is that 'disproportionate attention is usually given to the early events and insufficient attention to the later ones'. However, he did not draft the cases backward as suggested, but rather avoided this pitfall and spent the appropriate time on the major parts of the case studies.

In order to align the empirically observed auction process models with the literature of technology transactions (exploitation and acquisition process models), I applied the technology transaction process model by Lichtenthaler (2006a), which is presented in Chapter 3, to categorize the different phases of the transaction process. Investigating the bidding procedure as an essential step in the auction process, this part of the case descriptions was supported by Kumar and Feldman's (1998) auction process model presented in Chapter 5. Compared to Lichtenthaler's (2006a) model, the first and second stages of the auction model are part of the planning phase. Stages three and four represent the negotiation phase and the fifth stage represents the realization phase. Thus, the individual case descriptions follow a chronological order along the three core phases of Lichtenthaler's (2006a) model. These are presented in Section 9.1.

In the *second stage* of the analysis, I conducted a cross-case analysis similar to the fifth specific case analysis technique recommended by Yin (2003).[9] According to Bourgeois III and Eisenhardt (1988), the cross-case analysis is particularly applicable if one aims to identify the similarities and differences between multiple cases. Yin (2003) argues that this method is directly analogous to cross-experiment interpretations, which also have no numeric properties when there is only a small number of experiments.

To conduct a cross-case analysis, Eisenhardt (1989: 541) suggests three tactics that 'force investigators to go beyond initial impressions, especially through the use of structured and diverse lenses on the data... [and] ...enhance the probability that the investigators will capture the novel results which may exist in the data'. I followed the second tactic and selected specific aspects of the cases for which he then listed the similarities and differences between each pair. To extract the differences and similarities of the technology auction processes, I compiled tables for the different aspects of the study as recommended by Eisenhardt (1989). I aimed to reach external validity by conducting multiple case studies. According to Eisenhardt multiple case studies on a given topic have more external validity, that is, generalizability, than a single case, while construct validity was ensured through the use of multiple sources of evidence.

A criticism often raised against cross-case analyses with a limited number of cases (for example, mentioned by Yin (2003)) is that the examination relies strongly on an argumentative interpretation and not on numeric tallies. However, my case studies serve a revelatory purpose and do not claim any statistically significant results. I took measures to avoid the cross-case conclusions from being 'biased by undue attention to one or a few of the entire array of cases' as warned by Yin (2003: 165). Guided by the theoretical frameworks of Kumar and Feldman (1998) and Lichtenthaler (2006a), I had a thorough theoretical basis for the analysis of the auction process structures.

8.1.2 Case Studies of Successful Transactions

The second pre-study complements the first one since it also addresses the first specific research question presented in Chapter 2. Following the recommendation of Punch (2005), the second pre-study was designed to ensure internal consistency and a coherent logic across the components of the study, including the sampling. The general procedure to prepare and conduct the case studies are, again, based on Yin (2003). Similar to the description of the research approach of the first pre-study, I elaborate mainly on three methodological aspects in the following section. Again, these include the selection of the case studies, the data collection, and the data analysis.

Case Selection

The purpose of the second pre-study is to complement the results of the first pre-study, deepening the understanding of the governance structure of transactions via technology auctions. Complementary to the process analysis, additional insights into the organization and process structure of technology auctions were collected from the sellers of patented technologies. Case studies were selected from the set of transaction data compiled for the quantitative analysis. The dataset comprises 393 patented technologies that were offered at six technology auctions between spring 2006 and autumn 2007.

To identify cases that best serve the purpose of this study, providing rigorous insights into the governance structure of technology auctions, I applied a deliberate sampling approach ('purposive sampling') following Punch (2005). Miles and Huberman's (1994) framework, which outlines 16 sampling strategies was useful to ensure the best fit for the purpose of this study. Three of these strategies suited the purpose of this study. Accordingly, the sampling strategy can be regarded as a mix of criterion-based, stratified, and extreme case selection.

To ensure a maximum level of reliability, all case studies had to fulfill two basic criteria. First, the patented technologies had to have been sold in order to guarantee a certain level of illustrative power. This filter reduced the number of possible case studies to 156. Second, I selected only those patented technologies that were sold by firms, since this study is situated in the management science field, it is mainly interested in understanding firm behavior.[10] This filter revealed 66 possible case studies. To raise the reliability of the study's results, I also employed a *stratified selection* filter. The firms that had sold patented technologies were labeled as SME or large firms. According to Yin (2003), the comparison between case study results across two sub-groups (contrast groups) allows for deeper insights and increases the reliability of the results. Thus, I ensured that SME as well as large firms were equally present among the case studies. Finally, I attempted to design a filter to select those cases from which the most insightful results could be expected. In such situations, Yin (2003) recommends sampling *extreme cases* that often have the potential to provide the richest insights. In this study I therefore had to identify extreme cases from which it can be assumed that the sellers had gained considerable experience from the transaction ('best practice cases').

To identify best practice cases, one can argue that patented technologies that sold for particularly high prices did not reach these prices due to serendipity. Most likely, their sellers possess certain knowledge indicating that those technologies are particularly valuable. In contrast to patented technologies that sold for moderate prices, it appears reasonable to assume that the sellers of patented technologies with high price expectations had carefully prepared the transactions. Thus, one can expect to derive the most rich and solid insights from sellers of patented technologies that sold their technologies for the high prices.[11]

To identify these best practice cases in the dataset, I applied a monetary, but relative measure. From each of the six technology auctions, I selected the two transactions that sold for the highest and second highest prices. Applying this filter on each auction separately, I took into account that the different auctions reached different price levels (for example, the first European auction reached considerably lower average sales prices than the third US auction). Thus, I avoided that cases from certain auctions were discriminated from the sample.[12] Applying this filter revealed 12 potential case studies. Among them, eight patented technologies were sold by SME and four by large firms. The sales prices varied from €25 thousand to €3.7 million. Four of the 12 potential case studies agreed to participate in the study.

Data Collection

Among the six sources of evidence discussed by Yin (2003), for this pre-study I used interviews in order to collect first-hand primary data. Minichiello et al. (1990) provide a continuum model for interviews from which Punch (2005) derives three types of interviews (structured, group, and unstructured interviews) from which the first interview type 'entails more structured questions, along the lines of a formal survey' according to Yin (2003: 91). Punch (2005: 170) argues that, with structured interviews, 'flexibility and variation are minimized, while standardization is maximized'. Since I aimed to achieve the highest possible degree of comparability among the cases studies, I decided to use a structured interview approach for the collection of primary data. Having developed the interview guideline, structured interviews further allow for straightforward data coding. Furthermore, according to Larsson (1993: 1518), structured interviews increase the 'number of case studies that could be done within a certain resource limitation', although using a standardized interview guideline 'sacrifices the depth of case data'. However, the nature of this study is descriptive rather than explorative.

In order to best complement the first pre-study, three aspects were selected that should be addressed throughout the interviews. These include the 'organization of the transaction process', 'advantages and drawbacks of the technology auction process', and 'the sellers' perception of the transaction success'. For each of these aspects, questions had to be developed for the interview guidelines from prior literature from related research fields.[13] These include, in particular, the transaction process model proposed by Lichtenthaler (2006a), which is relevant for questions regarding the resource requirements in the transaction process phases, by Gambardella et al.

(2007) regarding the advantages and disadvantages of using intermediaries for technology sales, and by Milgrom (2005) for questions related to specific advantages and disadvantages of the auction design. Additionally, Rieck (1993), Kaufmann (2001), and Thiele (2003) provide helpful insights for questions about the firms' experience with traditional bargaining negotiations for technology transactions. To raise the validity of the questionnaire, I reused questions from prior questionnaires as far as possible, including the fourth version of the EU Community Innovation Survey (CIS4 2006), Granstrand (2000a), and Granstrand (forthcoming)). Throughout the development of the interview guide I aimed to fulfill the criteria regarding accuracy, unambiguousness, and complexity as specified by Atteslander and Cromm (2006) and Punch (2005).[14]

Having finished a first literature-based version of the interview guideline, to increase the questionnaire's validity, its content as well as its usability (that is understandability of the questions) were tested with three experts, who provided valuable comments and suggestions on the chosen phrases as well as the content and its answerability.[15] Having taken these comments into account, a second version of the guideline was piloted with four selected firms.[16] After some minor revisions, the final version of the interview guide comprised three parts, including 23 overall questions. The majority of the questions were of a closed nature in order to ensure that the respondents could answer the questions in due time as well as to ensure the comparability of the answers and thus the case studies in the cross-case analysis (Atteslander and Cromm, 2006).

In the first part of the interview guideline, 15 questions address the properties of the patented technology for sale, subsequently addressing the intra-firm organization of the transaction, the motivation of the seller to exploit this particular technology, as well as the selection process for that technology. Question types on this part include yes / no questions, alternative questions with multiple answer possibilities and closed questions with a five point Likert scale structure.[17] One open question was included regarding cost drivers throughout the transaction. In the second part of the interview guide, I investigated the advantages and disadvantages of transactions via technology auctions compared to classical bargaining negotiations. These questions have particular relevance for the analysis in terms of the success of a transaction from the sellers' perspective. This part comprises four closed questions with a five point Likert scale, plus one open question to identify further potential for improving technology auctions. In the final part, three control questions addressed the firms' particular experience regarding technology and IP asset transactions as well as their economic performance in the previous years.

In December 2007, I addressed the questions specified in the interview guideline to the 12 selected firms.[18] In order to raise the reliability of the results, I conducted interviews with the persons who were most involved in the transactions, preferably the former project manager responsible for the respective transaction.[19] Contact details of those persons were collected upfront through telephone enquiries, whereby I also wanted to raise the likelihood that firms participate in the study following the advice of Atteslander and Cromm (2006).[20]

The primary data collected from the interviews were complemented by additional data compiled from secondary sources. I collected secondary data from archival records (records of the traded patents from online databases incl. mainly the EPO and USPTO) and available documentation. From these patent documents, I acquired specific information about the patents, including their age, the number of claims, etc. Auction catalogues provided data on the type of technology sold, the number of patents included in a bundle, as well as the application area of the technology and the exact transaction volume. The corporate website provides data on the organizational structures of the seller as well as on business data from previous years.[21]

In order to raise the reliability and validity of the case studies, Morse et al. (2002: 1) suggest 'implementing verification strategies, integral and self-correcting'. Among different strategies he essentially suggests collecting data from various sources to cross-validate the results (for example, by interviewing different persons involved in a project) to avoid single response bias. Within this study's approach, the data were collected through a key respondent using the interview guide, which appeared to be sufficient to fulfill the descriptive purpose of this study. Thus, in contrast to what Morse et al. (2002) suggest, I collected no additional data from other persons involved in the transactions to correct for eventual respondent bias.

Case Analysis

Similar to the analysis of the case studies of the first pre-study, the analysis of technology transactions was carried out in a two-step procedure. Following, for instance, Punch (2005) and Yin (2003), I first prepared individual case study descriptions, which were then examined through a cross-case analysis in a second step.

In contrast to the case study descriptions of the first pre-study, the case descriptions of this study do not follow a chronological but rather a topic-focused order according to the three main aspects investigated, that is the intra-firm organization of the transaction, the advantages and drawbacks of the transaction, and the seller's judgment of the success of the transaction. The individual case descriptions are presented in Section 9.2.

The individual case descriptions were further subject to a cross-case analysis in a second step. The case studies are also split into two categories, that is SME and large firms. Following Eisenhardt's (1989) suggestion, this tactic allows for a deeper understanding of differences and similarities between the results and thus raises their reliability and generalizability. In order to cope with the rich data, followed the advice of Leonard-Barton (1988) and used tables to support the cross-case analysis.

In order to compare the individual case results I developed a quantitative approach, which however is not meant to inappropriately apply statistical analysis techniques to qualitative studies. Despite the low number of cases, the conservatively designed measures allowed me to interpret the results with respect to the homogeneity of the respondent's answers. The individual case study's results are compared using a triple-measure (M_x). This measure is a result of three independent measures (M_1; M_2; M_3). M_x has a value of 1 for any item if all three independent measures have a value of 1. M_1 takes into account the mean scores for each item (m_c) of each case study. M_1 has a value of 1 if any of the two extreme conditions is fulfilled: m_c is (i) either particularly high, that is equal to or above 2.5 (mean value of the Likert scale ranging from 1 to 5) or (ii) particularly low, that is equal to or below 1.0. M_2 takes into account the mean scoring behavior of each case study for each item and across all items (m_i). M_2 takes on a value of 1, if three out of four firms scored an item higher than the respective m_i. M_3 assesses the divergence of answers given by the firms for any item. Hence, M_3 is a measure of the homogeneity of the respondents' answers. For each item, M_3 has a value of 1 only if the difference between the highest and the lowest values given by a respondent are smaller than 2.5. An item is considered to have an impact only if the triple measure M_x has a value of 1. Thus, although quantitatively measured, M_x gives a qualitative indication of respondents' agreement on a certain extreme item value, but does not indicate whether the item has a high or low tendency. This must be interpreted from the respective m_c value.

8.2 A QUANTITATIVE STUDY OF AUCTIONED TECHNOLOGIES

The quantitative study addresses the second specific research question of this study as specified in Chapter 2. The objective of this study is to explain how auctioned technologies are characterized and whether they share certain properties in order to derive recommendations for technology owners.

This study should thus contribute to the understanding of technology properties that favor auctioning when firms have to choose a preferred governance structure for a transaction on the MfTI.

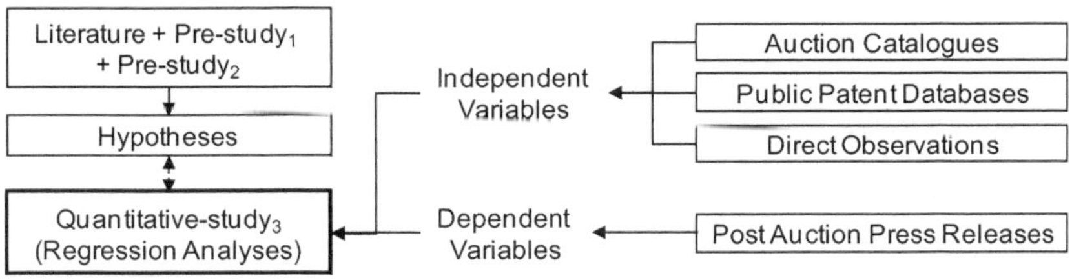

Figure 8.1 - Conceptual framework of the quantitative study

If this turns out to be the case, one can conclude that such technologies can be traded preferably via auctions. In contrast to the first two pre-studies, which are of an explorative and descriptive nature, this quantitative study builds on the first and second pre-studies and is of a confirmatory nature. As illustrated in Figure 8.1, the inductive pre-studies were complemented by the literature (on auctions and technology transactions) to develop hypotheses in Section 9.3 that were tested in this quantitative study, which is of a deductive nature.

In the following, I present relevant methodological aspects focusing on the data collection (with a special emphasis on data preparation) as well as the data analysis. The methodological framework illustrated in Figure 8.1 reveals the data sources of this quantitative study. The results from the regression analyses are presented in Chapter 10.

8.2.1 Data Collection

The quantitative study is based on patent data that were collected for technologies offered and sold at six auctions that took place between spring 2006 and autumn 2007 (see Table 8.1).[22] Five of the six auctions (four in the US and one in the UK) were organized by OT and one in Germany by IPA.[23]

Data was gathered and combined from four different sources as illustrated in Figure 8.1 and Table 8.2. Three data sources were provided by the auction firms (catalogues, direct observations, and post-auction press releases) which were complemented with data from official, public patent databases for the patents related to the auctioned technologies.

Table 8.1 - Overview of auctions covered in the dataset

Auction ID	Auction Date	Auction Firm	Location	Pulication Date of Press Release	Δ Auction Date - Publ. Date
A_1	6.4.2006	OT	San Francisco, USA	11.5.2006	35 days
A_2	26.10.2006	OT	New York, USA	2.11.2006	6 days
A_3	19.4.2007	OT	Chicago, USA	23.4.2007	4 days
A_4	15.5.2007	IPA	Munich, Germany	23.5.2007	8 days
A_5	1.6.2007	OT	London, UK	4.6.2007	3 days
A_6	25.10.2007	OT	Chicago, USA	30.10.2007	5 days

Notes: OT (Ocean Tomo, LLC); IPA (IP Auctions GmbH)

For each technology the auction catalogue provided data on the seller, its application area (for example, electronic commerce, microelectronics / IT or green technology), the sales price the sellers expected to achieve at the auction, and data with regard to the specific properties of each transaction.[24] These data relate to the transaction type (in most cases the technologies were offered for sale; only a few technology licenses were offered for sale), and the auction type (the majority of the technologies were subject to English auctions; a few were offered as Dutch auctions).[25]

Table 8.2 - Data compiled from different sources

Catalogues	Press releases	Observations	Patent databases
Application area (e.g. electronic commerce) [a]	Sales price	Opening bid	Application date
Auction type (e.g. English, Dutch) [b]	Sales status		Granting date
Country designations [a]			Inventor names
Expected sales price [b]			IPC classes
Legal status (e.g. granted, applied) [c]			Original assignee names
Count of claims [d]			Title of patent
Patent application numbers [a]			Patent abstract
Selected forward citations [a]			
Seller name [a]			
Transaction type (e.g. sales, license) [e]			

Notes:
[a] Available for all technologies
[b] Fully available for IPA auction, OT auctions are only English auctions
[c] Available for most OT technologies, not available for IPA technologies
[d] Partly available for OT and IPA auctions
[e] OT offered technologies only for sale, available for IPA technologies

Furthermore, the catalogue provided data on the patents embedded in each technology. Specific patent data includes the application numbers, the count of patent claims, selected forward citations, and country designations. Through direct observations at two technology auctions that I attended (A_4 and A_5 in Table 8.1), additional data were collected on opening bids. I complemented the data from the catalogues with data provided in the press releases published by the auction firm after each auction. For each technology, the press releases indicated the sales status or the sales price for each technology.

For each of the patents included in the technologies, I also collected data from publicly available patent databases, based on the patent application numbers listed in the catalogues (including Depatisnet by the DPMA, esp@cenet by the EPO, Patentscope by WIPO, PATFT and AppFT by USPTO, and databases from other national PTOs). The data I derived includes the application and granting dates, the names of the inventors, the four digit IPC classes, the original assignee names, the exact title of the patent, and the patent's abstract.[26]

8.2.2 Data Preparation

With regard to the preparation of the data, further issues need mentioning. First, I had to harmonize certain data stemming from different data sources. Second, I had to check for and handle missing data. Third, I had to consolidate the data for the individual patents collected from the patent database so that they could be applied in the analysis of the technologies.[27] Details are provided in Chapter 8 for the measures applied in the regression analyses.

In the following section, I elaborate on these three issues. However, this is not done exhaustively for the full dataset; instead, I focus on those measures that were used for the regression analyses. After the data collection, a few minor issues also raised questions. The catalogues of the different auction firms differed somewhat in their terminology. For instance, OT used the term 'issue date,' which IPA labeled 'granting date.' Moreover, OT had adjusted their catalogues slightly from auction to auction. These differences were clarified in dialogues with OT and IPA.

Data harmonization

Data harmonization was necessary in two cases with regard to the application areas collected from the catalogues and the assignee names (used to measure the technology source type) as collected from the patent databases. Although the catalogues from OT and IPA appeared to be fairly similar in general[28], as indicated in the notes of Table 8.2, the categorization of application areas differed across the catalogues. Furthermore, spelling formats of assignees differed in the data derived from the patent databases. This is a commonly known problem.[29]

Application areas

The auction catalogues provided data on the application area for each technology. This appeared particularly helpful, since, generally, 'patent data cannot be directly attributed to particular industries, as patent documents do not explicitly include the information that makes it possible to identify the economic sector to which the technology embodied in the patent is associated' (OECD, 2009: 90). However, the auction firms along with the sellers, had assigned each technology to an application area.[30] Nevertheless, the application areas were not standardized across the auctions, which had to be accounted for in the analysis. The classes OT had chosen in one catalogue were not exactly the same as those in other OT catalogues or in IPA catalogue.[31] In total, the technologies were assigned to 71 classes. Such an amount would not have allowed for solid statistical analyses due to the high fragmentation of the sample into too small groups that would include an average of just six cases from this database.

Table 8.3 - Classification of application areas

Application area	ID #
Automotive Systems & Devices / Industrial Machinery	1
Aviation & Aerospace	2
Business Methods / Financial Services	3
Chemical Engineering / Applied Sciences / Advanced Materials / Nanotechnology	4
Consumer Products / Electronics	5
Digital & Home Media / Entertainment	6
Display / Multimedia / User Interface Technologies	7
Electrical / Mechanical Engineering / Process Automation	8
Energy / Utilities / Power Technologies / Oil & Gas	9
Green Energy / Technology / Industrial Systems	10
Internet / Web Services / E-Commerce / Internet TV	11
Location-based Systems / Applications / Logistics	12
Medical & Life Sci / Devices / Telemedicine / Bio-chip / Bio-tech	13
(Micro)Electronics / Computer Systems & Technologies	14
Telecommunications (Wireless, Cellular, Optical & Other)	15
Miscellaneous	16

Thus, I had to apply a manual harmonization approach. Luckily the classes showed a reasonable overlap, indicating that they could be harmonized.[32] Furthermore, to solve this problem, different approaches were available to code standardized application areas (for example, NACE) from the patent data. For instance, OECD (2009) suggests three ways in which patents can be attributed to industries. These ways include having patents attributed directly to them by having an expert examine each patent ad hoc, having the applicant's (company's) industry code assigned, and by a priori establishing correspondence between the IPC classes and industries with the help of experts and subsequently integrating these into a concordance table that can be used to translate patent IPC classes into industrial application areas. For example, Oriani and Sobrero (2001) or Zeebroeck et al. (2005) analyze large patent datasets using four digit IPC classes that are matched with NACE industry classes to form concordance tables.[33] Since the four digit IPC classes were available in this dataset, I attempted to use the classification scheme suggested by Schmoch et al. (2003). However, due to the two-level structure of the dataset and the requirement for this analysis to consolidate individual patent data on a technology level, this approach did not turn out to be very useful. For each patent, I would have needed to translate its IPC classes into industry classes, which would then have been consolidated on a technology level. As IPC classes differ across the different patents belonging to a technology, I would have obtained different industry classes for each lot with the unsolved problem of how these classes could have been consolidated in a meaningful manner.

Attempting to identify and consolidate related classes bore the risk of producing biased results. I tried to limit subjectivity bias by applying a separate code and control mechanism using two persons. Finally, I was able to consolidate the application areas into the 16 classes presented in Table 8.3. Although I trust that my approach ensures that all the technologies are related to proper application areas, the classification does not relate to any official industry classification (for example, NACE) and thus only allows for limited comparability of this study's results with those of other studies. To limit this effect, the study applied a concordance table, which was used for the consolidation exercise (See Annex II).

Assignees

From the data derived from the patent databases, I had to standardize assignees' heterogeneous spelling formats, which were used to code the technology source type.[34] OECD (2009) recommends two approaches to deal with the problem of spelling variations. The first rule-based approach involves the definition of rules to compare the similarity of names. The second approach relies on the use of dictionaries, large collections of names which serve as examples for a specific entity class. Two examples are the USPTO and EPO standard assignee names file and the Derwent Patentee Codes. According to Magerman et al. (2006), it is also possible to build own dictionaries by means of a harmonization procedure.

Since, for the purpose of this study, it was only necessary to identify similar firms in order to assign them correctly to different organizational types, I decided to follow a less ambiguous approach, and consequently applied a suitable manual, two-step approach. First, I delimitated abbreviations indicating the legal structures of a firm (for example, Ltd., GmbH, AG) and assigned business units to corporations (for example, ABB Research Ltd and the ABB Group were harmonized as ABB). Examples of the results of this harmonization exercise are documented in Table 8.4.

Table 8.4 - Examples of consolidated assignee names

Original Assignee Name	Homogenized / Consolidated Original Assignee Name	OATY*
Li Ming Chiang, US	Ming-Chiang Li	1
Li; Ming-Chiang	Ming-Chiang Li	1
Min-Chiang Li	Ming-Chiang Li	1
Arco Chem Tech, US CPC International Inc, US	Unilever	4
Arco Chem Tech, US	Unilever	4
CPC International Inc.	Unilever	4
CPC International Inc, US Arco Chem Tech, US	Unilever	4
CPC International, Inc. Arco Chemical Technology, Inc.	Unilever	4
Bestfoods	Unilever	4

Note: * 1 (individual inventor); 4 (large firm)

From the names of the original assignees, I compiled a variable, which allowed me to determine the type of institution from which the technologies originated (technology source). I coded the harmonized list of original assignees into a nominal variable, reflecting five different organizational types (large firm, SME, research institute/university, individual inventor, and other). Hardly any difficulties were encountered during the coding of the two groups of independent inventors and universities / research institutes. An assignee's name was classified as an independent inventor if a natural person appeared as an assignee on the patent document. Patents assigned to universities, their TTO and other research institutes were further obvious and simple to identify.[35] To codify the firms in general would have been a simple process, however, I further differentiated firms into SME and large firms based on the official EU definitions for micro, small, and medium-sized enterprises (European Commission, 2003: 39):

> The category of micro, small and medium-sized enterprises (SME) is made up of enterprises which employ fewer than 250 persons and which have an annual turnover not exceeding EUR 50 million and / or an annual balance sheet total not exceeding EUR 43 million.

In order to apply this definition, I collected business figures for the firms in the dataset. However, business figures were not collected for all firms in the dataset. Through the harmonization exercise, I had already consolidated the different firm names. Furthermore, particularly well-known large corporations were easily identified and coded as large firms (for example, Siemens AG and Ford Corporation). After a first round of codification, I had identified 62 firms which remained unclassified and for which I had to collect business figures in order to assign them to the appropriate class. Data were collected from the firms' annual reports, balance sheets and other business reports on the firms' websites as well as from databases providing the financial figures of firms (including Google finance, Answers, Manta). As far as possible, data were gathered for the last available business year prior to the auction (preferably 2006 or 2005). However, I was not able to obtain data for all 62 firms. If no firm information was found through Google (not even a firm website), it was assumed that the firm was too small to provide this information, thus the firm was classified as an SME.

In addition, I needed to create the category 'others' for firms that did not fit into any of the two firm categories. While analyzing the data, it appeared that some of the firms were not manufacturing firms but rather PLF or other types of TMI, for instance, Competitive Technologies, Inc. is a technology market intermediary based in the US. Nevertheless, it is very unusual for a market intermediary to pursue its own R&D to develop patents. Since it usually acts as broker to support another firm to acquire or exploit technologies, this intermediary applied for patents under its own name.

Further data on the sellers were available from which I could have coded a variable indicating that the type of seller is similar to the type of assignee. Sellers do not necessarily have to be the same as the original assignees if patents have been traded after the filing and prior to the auction or if the sellers engage, for instance, a patent law firm to auction a technology on behalf of the patent owner. I performed various random cross-checks to detect whether coding such a variable would be meaningful, that is, be substantially different from the assignee. However, the cross-checks revealed that, in almost all cases, the assignee type would have been assigned identically to the seller type.[36]

Missing data

As cases with missing values pose a challenge to the data analysis, since typical modeling procedures simply discard these cases from the analysis (SPSS, 2007), it was necessary to maximize the number of cases with complete data. Thus, in a first step, I identified cases with missing data and (if possible) completed them in a second step. As a result from the first step, a missing data analysis (see Table 8.5) revealed some variables with few missing cases (for example, forward citations (FORWC) and number of inventors (ANNI)). However, two variables showed a considerable share of missing data. Excluding these variables was not an option, since it would have negatively impacted the explanatory power of the study's results through a smaller dataset. Therefore, to address the missing data, I turned to techniques discussed in Horton and Kleinman (2007), and Hair (2006).

Table 8.5 - Missing data overview

Univariate Statistics			
	N	Missing	
		Count	Per cent
BIOP	125	269	68,27
VAEX	299	95	24,11
FORWC	391	3	0,76
NPLO	394	0	0,00
IndTeam	390	4	1,02
ANNI	392	2	0,51
Remaining Protection Time (RPT)			
- ARPT	391	3	0,76
- MaxRPT	391	3	0,76
- MinRPT	391	3	0,76

First, the high number of missing cases (about 2/3) of the opening bid data (BIOP) was due to the fact that this data was collected through direct observations (see Table 8.2), and thus missing cases were only prevalent for two auctions (A_4, A_5). The share of missing data was too large to allow for a reliable use of imputation techniques. Therefore, the opening bid data were dismissed from the analysis. However, the use of both the opening bid data and the expected sales price data in one model would have caused difficulties regardless, since bivariate analyses revealed that they are highly correlated.

Second, the data for the expected sales price (VAEX) were also incomplete, although less data were missing, that is less than 25 per cent. Data on the expected sales price were not available for the fourth auction in IPA auction catalogue, or for a few technologies listed in the OT catalogues. Thus, to a large extent, the expected sales price data were missing for cases where opening bid data were available, which appeared correlated. Thus, to correct for the missing expected sales price data, the use of multivariate imputation techniques appeared favorable. Following Horton and Kleinman (2007), I imputed the missing data using the Expectation-Maximization algorithm (EM), one of the likelihood approaches available in SPSS.

To summarize, to develop the dataset for the quantitative study, I collected data from four different official and reliable sources in a mainly automatized process.[37] Following the data collecting, I prepared the dataset (harmonized and consolidated data and handled missing data) and conducted coherence tests. I randomly chose data records and compared them with the original raw data from the different sources to validate their accuracy. Moreover, I dealt with the problem that few technologies were re-offered, that is offered but not sold at an auction and then offered again at a subsequent auction (this case appeared for 12 technologies, from which three were sold later on). To avoid biasing the study's results, those technologies were eliminated from the analysis. Finally, I was able to compile a dataset that included information for 99 per cent of all data entries. The final dataset includes data on 393 technologies (1,461 patents), from which 39.6 per cent (156 technologies) were auctioned successfully.

8.2.3 Regression Analysis

The analysis of the dataset compiled for the quantitative study follows a two-step procedure. In the first part of the regression analysis, as is common in quantitative studies, I provide descriptive results (Section 10.2). In a second part, I provide the results from multivariate regression analyses (Section 10.3). The choice of the specific regression techniques (logistical and log-linear OLS regression) was motivated in Chapter 2. In the following, I explain how the descriptive and regression analyses were conducted and place particular emphasis on the assumptions that need to be tested so that the regression analyses produce reliable results.[38]

In the first part of the data analysis, the descriptive analysis was conducted using general analytical procedures. As most of these methods have been applied rather often in earlier works in this field, I do not describe them in detail. In preparing this section, I refer mainly to Rasch (2008), Luderer et al. (2007), Köhler et al. (2007), and Bortz and Döring (2006). Measures of central tendency (minimum, maximum, mean, and median) were employed as the main types of descriptive statistics encountered in research papers (University of the West of England, 2007).

To report descriptive statistics of individual variables, I employed contingency tables (also referred to as cross tabulation or crosstabs) in a similar manner for each variable. According to Powers and Xie (2008), contingency tables present (multivariate) joint frequency distributions of variables in a matrix format. They are most widely used to present the relationship between two or more categorical variables (Agresti, 2010). They are a flexible instrument for presenting data in an easily accessible way, which can be applied in a wide range of situations. Contingency tables are often used as a foundation for more advanced analyses, and sometimes they are used solely for sophisticated analysis (Punch (2005), Rosenberg (1968)).

The contingency tables employed in this study are referred to as 'two-way tables,' where each variable is presented across two categorical variables. From the three types of two-way contingency tables proposed by Goodman (1981), this study employs the second type, which presents a causal relationship of an outcome variable's dependence on an explanatory variable. For each variable, two contingency tables are plotted across the dichotomous (two categorical) outcome variable, 'sales status' (sold or unsold).[39] In a first 2 x 6 contingency table, the explanatory variable is the 'auction number' with six categories (A_1 to A_6) and, in the second 2 x 4 contingency table it is 'source of invention' with four categories (individual inventor, university / research institute, SME, or large firm). While 'sales status' represents the dependent variable used in the logistic regression models, the two explanatory variables are those which were used as control variables in the regression models. In each contingency table, I report the measures of central tendency. Moreover, the table adds a bottom row and right column presenting the total numbers for each category. In addition to the contingency tables, frequency tables or histograms are used for selected variables.

Throughout the second part of the data analysis, I employed regression analyses to 'estimate how much of the variance in a phenomenon I can account for, using a particular set of independent variables... [and to] ...determine the effects of the different independent variables on the dependent variable' (Punch, 2005: 79). To develop the models, the approach of Hair (2006) for multivariate model building was followed. In order to validate the assumptions of the regressions, I mainly refer to Hair (2006), Gujarati (2003), and Garson (2010).[40]

OLS regression

To test the effects of independent variables on the metric-scaled sales price, I employed a multivariate OLS regression procedure as presented in Equation 1. Following Hair (2006) and Gujarati (2003), OLS regressions need to comply with certain assumptions to return reliable results. In order to test the regression assumptions, I examined individual variables (dependent and independent) prior to computing the models and those referring to the relationship between the variables after the model was computed (the variate).

$$Y_i = b_0 + b_1X_1 + \cdots + b_nX_n$$

Equation 1 – Generic form of OLS regression models

For the individual variables, I tested for *normality*, using the skewness and kurtosis as well as graphical analyses, particularly normal probability and box plots. According to Garson (2010), skew should be within the ± 2 range when the data are normally distributed and kurtosis should be within the ± 2 range. The tests revealed a positive (left) skew for almost all metric-scaled variables (with the tail to the right). These results, however, were not surprising. Patent properties (for example, forward and backward citations, claims, and number of inventors) are well known to be skewed due to their count data nature (Winkelmann, 2008). Moreover, patent values are known to be skewed (Hall et al. (2005), Harhoff and Scherer (2000), Trajtenberg (1990), and Hausman et al. (1984)).

Table 8.6 - Normality of original / transformed variables

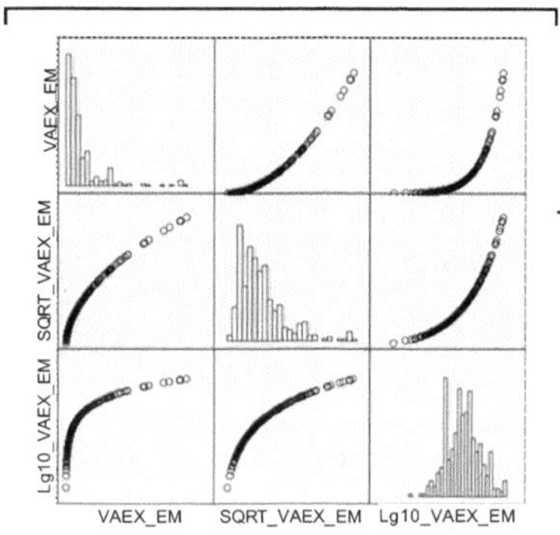

	Skewness Statistic	Kurtosis Statistic
BICL	5,343	35,447
Lg10_BICL	,135	-,663
SQRT_BICL	2,320	7,737
VAEX	2,963	9,992
Lg10_VAEX_EM	,086	-,071
SQRT_VAEX_EM	1,624	3,153
NPLO	4,587	33,017
Lg10_NPLO	,957	,055
SQRT_NPLO	2,101	5,858
FORWC	3,105	13,870
SQRT_FORWC	,908	,662
Lg10_1_FORWC	,341	-1,041
ANNI	2,595	12,729
Lg10_ANNI	,546	-,468
SQRT_ANNI	1,239	2,508
ARPT	-,851	,463
Lg10_ARPT	-2,418	8,122
SQRT_ARPT	-1,415	2,202

As remedies for non-normality, Hair (2006) recommends correcting positively (left) skewed distributions with Log, Ln or Square root transformations. This approach is well known and common in patent studies.[41] For instance, Gambardella et al. (2008) employ log-transformations to achieve normality for OLS regression. The transformations reveal much improved normality measures as can be seen in Table 8.6, with a graphical example provided for the expected value. Among the three transformations – across almost all variables – the log-transformation achieved the best results and was thus applied.[42]

Furthermore, I tested for the presence of correlations between two or more independent variables (multicollinearity). While a simple correlation matrix of the independent variables indicates correlations between two variables only[43], multicollinearity might be due to the combined effects of variables. To test this effect, I employed the tolerance measure as suggested by Hair (2006) and Gujarati (2003). According to Brosius (1998), values below 0.1 indicate collinearity and values below 0.01 are a clear sign of collinearity. The collinearity statistics of this study's models shows values close to 1 for all variables. Therefore, it can be concluded that multicollinearity is not present.[44]

Furthermore, Hair (2006) and Gujarati (2003) demand that the model should be linear in the parameters. *Linearity* represents the degree to which the change in the dependent variables is associated with independent variables. Examining scatter plots of the variables did not reveal any pattern indicating non-linearity.

With regard to the variate, I tested for *homoscedasticity*. According to Hair (2006: 83), 'the dependent variable should exhibit equal levels of variance across the range of predictor variables'. Thus, for each given value of any independent variable, the variance of the residual should be constant for all observations.

To test this assumption, Backhaus et al. (2006) recommend using scatter diagrams, plotting the standardized predicted values against the standardized residuals. The values in these plots should be randomly distributed and the plotted data should not show any pattern. The scatter plot of the untransformed model indicates an increasing distribution of residuals, which signifies heteroscedasticity. However, showing randomly distributed values, the scatter plot of the log-transformed model does not indicate heteroscedasticity.

I also tested the *independence of residuals* (Hair, 2006) or whether the residuals show no autocorrelation (Gujarati, 2003). A violation of this assumption would indicate that unexplained systematic relationships exist in the dependent variable. Thus, some factors affecting the results would have been omitted from the analysis. I employed the Durbin-Watson statistic as a common measure to detect such correlations. The Durbin-Watson Statistic's value ranges from 0 to 4 with an ideal value of 2 indicating that errors are not correlated. A value significantly below 2 would indicate a positive correlation and a value significantly greater than 2 a negative correlation. Values ranging from 1.5 to 2.5 are considered to indicate no autocorrelation (Brosius, 1998). All the models used in this study show a Durbin-Watson statistic value close to 2.0.

I tested further whether the *mean values of residuals are zero* (Gujarati (2003), Hair (2006)). The residual statistics of this study's models show mean values of 0. Thus, I could not detect any violations of this assumption. Regression models also demand that the *covariance between the residuals and* X_i *is zero* (Gujarati, 2003). However, according to Backhaus et al. (2006), the significance test is valid regardless of the distribution if the number of cases / observations is high (that is larger than 40). The number of cases in this study's models fulfills this criterion. Therefore, I assume that this assumption is valid. For the log-linear OLS regression, I employed the coefficient of determination R^2 as a *'Goodness of fit'* measure.

Logistic regression

A logistic regression model is distinguished from a linear regression model in that the dependent variable is not metric but dichotomous (bivariate). Therefore, the logistic regression applies a logistic transformation to the dependent variable (a logit link function) as shown in Equation 2. It uses a specific form of the logistic curve, which is s-shaped for the dependent values to stay within the range between 0 and 1. To estimate the model, the curve of the predicted values is fitted to the actual data. For each observation, the regression then predicts a probability value between 0 and 1. If the predicted probability is greater than 0.5, then the outcome is estimated at 1; otherwise, the outcomes are predicted to be 0 (Hair, 2006). While OLS regression employs the method of ordinary least squares, the nonlinear nature of the logistic transformation requires the maximum likelihood procedure, which should be used in an iterative manner to find the most likely estimates for the coefficients. Instead of minimizing the squared deviations (least squares), the logistic regression maximizes the likelihood that an event will occur (Hair, 2006).

$$Logit_i = \frac{prob_{event}}{1 - prob_{event}} = b_0 + b_1X_1 + \cdots + b_nX_n$$

Equation 2 - Generic form of logistic regression model

In other respects, the logistic regression follows the same principles and rules used in OLS regressions. However, according to Statistics Solutions (2010), logistic regression does not make many of the key assumptions of linear regression and general linear models that are based in the OLS algorithms – particularly regarding linearity (although linearity of independent variables and log odds should be present), normality, homoscedasticity, and measurement level (Garson, 2010). However, according to Garson logistic regression still requires two assumptions: The residuals need to be independent and the model should have little or no multicollinearity. How these assumptions were tested is discussed above.

For logistic regressions, individual coefficients of determination (that is *Goodness of fit measures*) are not available. Instead, I applied four common measures from the three categories (statistical, pseudo R^2, and classification accuracy) (Garson (2010), Hair (2006)).[45] I employed two measures from the category of statistical measures. The *-2 times log likelihood (-2LL)* is a common statistical measure for the model fit. Good results of the observed results are indicated by high likelihood values, which translate into small -2LL values. If a model fits perfectly, the likelihood is 1. Thus, -2LL is equal to 0. However, since there is no upper boundary for -2LL, it is difficult to make a statement about the meaning of the score. It is more often used relatively to observe how the model fit develops if additional variables are added to the model. The fit increases if additional variables lead to reduced -2LL values. As a second measure, I employed the *'Hosmer and Lemeshow chi-square'* test, which assesses the correspondence of the actual and predicted values of the dependent variable. A better model fit is indicated by a smaller difference in the observed and predicted classification. Thus, well-fitting models show non-significance (Hair, 2006).

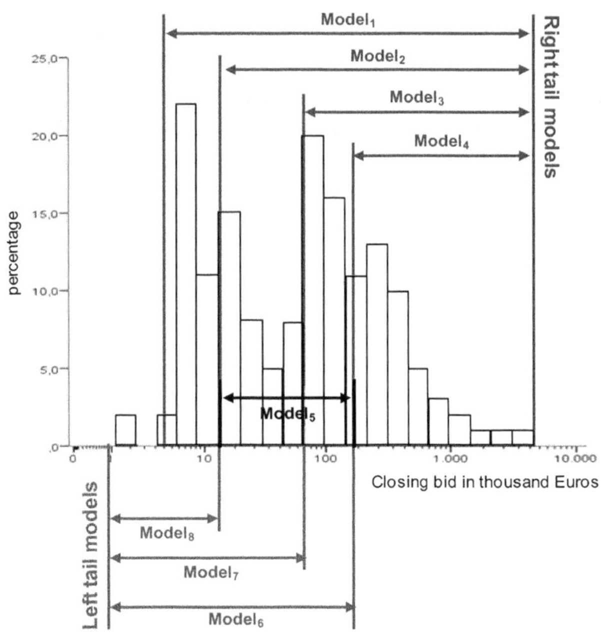

Figure 8.2 - Sensitivity analysis of value ranges

Among various pseudo R^2 measures, I employed the *Nagelkerke's R^2*, which is a modification of the Cox-Snell coefficient. While the Cox-Snell coefficient can hardly have a value of 1, Nagelkerke's R^2 is adjusted to compensate for this problem. These 'Pseudo R^2' measures are interpreted as 'reflecting the amount of variation accounted for by the logistics model, with 1.0 indicating a perfect model fit' according to Hair (2006: 362). According to Hosmer and Lemeshow (2000: 167), 'low R^2 values in logistic regression are the norm, which presents a problem when reporting their values to an audience accustomed to seeing (classical) linear regression values'. Models with values between 0.2 and 0.4 are thus considered to have 'very good' explanatory power (Ludwig-Mayerhofer et al., 2010). The third category of measures to assess the overall model fit draws on the classification accuracy. The *classification accuracy* presents the levels of predictive accuracy achieved by the logistic model. The measure of classification accuracy I employed is the 'hit ratio', i.e. the share of correctly classified cases Hair (2006: 362).

To summarize, having tested assumptions for the individual variables and the variate, I was able to judge whether the regression models could be applied to analyze the dataset. In accordance with Gujarati (2003), for the OLS (ordinary least square) regression, I had to decide whether to apply a linear or a log-linear regression model employing the transformed variables (the log of the regressand is a function of the logs of the regressors). The problems of non-normally distributed variables and heteroscedasticity did not occur in the model using transformed variables. Therefore, I decided on the log-linear regression model. Since the logistical regression did not violate any assumptions, it appeared to be applicable.

To validate the robustness of the results of the multivariate regression analyses, I conducted numerous sensitivity analyses (Forrester, 2001).[46] In particular, models were computed for eight different sales price ranges calculated based on the 25 per cent, 40 per cent, and 75 per cent quartiles of the sold technologies as, for example, proposed by Alcácer et al. (2009). These left (right) tail models take the low (high) value technologies into account as illustrated in Figure 8.2. Furthermore, I tested the models for different auction groups. In particular, I differentiated between the US and EU auctions as well as the first auctions in each country (in the US, Germany, and UK) and the subsequent auctions.[47]

NOTES

1 A broad discussion is ongoing in the literature on the quality of case study research. Aside many critics, Eisenhardt and Graebner (2007: 25) claim that 'papers that build theory from cases are often regarded as the "most interesting" research and are among the most highly cited pieces in AMJ, with impact disproportionate to their numbers'. A review of the methodology for case study research can be found in Larsson (1993).

2 For this pre-study I would like to acknowledge the support of Antoneta Hristova and Sascha Havestadt.

3 Although some arguments would have actually favored conducting a single case study, I have already provided arguments for why I chose to conduct multiple case studies in Chapter 2.

Furthermore, due to the limited availability of cases, I did not conduct any pilot cases, although this is recommended by Yin (2003).

4 Another selection criterion would have been to choose a performance measure (for example, total returns of the auction). However I dismissed this criterion, since the results from the auctions (particular the US and German auctions) were not equally distributed. Thus, transactions from certain auctions would have been filtered out entirely.

5 Unless otherwise specified, the sources used are the OT auction catalogues from 2006 (April, October) and 2007 (April, June, October) and the IPA auction catalogue from May 2007.

6 The IPA auction was held in Munich in May 2007 and the OT auction in London in June 2007.

7 The participants of the IPA auction actually received a questionnaire that was jointly developed with IPA and should have served as a feedback form as well. However, too few participants returned the questionnaire to allow for a solid analysis.

8 Personal and telephone interviews were held with Andrew Ramer and Justin Basara (OT; 14.05.2007), Boris Peters (IP Bewertungs AG; multiple times in 2006-2008), Ron Laurie (Inflexion Point; multiple times but also on 15.05.2007), Ron Epstein (IPotential; 14.05.2007), Richard Ehrlickman and Radu Christian (TAEUS; 14.05.2007), Dr. Sven Wehser (Siemens AG; 14.05.2007), and Mikko Väänänen (Suinno OY; 1.6.2007). Additional interviews were conducted during the course of the pre-study reported in Tietze (2008).

9 This cross-case analysis can be interpreted, as well as a meta-analysis of individual case studies. However, these approaches usually use much more cases (that is the review in Larsson (1993) reports between 25 and 313 cases) in order to allow for statistical analysis, while the cases can come from different prior studies. According to Larsson (1993), there have only been a few attempts to perform a meta-analysis of case studies, which have been labeled the 'case survey' (Lucas (1974); Yin and Heald (1975)), 'case meta-analysis' (Bullock and Tubbs, 1987), and 'structured content analysis of cases' (Jauch et al., 1980); while there are also various nameless approaches (for example, Miller and Peter (1977)).

10 Consequently, transactions for which the sellers were universities or research institutes (for example, technology transfer offices), independent inventors or others (for example, law firms) were excluded.

11 I am aware that, due to this selection criterion, the perceived success of the transaction by the sellers might be positively biased, that is assuming that those who achieved particularly high sales results are most satisfied with the auction model.

12 In a first approach, I selected transactions with sales prices above the overall average value across all auctions. However, this approach was dismissed, since it would have discriminated against the German auction. None of the transactions completed at the IPA auction had reached sales prices above the overall average value.

13 With regard to the order and phrasing of the questions of this study's interview guidelines, I followed the advice provided mainly by Schaeffer and Presser (2003), Fielding (2001), and Wilson (1996).

14 For the preparation of the questionnaire, I would like to acknowledge the support of Henning Krenz.

15 I would like to acknowledge the support, the valuable comments, and suggestions from Thomas Ewing (IP ValueADDED), and Boris Peters (IP Bewertungs AG).

16 The support of Diehl Stiftung & Co. KG (Nürnberg), Fraunhofer-Gesellschaft (Munich), HILF! GmbH (Oberhaching), and TuTech Innovations GmbH (Hamburg) is acknowledged.

17 Likert scale questions, that is questions designed to collect quantitative data, are not controversial for case studies. According to Yin (2003: 14f), 'case studies can include and even be limited to quantitative evidence … and they can be based on any mix of quantitative and qualitative evidence'.

18 Upon request, firms received additional paper copies of the questionnaire via regular mail as well as electronic PDF documents by e-mail which were used by two firms.

[19] Following Punch (2005), all respondents received the same questions in the same order, delivered in a standardized manner, where the interviewer takes a neutral role thereby stressing rational rather than emotional responses.

[20] I was able to identify specific contact persons of eight firms. The remaining four firms received a written request for an interview directed to the head of the patent or licensing department. The auction firms denied providing access to the specific contact data of the responsible persons in the sellers' firms.

[21] In order to draw a more complete picture of the transactions, it would have been valuable to take the opinions of the buyers of the technologies into account. However, their contact data were almost impossible to access for reasons explained above.

[22] Patent documents represent a large body of information and adequate quantities of comparable units suitable for detailed statistical analyses of technological developments, although they have inherent weaknesses (Tijssen, 2002). Not all innovations are patented (for example, Arundel and Kabla (1998)), nor are patents equivalent to innovations. Basically, inventions and patents represent ideas; innovations convey those ideas into the market. Moreover, various patented inventions are never fully developed or adjusted to market demands. The knowledge codified and protected in a single broad 'generic' patent may be used in several innovations, whereas a complex technical system may cover a range of interrelated and incremental patents.

[23] In Chapter 2, I explain why I have chosen to combine existing data instead of collecting own primary data, for example, by means of a survey. In contrast to secondary data, survey data has the major disadvantage of being naturally subject to the general limitations that the data are based on respondents' stated beliefs rather than non-subjective valuations. Moreover, secondary data offer shorter collection times and larger sample sizes but can be less refined or tailored to the object of study.

[24] For some auctions close to the auction date, the firms published supplement catalogues, which included some updated information. These supplements were also included in this study's dataset.

[25] Information on different auction types is provided in Chapter 5.

[26] Since I only had access to publicly available patent databases, I was unable to collect certain data. For instance, this study lacks opposition data although oppositions have been shown to have an impact on patent values. Certain patent offices offer the possibility for third parties to oppose granted patents that they deem invalid. As opposing a patent is costly, it can be inferred that only patents with some damaging effects on competition, and thus some economic value, are opposed. Hence the fact that a patent is opposed can be interpreted as a signal of value (OECD, 2009: 146).

[27] I considered conducting an outlier analysis. However, according to Osborne and Overbay (2004), the causes of outliers appeared unlikely. Owing to the automated process of retrieving data from official patent databases, which can be expected to have a high data quality, I did not expect any 'outliers from data errors' caused by human coding. In contrast, however, the coding of the data from the other sources could have been subject to human coding error. To verify whether this was the case I cross-checked the data, drawing random samples that were then compared to the raw data. Furthermore, the dataset could not be expected to suffer from 'intentional or motivated mis-reporting', which is rather common in surveys or experiments, but not in secondary data studies.

[28] The catalogues appeared very similar. Consequently, one can assume that the IPA catalogue was inspired by the OT catalogue, especially since three OT events had already taken place before IPA launched its first auction.

[29] Reasons for this problem are manifold. According to OECD (2009), firms are known under several different names (for example, acronyms: IBM, International Business Machines), some qualifications can be added to the name (for example, Siemens, Siemens AG), and patents can be taken by affiliates, some of which are easily identified (for example, Sony US is an affiliate of Sony) whereas others are more difficult (Citroën is part of the PSA group).

Changes in the company's legal status as well as changes in company names, affiliations, as well as mergers and acquisitions often make the use of patentees' names in patent data difficult.

[30] This approach can be assumed to be similar to the first approach mentioned by OECD (2009), using experts to study patents and assign them to application areas. The auction firms, together with the sellers, who can be assumed to possess detailed knowledge about the offered technologies and their application areas, had assigned the technologies to specific application areas in which they were presented in the auction catalogues. Although I cannot judge the quality of this allocation, I am reasonably confident that this categorization was carried out in a reliable manner. I consider it a critical factor for auctions' success that the technologies have to be assigned to the appropriate application areas so that the prospective buyers can find what they are looking for.

[31] These differences result from the fact that the information on the application area was provided by the sellers. The catalogues state that 'all descriptions of the Lots enclosed in the Catalogue have been provided to OT by the Sellers', and by IPA that 'all information relating to and in connection with the auction of the intellectual property right are deemed to be information from the Authorizing Principal'.

[32] Examples are Aerospace and Aviation and Aerospace, which were consolidated into the class, Aviation and Aerospace. Furthermore, I consolidated the classes 'Bar code/RFID technologies/Smartcard', 'Communications', 'Telecommunications', and 'Wireless / Cellular / Optical and other telecom' into the category 'Telecommunications (wireless, cellular, optical & other)'.

[33] An early attempt to build an industry concordance table for patents was the 'Yale Concordance', which was developed by Evenson et al. (1991). Another attempt was the 'OTAF Concordance,' the USPTO concordance between the US Patent Classification (USPC) system and the US Standard Industrial Classification (SIC) system, created in 1974. Other work in this field includes the concordance proposed by Johnson (2002), based on data from the Canadian Patent Office. A more recent concordance table was designed by Schmoch et al. (2003) from the Fraunhofer Institute for Systems and Innovation Research, the Observatoire des Sciences et des Techniques (OST), and the University of Sussex's, Science and Policy Research Unit (SPRU).

[34] Delphion (2004) defines assignees as the person(s) or corporate body to whom all or limited rights under a patent are legally transferred.

[35] A small bias might have remained in the data because technologies developed at universities might have been assigned to a faculty member personally instead of being assigned to the university. However, it appears reasonable to assume that this case applies, since US patents constitute the largest share in this study's dataset. US universities commonly do not apply the 'teacher's exemption' as has been rather common at European universities (Lang, 1998).

[36] Compiling a variable based on the seller's name and comparing this variable with the original assignee could have provided insights into the share of patents that had been traded prior to the auctions. However, the way in which both nominally scaled variables would have been coded would have only allowed for detecting trade if it had occurred between different organizational types (for example, when a patent had been invented by a university and then been sold to a large firm). However, this was not of primary interest in this study.

[37] The auction catalogues are the only non-official sources that could be biased as they can be perceived as marketing material. However, I am confident that the data used from the catalogues had no substantial bias. The auction firms operate in an emerging market. Although they would have an incentive to sugarcoat the data in order to attract new customers, they needed to create trust (reputation) among their customers, persuading them to repeatedly participate in upcoming auctions. The market does not appear robust enough to risk destroying trust before having created a robust customer basis. It is certainly not of interest to the auction firms to create expectations that they cannot fulfill at the event by publishing cata-

logues that, for example, include certain high profile technologies that would not have been auctioned.

[38] All analyses were computed with the SPSS software package, version 17.0.

[39] The Chi²-Test following Pearson could be employed to test the significance of the value difference in the categories. As the variables were subject to tests in multivariate regression analyses, I do not report results from bivariate analyses (Punch, 2005).

[40] The order of presentation from both regression techniques in the following has been changed for didactical reasons only.

[41] Log-transformations appear problematic if the variable contains zero values, since log(0) is undefined. In the case of forward citation data, I thus added a constant with a value of one to the original value, thereby avoiding the problem of undefined values (De Smith et al. (2007), BBN Corporation (1997)).

[42] Multivariate normality was not tested separately. According to Hair (2006: 80) 'assessing and achieving univariate normality for all variables is sufficient'.

[43] As mentioned above, the correlation matrix revealed a strong correlation between the opening bid and the sellers' expected sales values. In addition to the missing data (see above), the opening bid was omitted from the models.

[44] High standard errors also indicate possible multicollinearity, however, none were observed in this study's models.

[45] See Hosmer et al. (1997) and Hosmer et al. (1991) for discussions of different goodness of fit measures.

[46] Further approaches to sensitivity analyses are reported, for instance, by Severini (1996), Chatterjee and Hadi (1988), and Hampel (1971).

[47] Owing to the low number of cases for each auction (on average 65 technologies), it was not possible to compute the multivariate model for the individual auctions. The results showed symptoms of overfitting (that is unexpectedly high goodness of fit measures or standard errors).

9. Auction Governance Structures

9.1 COMPARATIVE ANALYSIS OF TECHNOLOGY AUCTION PROCESSES STRUCTURES

Since the literature did not provide sufficient insight into the transaction governance structure (TGS) of auctions for patented technologies, I conducted a first, qualitative pre-study to complement the insights from the auction and management literature presented in Part II. For this study, empirical data were collected by means of interviews, live observations of two auctions, and from auction catalogues.

The presentation of the study's results is in accordance with the two stages of the analysis. First, following the three core phases of Lichtenthaler's (2006b) model, I present individual case descriptions for each of the two auction designs as results for the within-case analyses. The case descriptions were prepared by combining the transaction process model proposed by Lichtenthaler (2006b) (see Chapter 3) and the auction process model proposed by Kumar and Feldman (1998) (see Chapter 5). Second, I present the results of the cross-case analysis that aimed to derive similarities and differences from the process structures.

9.1.1 The Auction Process of OT

The *planning phase* starts with initial preparations of the auction. All activities during this phase relate to three areas, including the definition of rules, the announcement of the auction date and invitations to sellers and interested bidders to register for the auction. Defining the auction's terms and conditions, which govern the transactions, is one of the very first preparatory steps. The rules define the requirements for the selection of the technologies to be offered at the auction, the registration requirements for participants (sellers and bidders), how disputes would be resolved, the conditions of sale, the protection of bidders' anonymity, the terms for the due diligence, and the bidding procedures allowed at the live auction event (for example, English, Dutch auctions).

Following the internal preparation activities, the upcoming auction is announced to the public via various media sources, including the auction firm's webpage, online newspapers, the auction catalogues (except for the 1st auction), as well as throughout seminars and other community events.[1] The first live auction conducted by OT was announced about ten months prior to the live auction event.[2] The subsequent auctions were announced approximately six months prior to the upcoming auctions in the auction catalogues of the prior auction (in the form of open call for technologies) or in the post-auction press releases.

As a result of the auction announcement, the auction firm starts to collect suggestions for technologies that sellers propose to offer at the upcoming auction. Upon announcing the auction, OT called primarily for technologies from pre-determined areas. In addition, OT announced that the 'most lucrative and appropriate for auctioning are patents that cover existing technologies... [that are]... not used by their owners'.[3] For this purpose, the firm requires potential sellers to register for the event by signing a Seller's Consignment Agreement.

Sellers have to pay a listing fee for each technology they submit and are required to take part in an internal evaluation of the auction firm. The fee varies depending on the type of the asset and whether there is a minimum reserve price (MRP) as indicated in Table 9.1.

Table 9.1 - Listing fees for different asset types of an OT auction

IP Type	Conditions	Fee
Patent		
	At Minimum Reserve Amount	$1,000
	With a Consignor-specified Reserve Amount	$3,000
Patent Pool		
	At Minimum Reserve Amount	$1,000
	With a Consignor-specified Reserve Amount	$6,000
Trademark		
	At Minimum Reserve Amount	$1,000
	With a Consignor-specified Reserve Amount	$2,000
Domain Name		
	At Minimum Reserve Amount	$250
	With a Consignor-specified Reserve Amount	$500

Sellers can either accept the MRP as predetermined by OT or can specify an own MRP. For specifying an own MRP, the listing fees were at least double the price of the listings when using the pre-determined MRP. For auctioning, OT accepted different technologies or IP asset types, including patents, trademarks, and domain names.[4] However, in the case of patents, OT further differentiated between technologies covered by a single patent only and technologies covering various patents.

For each registered asset, sellers are requested to provide information to enable the auction firm to conduct a first internal evaluation. The information requested from the sellers includes: specific IPR registration numbers; the chain of title; a brief description of the asset; a list of the rights to the IPR being sold; a list of licenses or encumbrances of any kind or nature that have been granted under the specific asset; indications of involvement by standards bodies, universities, government agencies, assistance groups or other non-profit institutions; the identity of any existing or potential secured or unsecured claims of ownership by third party creditors or license-holders; the identity of any claims by a third-party, which could potentially affect ownership, licensing, infringement, validity and enforcement or other issues; knowledge of any usage or suspected usage of or infringement of the asset; information on litigation (resolved, pending or threatened) involving the asset; due diligence materials as prototypes or tertiary data, technology or know-how; as well as information on the availability of personnel to assist the commercialization.

OT reserves a right to reject technologies if they do not comply with the internal evaluation criteria. To conduct the internal evaluation in the first half of 2005 OT had acquired the web-based firm Patentrating.com, a patent rating boutique from California, US, in order to in-source the necessary knowledge and expertise to perform the internal evaluations (IP Law Bulletin, 2005). According to Patent Ratings (2007), the internal evaluation is conducted using a 'confidential method for studying inter-relationships of about 50 discrete attributes associated with a single patent based on the entire 7 million US issued patents'.

Following the internal evaluation, the auction catalogue is published, making the selected technologies available to prospective buyers, inviting them to register and announce their interests. With prospective buyers becoming involved in the process, the *negotiation phase* is initiated. The catalogue serves as a basis for prospective buyers, thus the auction firm aims to distribute it widely. It is published electronically on the OT website and, in printed form, it is distributed across a network of selected firms. In addition to the catalogue, a catalogue supplement is usually published a few weeks prior to the auction and includes similar information about additional technologies to be offered at the auction.

The catalogue is split into various sections of different technology application areas (for example, electronics & handheld devices and display technology & digital imaging).[5] Within each section, the technologies are presented as lots. A lot represents one technology with a minimum of one patent (it can include several patents). Each lot is described by a lot summary, complemented by specific details about each patent included in the technology. The summary includes the name of the seller, the title of the technology, the official registration numbers of all the related patents, and a list of selected forward citations of other firms that cite or use the technology. Furthermore, the summary suggests licensees for whom the technology might be of interest and gives an indication of the opening bid or expected sales price (commonly ranging from a lower to a higher bound). Following the summary, the catalogue provides specific additional information on each patent offered as part of the technology. OT provides mainly bibliographic information about the patents derived from the legal patent documents. This information includes the legal registration number, the name of the inventor(s), the title of the patent, the issue date, the number of claims, and the abstract of the patent. In some cases this information is complemented by an illustration, usually taken from the patent documents.

Immediately following the publication of the catalogue, prospective buyers are invited to register on OT's corporate website. Similar to the seller registration, prospective buyers have to purchase a bidder registration package for a registration fee of US$1.5 thousand. Prospective buyers have to sign a bidder agreement and provide a bank warranty to ensure coverage of the auction costs. Following the successful registration, OT provides the prospective buyers with instructions to access the data rooms, enabling them to perform due diligences for technologies they are interested in, and thus engage in negotiations with the sellers.

Throughout the due diligence, prospective buyers have the possibility to evaluate the technological feasibility, verify the legal status of the patents, and the technology's economic potential within a virtual due diligence room. To enhance the due diligence and support the prospective buyers, OT provides different tools. Aside from the auction catalogues, which include mainly public bibliographic data, OT provides a secure online data room where prospective buyers can examine technologies in detail with regard to confidential information provided by the sellers. To ensure the anonymity of the prospective buyers, the data room is maintained by an independent third party. Additionally, OT provides the possibility for personal meetings with the sellers, either prior to or on the date of the live auction event. Throughout the five OT auctions, the period to conduct the due diligence (measured from the publication of the catalogue until the auction date) varied between five and 12 weeks, with an average of about seven weeks.

Following the due diligence phase, the live auction event takes place, where prospective buyers have the possibility to participate in person, via telephone, via an OT representative, via the internet or in absence by providing written instructions beforehand.[6] The auction is run by a third party auctioneer appointed by OT. When the auctioneer presents a technology for sale, each registered bidder can submit bids within her or his predetermined financial limits. Throughout the six auctions, OT used solely English bidding procedures (see Chapter 5). As specified in the terms and conditions, a minimum reserve price (MRP) property was applied in all auctions, defined by OT as the 'lowest price for which the lot can be sold'. The MRP was confidential in the first two auctions held in April 2006 and October 2006. In the later auctions, OT published the MRP.[7]

OT claims to protect the anonymity of the bidders throughout the whole auction process. In the due diligence phase, the online secure data room is maintained by an independent third party. During the auction event, bidders are identified by 'paddle numbers' only. For those bidders who prefer double-blind anonymity, an OT representative executes bids on the bidder's behalf in accordance with written instructions. Technologies are successfully sold if the final bid is higher than the MRP. If the MRP is not reached, the technology is withdrawn from the auction. In the case of a successful sale, the seller enters into a consignment contract. As a condition of participation, each seller and bidder has to agree not to use the participation of any registered seller or bidder as evidence in any action seeking to enforce a patent, copyright or trademark for any purpose.

Having concluded the due diligence and finished the price negotiations (bidding) of all technologies, the *realization phase* of an auction starts. However, this phase differs for technologies that were sold and for those that remained unsold. For successfully sold technologies, the ownership rights of the technology (and necessary complementary tacit knowledge components) are transferred to the new owner. In exchange, the new owner pays the final bid price to the seller. Additionally, both the seller and the buyer have to pay the commission fees to the auction firm. These include a 10 per cent buyer's premium and a 15 per cent seller's premium of the final bid price. If requested, buyers can request additional knowledge transfer for a technology from the seller. Technologies that remain unsold and which were thus withdrawn from the auction are available for post-auction negotiations, thus a re-negotiation phase. For instance, if prospective buyers showed interest and placed bids for a particular technology, but the bids did not reach the (at some auctions unknown) MRP, the seller might however be willing to lower the MRP in post-auction negotiations. OT reserves the right to benefit from any transaction closed at 'any period of time after the auction', even without OT mediation. However the firm does not provide insights into how OT controls and monitors the correctness of this post-auction process.

9.1.2 The Auction Process of IPA

To initiate the *planning phase*, IPA developed a set of auction terms and conditions with the support of an independent law firm. In addition, this phase covers the sellers' registration, a first internal evaluation and selection of technologies for the auction, followed by the publication of an auction catalogue.

Approximately five months prior to the live auction event, IPA announces the auction date to the public through their corporate website and various types of online and offline publications, including a newsletter targeted for experts in the community.[8] Parties willing to sell technologies on the auction are invited to register on IPA's corporate website. Sellers are allowed to offer patented technologies as well as licenses to technologies but also other IP assets, including domain names and copyrights. For selling licenses, specific rules apply. Only licenses are accepted for which the license rights could be acquired through a lump sum payment or by means of a down payment with following royalty payments. Sellers of licenses, furthermore, have to provide prospective buyers with a license contract in the data room.

Following the auction announcement, interested sellers have to complete an online questionnaire, in which they have to provide details about the technologies they want to offer. The technologies are then evaluated by IPA's mother firm, Intellectual Property Bewertungs AG[9] using cost-, market- and income-approaches (IPB, 2005). Following the internal evaluation, IPA informs the sellers of which technologies they accepted for the auction. Sellers, whose technologies are accepted for the auction then have to provide IPA with a signed copy of the terms and conditions for authorizing auctions and a complete and signed list of all the related IP assets they want to submit for this auction. Registered sellers then receive access to an online data room so as to provide 'all the documents necessary in order to provide sufficient first round information to a potential bidder'. Sellers have to pay an auction authorization fee for each technology accepted for the auction. The fee for the auction is one thousand for a technology with one associated patent or license agreement and two thousand euros for technologies with two or more associated patents. By paying these fees, sellers accept IPA's predetermined MRP of €25 thousand. However, sellers are allowed to specify an own MRP for an extra fee. The 'reserve price supplement charge' is three per cent of the sales price awarded in the case of a successful sale.

Following the internal selection procedure, IPA compiles and publishes the auction catalogue on IPA's corporate website and distributes it to prospective buyers.[10] Following an introductory section, IPA lists the terms and conditions of the auction as well as some further organizational information (for example, the location of the auction). In the main section, the auction catalogue provides information on the available technologies. This section is split into chapters for separate application areas (for example, mechanical engineering, and automotive). Throughout the different chapters, the technologies are presented in lots, while each lot can include one or more patents. Each lot presents a summary followed by specific information on the related patent. The summary provides information on the title of the technology, the name of the seller, the type of transaction (sale or license), the type of the auction (English or Dutch), the amount of related patents, the technology field, as well as a list of firms that cite one or more patents included in the lot (forward citations). For few lots, for illustrative purposes, the summary presents a figure (for example, process diagram taken from the patent documents).[11] In addition to the lot summary, for each patent, the catalogue provides specific information on the title of the patented invention, the priority date, the name of the owner, the name of the inventor(s), the official patent document numbers, and the abstract of the description extracted from a legal document.

Following the publication of the auction catalogue, the *negotiation phase* starts with a request to parties who are interested in acquiring technologies to complete a bidder registration form and to provide a financial guarantee from their bank. Prospective buyers have to pay a bidder registration fee, of which the amount depends on how much they were planning to bid. For live bidding, IPA charges €1.5 thousand, for outside bidding IPA further differentiates between telephone bidding (€750) and written absentee bidding (€500). Successfully registered bidders receive a bidding number (for the live auction event) and access to the online data room in order to perform due diligences on particular technologies they are interested in. The virtual data room is operated by an independent third party in order to guarantee the anonymity of prospective buyers. For the first auction, the online data room was available from 01 March 2007 until the live auction event, thus for about ten weeks. In addition to the information on technologies and related patents provided in the auction catalogue, prospective buyers have the opportunity to review classified information on the technologies' or patents' histories. Furthermore, prospective buyers have the opportunity to contact the sellers via the data room and request additional information. Moreover, IPA offers the possibility to organize meetings or phone conferences. In addition to the information provided by the sellers, prospective buyers may request Patent Analysis Reports prepared by a third party intermediary.[12] The data room closes on the date of the live auction event.

The bidding event, which was held for the first auction in Munich, Germany on 15 May 2007, is governed by the auction's terms and conditions. The bidding proceeds anonymously and is led by an independent auctioneer. To place bids, bidders raise bidding numbers when the auctioneer calls a specific lot. For outside bidders, in the case of telephone bidding, an IPA representative holds direct contact via the phone with the prospective buyers and places open bids for them. For absentee bids, the prospective buyers have to hand in their maximum bid prior to the live auction event. This maximum bid is known to the auction firm during the auction, but not to the other bidders. Either English or Dutch bidding procedures are applied. For English auctions, the opening bid was €25 thousand, if the seller had not specified an own MRP. As specified in the terms and conditions, calls for bids are usually raised by about 10 per cent above the previous bid. A bid is considered successful if it is above the MRP. For Dutch auctions, the bidding starts with a price set by the authorizing principle. This price is systematically reduced by the auctioneer and the technology is awarded to the first bidder.

Following the live auction event, the *realization phase* starts. Successful bids are reserved for a period of one month after the event. During this period buyers are obliged to pay the bidding price to the seller and the commission fee to IPA, plus the statutory value added tax. Accordingly commission fees are to be paid only on 'successful bid prices on the lump sums or on the upfront payment'. For successful bids, IPA collects commission fees of ten per cent of the sales price from the buyer. Future royalties from any acquired technologies are not subject to any payment. The sellers' premium is 15 per cent of the successful bid price. For unsuccessful bidders and sellers, IPA provides the opportunity to continue negotiations in separate rooms. In case a technology is not sold during the live auction event, IPA offers the possibility for follow-up sales, which begin on 'the day following the respective auction and lasts six months'. During this period, IPA reserves a right to still commission the fees by the sellers and buyers, even if IPA is not directly engaged in the transaction.

9.1.3 Results from the Comparative Analysis

Having provided individual case descriptions of auction designs developed by OT and IPA, I subsequently present the results from a comparative analysis in order to identify similarities and differences between the two process structures. I proceed along the three phases of the auction process with particular emphasis on the auction fees towards the end.

Figure 9.1 illustrates the generic auction process structure as synthesized from the analysis of the two auction processes highlighting the three core phases of the Lichtenthaler (2006a) model and also with reference to the auction process model proposed by Kumar and Feldman (1998). The conceptual process model presented in Figure 9.1 also indicates the different actors involved in each phase.

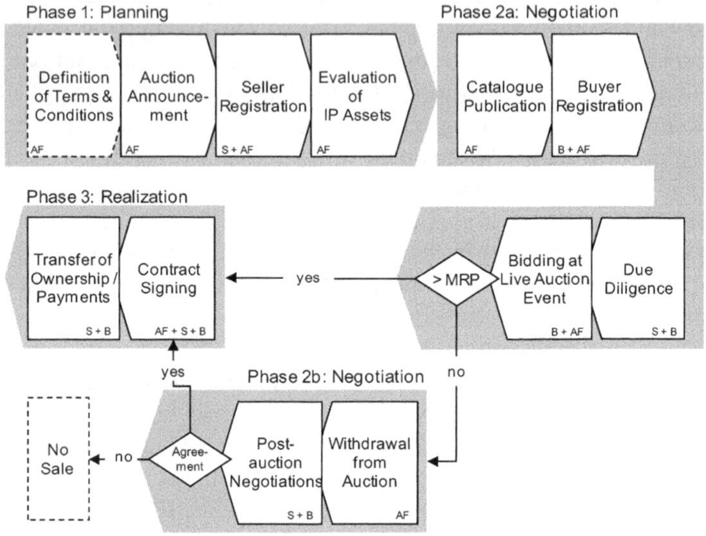

Notes: S (Seller); B ((Prospective)Buyer); AF (Auction Firm); MRP(Minimum Reserve Price)

Figure 9.1 - Technology auction process structure

Comparing the structure of both auction processes developed by OT and IPA, I found that they appear similar to a large extent.[13] Moreover both processes conform to the three core phases of the Lichtenthaler (2006b) model. Both processes start with the announcement of the auction[14], followed by a seller registration and an internal evaluation of the submitted technologies. The second, phase starts with the publication of the auction catalogue, initiating the negotiation phase that terminates after the live auction event. Following the live auction event, during the realization phase, technologies and IP asset ownership rights as well as additional (tacit) knowledge are transferred in exchange for the payments (including the commission fees to the auction firm). However, following the auction event, unsold technologies are made available for post-auction negotiations.

Both firms initiated the *planning phase* with the announcement of the live auction event to the public through their corporate websites and other publications, including IP expert journals, newsletters targeted to the community, as well as regular newspapers.[15] OT, having already organized several auctions, additionally used the auction catalogues to promote the next upcoming auctions. Both auction firms invited sellers to submit technologies for sale. For both auction processes, the sellers had to complete a registration procedure, which differs slightly from those of other auction firms, and provide specific documents for an internal evaluation. For instance, for technologies delivered to IPA, the sellers are required to complete a specific questionnaire to deliver particular information.

Both firms accept various types of assets for the auctions. These mainly include patented technologies (which are the focus of this study), but also trademarks, copyrights, and domain names. With regard to the technologies, both firms offered each technology sequentially and independently in a lot. A lot could comprise one or more related patents bundled together. While these four IP asset types are offered at both OT and IPA auctions, IPA additionally offers licenses. However, only licenses are accepted for which the license rights could be acquired through a lump-sum payment or through a down payment with following royalty payments. If offering licenses, sellers have to provide a license contract in the data room to prospective buyers upfront.

Table 9.2 - Bidding processes and MRPs of OT and IPA auctions

	OT	IPA
IP asset type	Patent, trademark, copyright, domain name	Patent, trademark, copyright, domain name, license
Bidding procedure	English	English, Dutch
Minimum reserve price	A_1: Confidential (not published) A_2: Not published (assumed to be confidential) A_3: US$ 10,000 A_5: £ 5,000 A_6: US$ 10,000	A_4: € 25,000

As shown in Table 9.2, both firms employ the MRP concept in their auction designs. Sellers have the possibility to either accept the MRP pre-determined by the auction firm or to determine an individual MRP for an extra fee. The amounts of the pre-determined MRP differ from auction to auction. For the first and second auctions OT kept the MRP confidential, thus the MRP remained unknown to bidders. OT changed this strategy after its second auction. In the following US auctions (A_3 and A_6), the MRP was announced to be US$10 thousand, while the MRP was £5 thousand at the

UK auction (A_5). For their first auction, IPA set the MRP at €25 thousand, thus a considerably higher MRP than OT had applied at prior auctions.

Sellers as well as prospective buyers have to pay registration fees in order to be allowed to participate in an auction (see Table 9.3). Both auction firms charge different registration fees depending on different variables. For sellers, OT based the registration fee on three criteria that refer to the asset type, the number of related IP assets bundled together, and the MRP. Sellers are allowed to define an own MRP for any asset. In this case, the registration fee for this asset was considerably higher (see also Table 9.1). IPA based the seller registration fee on two criteria that refer to the number of assets and the MRP. In cases where the seller specified an own MRP, IPA charged a variable fee of three per cent on the sales price. This additional fee was thus only due in the case of a successful sale.

Table 9.3 - Fee structures of OT and IPA auctions

Auction Firm	**OT**	**IPA**
	Registration Fees	
Sellers	Depending on IP Asset Type, Numer of IP Assets and MRP	Depending on Number of IP Assets and MRP
	For one Patent*: US$ 1,000 with OT MRP US$ 3,000 with own MRP	Single IP Asset: €1,000 Two or more IP Assets: €2,000
	For a 'Patent Pool', i.e. Bundle: $1,000 with OT MRP $6,000 with own MRP	Own MRP: Plus 3% Fee on the Sales Price
Buyers	Fixed Fee of $1,500	Depending on Type of Bidding Live Bidding: €1,500 Outside Bidding: €750 (Telephone); €500 (Absentee)
	Commission Fees	
Sellers	15% on Sales Price	15% on Sales Price
Buyers	10% on Sales Price	10% on Sales Price

Note: * Analogue for other IP Asset Types

Following the submission of technologies, OT performs an internal evaluation through a specialized patent rating firm, which they have acquired solely for this purpose. IPA conducts the internal evaluation with support of its mother firm, IPB. Having evaluated the technologies, those that fulfilled the internal evaluation criteria are accepted, those that do not are rejected. Accepted technologies are then compiled in an auction catalogue, which is subsequently published.

With the publication of the catalogue, the firms initiate the *negotiation phase*. The information provided in the catalogues of both firms is similar. The catalogues include mainly public bibliographical data on the patents embedded in the technologies, thus providing first information for prospective buyers for a first screening. Prospective buyers are then invited to register on the firms' websites.

Prospective buyers have to provide a signed bidder registration form and a bidder agreement, for which a draft is published in the respective auction catalogue. In case prospective buyers are not able to attend the bidding session personally, they have to further provide a signed alternative bid form. In addition, prospective buyers have to provide a bank guarantee and, in case the buyers would send a bidding agent, they have to deliver a written authorization confirmation. Furthermore, prospective buyers have to pay a bidder registration fee that varies slightly between the two auction firms (see further below). In order to participate in an auction, OT charges prospective buyers a flat fee of US$1.5 thousand. IPA charges prospective buyers a registration fee. Depending on the type of bidding, IPA charged the highest fee of €1.5 thousand for live bidding (that is prospective buyers being personally present in the auction room). For outside bidding, the fee was half the price (for telephone bidding) and a third of the price for absentee bidding (see Table 9.3).

Having completed the registration procedure, prospective buyers receive access data to the online data room in order to conduct in-depth due diligences. The data rooms offer the possibility to review detailed data on specific technologies provided by the sellers but also to directly contact the seller to request additional data. Both auction firms use online data room services of third party providers.[16] In addition to the data rooms, both firms engaged a specific IP evaluation firm to produce detailed reports for buyers on certain technologies on request and for additional payments. Thus, throughout the auction processes, prospective buyers had four information sources to evaluate technologies: the catalogue, the data room, personal or telephone meetings with the sellers, and tailored reports from third party experts, which prospective buyers could request.

Table 9.4 shows the publication dates of the catalogues and the auction dates. The length of this period in which prospective buyers could conduct a due diligence through the virtual data rooms varied from five (A_5) to 12 weeks (A_2 and A_3). Whether the available time for the due diligence can be regarded as sufficient, appears difficult to judge. The available literature does not indicate nor provide any recommendations for a sufficient length of such a period. However, according to Niioka (2006), an average of seven weeks should be regarded as relatively insufficient, particularly because not all prospective buyers might be aware of the upcoming auction on the date of the publication of the catalogue.[17]

Following the due diligence phase, the live auction event is the final step in the negotiation phase. For the execution of the biddings, both auction firms employ third party auctioneers. OT and IPA employ open 'cry-out,' English (ascending) bidding procedures, while IPA also applies Dutch (descending) bidding procedures (see Chapter 5) where bids are increased not by a fixed increment but rather reduced through the auctioneer. Although increasingly discussed in the literature, the second-price sealed-bid auction (Vickrey-auction) was not used. Each technology is offered sequentially and independently in line with the arguments from the auction design literature for heterogeneous assets for which auctions are commonly designed as single-unit, although multi-lot (that is various assets are auctioned sequentially at one auction). The bidding starts with an opening bid determined by the auctioneer. Technologies are sold if a bid reaches prices above the MRP pre-determined by the seller's auction firm. If for a technology the bids do not reach the MRP, this technology is withdrawn from the auction and remains unsold.

Table 9.4 - Duration of due diligence period

Auction #	Catalogue Publication Date	Auction Date	Due Diligence Period (weeks)
A_1	14.02.2006	06.04.2006	7
A_2	August 2006	26.10.2006	8 - 12
A_3	25.01.2007	19.04.2007	12
A_4	n.a	15.05.2007	~10 *
A_5	25.04.2007	01.06.2007	5
A_6	21.08.2007	25.10.2007	9

Note: * following an interview with an IPA representative

Following the live auction event, the *realization phase* starts. Within this phase, the sellers transfer the technologies, their IP ownership rights, and additional knowledge to the new owners in exchange for the payment. The exchange happens as specified by a standardized auction contract, which both the seller and buyer have to accept. The contracts employed by the auction firms however allow only for very limited adjustments to any transaction. For instance, it is not possible to include grant back clauses. Furthermore, OT and IPA only allow lump-sum payments. Thus sellers and buyers are restricted and cannot negotiate any royalty schemes, which are quite common when trading technologies. In addition to the payment of the final bid price, buyers as well as sellers, have to pay two types of fees (see Table 9.3). Commission fees have to be paid for successful sales, in other words biddings where the final bid exceeds the MRP at the live auction event or during post-auction negotiations. The fees are similar for the OT and IPA auctions. Sellers have to pay a premium of 15 per cent on the sales price (disagio) and buyers are charged ten per cent on the sales price (agio).[18]

Following the live auction event, both firms also provide the opportunity to unsuccessful bidders and sellers to continue with further negotiations throughout a post-auction phase. Throughout this phase, the rules governing the post-auction sales differ between OT and IPA. Unsuccessful technologies auctioned by IPA are subject to post-auction sales up to six months after the live auction event, whereas OT requires the payment of commission fees if the offered technologies are sold at 'any time after the Live Auction'.

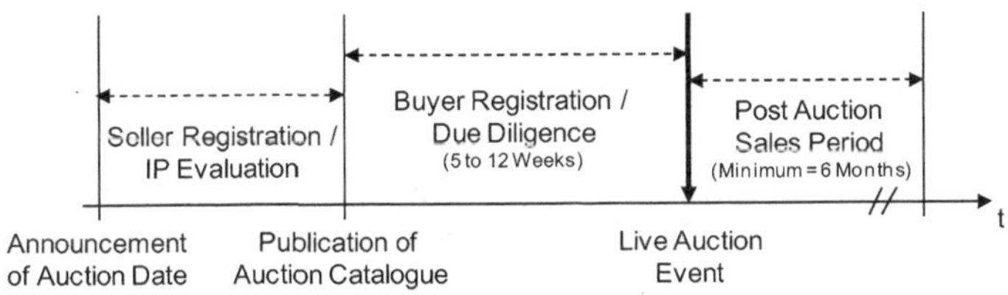

Figure 9.2 -Phases and distinguishing events in the auction process

To conclude, Figure 9.2 summarizes the three phases of the technology auction structure and the events that distinguish them. In short, the three phases are the:

- Pre-auction (planning) phase: Initiated by the auction announcement. It includes the registration of sellers and the submission of their technology proposals as well as the internal evaluation of technologies through the auction firm.
- Bidding (negotiation) phase: Initiated by the auction catalogue publication. It includes the registration of prospective buyers, the due diligence, and the bidding at the live auction event.
- Post-auction (realization) phase: Initiated when the live auction event is completed. It includes the formal transfer of the payments from the buyer to the seller in exchange for the ownership rights (that is intellectual property rights), eventually a transfer of complementary (tacit) knowledge from seller to buyer, the payment of commission fees to the auction firm by the buyer and seller, and eventual post-auction negotiations of unsold assets that were withdrawn from the auction.

9.2 CASE STUDIES OF SUCCESSFUL TRANSACTIONS

The second study (see Section 8.1 for methodological remarks) complements the qualitative insights from the comparative auction process analysis with empirical insights from a limited number of sellers (based on interviews, live observations, and documentation). Descriptive statistics of all the total transactions and the properties of the transacted technologies are provided in Chapter 10. The results from this second study are presented in two steps. First, I present individual case study descriptions illustrating how patented technologies can be auctioned. Although based on a key respondent approach, the case studies must be clearly understood to serve a primarily explorative and descriptive purpose. All case studies follow a similar structure. In an introductory section, I describe technology properties including the age of the technology, the number of related patents, and the geographical reach of the international patent protection. This section also includes information on the origin of the technology (for example, whether it was developed internally or jointly with another firm) and the technology's strategic and operational importance to the owner prior to the auction. In a second part, I describe the motives for the external exploitation and how the sellers have managed the transaction internally.[19] I emphasize on who was leading the exploitation project, the necessary resources throughout the various project phases, the patent strategy of the firm, and who had initiated the exploitation project. The third part of each case study discusses the firm's perception of the transaction's success as well as the firm's perception of the advantages and disadvantages of technology auctions compared to bargaining negotiations, that is where firms do not use TMI (for example, an auction firm) but rather make use of internal resources (for example, directly negotiated between two firms' patent departments). To evaluate the auction's success, the sellers were asked how they perceive the transaction in comparison to the resources that were necessary to complete the transaction, the total sales price and in comparison to the resources and costs the firm had incurred developing the technology. I present further insights into the perception of how the firms perceive the auction firm's role concerning different tasks (for example, marketing the technology). Each case study closes with a brief summary.

In the second step, I present results from the cross-case analysis in which I focus on two major aspects. These aspects refer to the internal transaction management and the transaction success, or how the firms perceived the advantages and drawbacks of auctions in comparison to bargaining negotiations as well as their perception of the auction firm's role within the transaction process. For the cross-case analysis, this study makes use of the triple measure developed in Chapter 8, which should be interpreted as a measure of agreement of the answers provided by the four case study firms. However, before presenting the individual case studies, a brief overview of the case study firms is presented.

9.2.1 Introduction to Case Study Firms

As illustrated in Table 9.5, the case studies were selected to represent a two-pair sample, including two SME (Comp1 and Comp2) and two large firms (Comp3 and Comp4), of which three are German firms and one is a Finnish firm. The four firms sold patented technologies at three auctions (two conducted by OT on 06 April 2006 and 01 June 2007; one conducted by IPA on 15 May 2007). The transaction volume ranges from a minimum of €25 thousand to a maximum of €650 thousand, with a mean value of €202 thousand. On average, both SME sold their technologies for a higher value, with a mean value of €350 thousand, than those sold by both large firms, with a mean value of €55 thousand.

The application areas of the technologies include 'surface / material science', 'signal transmission / telecommunications', 'measuring instruments', and 'office machinery and computers'. The age of the technologies varied from 5.5 to 8.0 years, with an overall average of 6.6 years. With an average age of 5.8 years, the technologies sold by the SME were 1.8 years younger than those sold by the large firms with an average age of 7.6 years. However, none of the technologies can be regarded as particularly young or particularly close to the expiration dates of related patents. The number of patented inventions that were bundled together in a technology varied more widely. The technology sold by Comp4 included only one US patent, while the technology sold by Comp3 included 12 patents. On average, the technologies consisted of 5.8 patents.

Comp1 is an SME, founded in 2001 and based in Germany. It specializes in molecular biotechnology, nanotechnology, and laboratory equipment. In August 2005, Comp1 sold all of its biotechnology-related assets to Qiagen, a company based in Venlo, the Netherlands that specializes in manufacturing products for the analysis on biological tests. The equity firm, Ventizz Capital Partners Advisory AG, currently runs four investment funds and holds a majority share in Comp1 via Ventizz Capital Font I. Before its liquidation started in 2005, Comp1 employed fewer than 50 people. Comp1 owned two major technologies for the preparation of peptide and protein tests for liquid chromotographic analytics and mass spectroscopy. Applying this type of preparation, testing structures can be modeled more precisely and the resources can be minimized. Prior to the transaction, Comp1 had employed a central patent manager, but did not have any particular experience with external technology transactions. In general, the external exploitation of technologies was of relevance to Comp1 in 2004, but rather irrelevant from 2007 to 2010, due to the liquidation of the firm.

Table 9.5 - Overview of case study firms and technologies

Firm Abbreviation	Comp1 [a]	Comp2	Comp3	Comp4
Seller Type		SME [b]		Large Firm
Nationality	German	Finnish	German	German
Main Industries	Bio- and Nanotechnology	Patent Management	Chemical and Pharmaceuticals	Energy, Electronics and Industry Technologies
# of Employees	≤ 50	1	Ca. 32,000	Ca. 400,000
Revenues (2007)	≤ 10 Mio. €	≤ 10 Mio. €	Ca. 7,000 Mio. €	Ca. 75,500 Mio. €
R&D Expenditures (2007)	n.A.	n.A.	Ca. 1,000 Mio. €	Ca. 3,500 Mio. €
Transaction Date	15.05.2007	01.06.2007	15.05.2007	06.04.2006
Auction Company	IPA	OT	IPA	OT
Transaction Volume (€)	50.000	649.044	25.000	85.178
Application Area	Surface/ Material Science	Signal Transmission, Telecommunications	Measuring Instrument	Office Machinery and Computers
# of Related Patents	7	3	12	1
Age of Technology Prior to Auction Date (years)	6,10	5,50	7,00	8,10
Experience with Patent Management	Low	Low - Medium	Medium - High	High
ETE Relevance (2004/2007/2010)	Medium / Low / Low	High / High / High	High / Medium / Medium	Low / Medium – High / Medium - High

Notes:
[a] In liquidation phase prior to the auction
[b] Following the EU definition for SMEs. For both firms revenues and R&D expenditures were not available. They were too small to be required publishing these figures.

Comp2 is a Finnish SME with annual revenues below €10 million. It specializes in the implementation of patent strategies for Finnish firms and inventors. The firm was founded primarily to capitalize on the inventions of the founder and CEO. Currently, Comp2 runs technology transactions in a range of industries, mainly telecommunications, computer implemented inventions, industrial automation, as well as chemistry, and biotechnology. Comp2 is run solely by the CEO who manages all patenting activities and external technology transactions. For necessary support, the firm works together with various sub-contractors. Thus, external technology transaction expertise was particularly relevant for the company's business model in 2004, 2007 and was expected to remain high in 2010.

Comp3 is a large firm based in Germany specializing in the pharmaceutical and chemical industries. In 2007, Comp3 had operations in about 80 countries with more than 30 thousand employees globally. Its total revenues accounted for approximately €7 billion, with the pharmaceutical business accounting for 70 per cent and chemistry for 30 per cent. In 2007, Comp3 spent approximately €1 billion on R&D activities. Comp3 operates an in-house patent department with a central patent manager and runs regular meetings of a committee responsible for intangible assets. Comp3 systematically evaluates its technology and IP portfolio using dedicated patent databases and portfolio management software. In 2004 and 2007, external technology exploitation was highly relevant for Comp3, however, this was expected to decrease slightly in 2010.

Comp4 is a German-based large firm that employed approximately 400 thousand people in 2007. In 2007, its revenues amounted to approximately €75 billion with the company spending €3.5 billion on R&D activities of which almost 50 per cent was for industrial technologies. Comp4 holds approximately 60 thousand patents with 50 per cent of them being in the energy and environment sector. To manage its patent portfolio, Comp4 employs several patent managers in a corporate IP unit with separate licensing departments for different regions. The patent management follows an explicit patent strategy which regulates the exploitation of technologies outside Comp4. A dedicated committee holds regular meetings to review the company's technology and IP portfolio. The firm makes use of various patent databases and portfolio management software tools. In 2004, the external exploitation of technologies outside the own boundaries were of no relevance for Comp4. In 2007 it was however of high relevance and expected to remain high until 2010.

A Special Surface Technology

Comp1 exploited a technology consisting of two inventions in the area of surface technologies. The first invention is a substrate with a particular light scattering and ultraphobic surface. The second invention covers a long-term robust water and oil rejecting surface. Both related patent families cover not only the structure of the surface, but also the manufacturing processes of and a wide range of application areas. The technology can be used to coat facades and solar panels and the insides of optical equipment and in the varnish industry.

Comp1 developed the technology in-house about nine years prior to the auction. The technology was of particular strategic relevance to Comp1 in 2002, although the technology never reached a positive return on investment. In 2007, the technologies were of neither strategic nor operational relevance to Comp1. Outside of the own business areas, Comp1 believed that both technologies would have a high number of potential applications in various business areas with a moderate revenue potential.

Internal Transaction Management

The project to exploit the technology externally was initiated and led by the company's CEO. Comp1's major motive for externally exploiting the technology was the firm's strategic decision not to develop the technology any further. Furthermore, the firm was in the process of liquidation. Thus, an auction was suitable as the firm was looking for less resource-intensive approaches to monetize its remaining assets. The technology was not selected internally by means of a specific mechanism. Instead, all related IP assets were sold jointly as a package.

To complete the transaction, Comp1 spent a total of 14 man days: ten man days for planning the transaction, one man day for the negotiation phase with prospective buyers, and three man days for realizing the transaction. Comp1 estimated the costs for the transaction in the range of €10 thousand to €50 thousand. The major cost drivers behind the transactions were the commission fees for the auction firm and the participation fee. Prior to the auction, Comp1 made a major attempt to exploit the technology only using internal resources, as well as in conjunction with another TMI. However, since these attempts were to no avail, the company decided to auction the technology. After completing the transaction, Comp1 reported that it had no knowledge on the buyer of the technology. Comp1 did not offer any post-auction knowledge transfer, nor did the buyer request it. Thus, the new and old owners do not intend to cooperate on this technology in the future.

Transaction Success

As a particular disadvantage of a public auction, Comp1 identified the price negotiations with prospective buyers, the costs of the whole transaction, and the disclosure of sensitive corporate information. Furthermore, the complexity of the transaction and the missing possibility to exclude particular buyers from the auction were considered problematic. Comp1 only experienced a few problems regarding communication with the auction firm, the identification of prospective buyers, and general organizational issues. The firm did not find the preparation of the due diligence and the limited information on the buyer problematic.

As a particular advantage of the auction, Comp1 identified the increased transparency in the transaction process, the possibility to exploit technologies with a limited value, and the fact that the firm could complete the transaction without needing to possess strong expertise on transactions. Furthermore, the firm had a positive opinion about not having to spend much in terms of internal resources, the standardized transaction process, the high quality of the technology valuation process, and the standardized contracts. Additionally, Comp1 valued the low total costs for the transaction, the high sales price, and the reduced buyer expectations of low prices. However Comp1 did not consider the reduction of costs, the standardized and fixed fees or the fact that after the auction no knowledge transfer was necessary particularly beneficial. Comp1 did not indicate that the duration of the transaction was significantly reduced by the auction process.

Regarding the role of the auction firm, Comp1 identified the closing of the transaction and the provision of a standardized legal framework as decisive aspects. Additionally, the auction firm provided some helpful support for selecting the technologies and supported the communication between the seller and prospective buyers as well as the organization of the transaction process by the auction firm. Furthermore, Comp1 considered the marketing activities of the auction firm in other countries helpful while the company perceived the technology valuation by the auction firm, the general marketing activities, and the reduction of overall transaction costs throughout the exploitation process somewhat helpful. Comp1 did not find the auction firm's marketing activities *in other industries* or its support to identify prospective buyers helpful.

Comp1 was very satisfied with regard to the overall success of the transaction. Moreover, the company was very pleased with the sales prices compared to the costs for the development of the technology and the simple and straightforward transaction. However, Comp1 was unsatisfied with the absolute sales price, that is had expected to reach a higher, absolute sales price for the technology. In general Comp1 was satisfied with the outcome and the transaction. Comp1 did not identify any particular potential for improving the auction design.

Summary

On 15 May 2007, Comp1 auctioned a nine years old surface / nanotechnology for €50 thousand through the IPA auction. Prior to the auction, the technology was not used internally and did not generate any revenues for the firm. Comp1 expected the technology to have strong potential for applications outside the firm's own boundaries, however with only a moderate revenue potential. The technology was externally exploited mainly because the firm was liquidating, that is due to the strategic decision of the firm not to develop the technology further.

Comp1 spent about 14 man days to complete the transaction, which translates to between €10 thousand and €50 thousand in monetary terms. As major cost drivers behind the transaction, Comp1 identified the auction fees, particularly the commission and the registration fees. Having no particular experience in transaction management, the firm considered the price negotiations with prospective buyers, the costs of the transaction and the disclosure of corporate information disadvantageous.

Comp1 considered the transparency of the process, the possibility to exploit even technologies with low and moderate values, and the fact that basically no internal expertise on the subject was necessary advantageous. The role of the auction firm appeared particularly valuable for the contract closing and the provision of a standard legal framework. Comp1 was generally satisfied with the overall transaction. Being in the liquidation stage, the firm aimed to externally exploit its remaining assets for the lowest possible costs. Completing this transaction, the firm reached this goal in a satisfactory ratio compared to the costs it had incurred to develop the technology. However, the firm had expected a higher absolute price for it.

A Wireless Communication Technology

The technology sold by Comp2 relates to applications for wireless, file-based multimedia data transfer. The technology enables multimedia messaging (MMS) as data files between mobile phones. In 2007, the technology was used worldwide in about 100 million high-end phones, for example, by the Finnish mobile phone manufacturer Nokia. In mid-2007, the firm ran a patent infringement suit (out of court settlement) with Nokia regarding this technology. Comp2 expects several billion mobile phones to implement its technology until the patent expires in 2020.

The technology was developed in-house in 2001 by the CEO of the firm as a combination of product and process inventions. From 2002 to 2007, the technology had a high strategic relevance for Comp2, although it was never used actively in the firm's operational business. Therefore, it was never of operational relevance. The firm expected the technology to have very high potential for various applications in various businesses and industries, with a particularly high revenue potential.

Internal Transaction Management

The transaction project was initiated not by Comp2 itself but rather by an independent consultant. The primary motive to engage in external exploitation activities was to monetize the technology. The technology was selected from the portfolio by an external consultant, while the transaction was managed by the CEO of Comp2.

For the total transaction, Comp2 spent seven man days: five days for the preparation and one day for the negotiations with prospective buyers, particularly the due diligence, and one day for the realization of the transaction. In terms of monetary resource requirements, Comp2 estimated the total costs of more than €100 thousand including the auction costs as well as the costs for the infringement suit. Thus, drivers for these costs were primarily the commission fees and legal costs for the infringement suit as well as interest fees for loans to pay the registration fee for the auction and to cover the costs of the infringement suit.

Prior to the auction, Comp2 had tried to exploit the technology with the help of a patent law firm and consequently developed a concept for a spin-off which would have further developed the technology. Prior to the auction, Comp2 had also attempted to make use of other TMI, but never seriously considered a joint venture with another firm.

The buyer of the technology remained unknown to Comp2. Although Comp2 was willing to provide additional knowledge transfer as a complement to the transfer of the ownership rights, and had even offered this to the buyer, the buyer did not request any additional knowledge transfer. Thus, the old and new owners do not intend to cooperate on further developments of this technology in the future.

Transaction Success

The patented technology for a wireless, multimedia file transfer was sold by Comp2 for €649 thousand on 1 June 2007 in London, UK on the auction organized by OT. Throughout the transaction, Comp2 was generally satisfied with the process. The due diligence, the identification of prospective buyers, the disclosure of sensitive information, limited information about the buyer, and the complexity of the transaction did not present any difficulties throughout the transaction. Comp2 identified only the high commission fees as a major negative issue of the auction.

In contrast to bargaining negotiations, Comp2 valued the acceleration of the transaction process, the high transparency of the process, the standardized legal framework, as well as the simple fee structure. Additionally, Comp2 experienced the standardized transaction process as partly advantageous and did not experience any particular advantages regarding low overall costs for the transaction, low external costs (for example, for a patent law firm), a higher quality of the technology valuation, a higher sales price compared to bargaining negotiations, reduced price expectations of the buyers as well as the knowledge transfer to the buyer. The firm did not see any advantages in a possibility to sell technologies with low and moderate values or that there is little need for internal resources and expertise.

Comp2 perceived the role of the auction firm as particularly advantageous with regard to the selection of the technology, marketing of the technology in general, as well as in other countries and businesses, the identification of prospective buyers, the support of the information exchange between prospective buyers and sellers, the closure of the transaction, the provision of a standardized legal framework, as well as the general administration process. Comp2 did not perceive the role of the auction firm as advantageous with regard to the reduction of total transaction costs or as valuation of the firm's technology portfolio.

Comp2 appeared completely satisfied with the sales price, even compared to the costs for developing the technology and patenting it, regarding the ease of the transaction and the duration. Comp2 was completely satisfied with the outcome of the transaction. However, for future auctions, Comp2 suggests that all related intangible assets should be made explicit to prospective buyers throughout the due diligence stage, especially all patents that might conflict with or infringe a technology.

Summary

Comp2 successfully completed a transaction for a wireless, multimedia file transfer technology for €649 044 via an auction organized by OT in the UK on 1 June 2007. Prior to the auction, Comp2 had tried to exploit the technology internally, with the help of a patent law firm, with another TMI, as well as developed an own spin-off solely dedicated to this technology. Although this technology had always had a high strategic relevance, Comp2 never made use of the technology in its own operational business.

The transaction was motivated solely to generate revenues. For completing the transaction, Comp2 spent seven man days of work. However, Comp2 claimed that the total costs including the auction fees were above €100 thousand. An additional cost driver was an infringement suit prior to the auction.

The accelerated transaction of the auction, the high transparency in the process, and the standardized legal framework (particularly the contracts) and fee structure were major advantages for Comp2. However, a major disadvantage for the firm was the high commission fees of the auction firm. However, despite the high commission fees, the firm was completely satisfied with the transaction.

An Analytic Technology for Splitting Molecules

Comp3 exploited a technology for a miniaturized analytical system for the splitting of molecules as well as a field generating membrane electrode for separating electro chemical reactions from ions, developed particularly for contact with ionized liquids. The technology – a combination of product and process inventions – can be applied in process technologies, particularly for applications in material sciences and nano technology.

From 2002 to 2007, the technology had a moderate strategic relevance for Comp3. Prior to the auction, it had almost no strategic relevance for Comp3. The technology was never implemented in any of Comp3's own products, and thus never contributed to the sales of the firm. Comp3 projected a moderate market potential for various applications in various businesses, as well as a moderate revenue potential.

The analytical system was developed about nine years prior to the auction together with another firm in a joint venture. A first patent application was filed on 16 June 1999. The field generating membrane electrode was developed prior to the auction and a patent was applied for on 26 November 2001. The technology is protected by two patent families, including four European, four Japanese and three German patents. The membrane electrode is protected separately by one European patent.

Internal Transaction Management

Comp3 decided to externally exploit the technology, since the firm was not planning to develop the technology any further. Moreover, the firm wanted to identify prospective buyers outside the firm's own boundaries. Furthermore, the firm intended to become more active on the MfTI, show presence in the market, and capitalize on the market potential of the technology outside the firm's own boundaries. Neither the lack of internal resources required nor the organizational barriers relevant for further developing the technology impacted the firm's choice to auction the technology.

The transaction was initiated by the head of Comp3's patent department. The technology was selected when the firm ran a regular screening and prepared a status report of the patent portfolio. The head of the patent department was responsible for the exploitation project. In total, 12 man days were necessary to complete the transaction. Ten man days were spent on planning and preparing the transaction and two days were spent on realizing the project. The negotiation phase did not require any significant work. The total costs for the transaction were estimated between €10 thousand and €50 thousand. The major cost drivers were the internal resources (for example, salaries) and the fees to the auction firm.

Prior to the auction, Comp3 had tried to exploit the technology externally, using internal resources, but did not try to exploit it with the help of an external partner, neither a patent law firm nor another TMI. Comp3 suspects that the buyer of the technology is a 'patent troll' (that is a non-practicing entity that acquires IP assets without planning to make use of it except for collecting royalties from other firms). After the auction, Comp3 was not willing to deliver additional knowledge to the new owner nor did the buyer request it. Thus, cooperation between the old and new owner appears very unlikely in the future.

Transaction Success

The technology was sold for €25 thousand on the auction organized by IPA on 15 May 2007. Throughout the transaction, Comp3 observed coordination and organizational problems with the auction firm. Furthermore, Comp3 experienced the efforts to prepare the auction as difficult. Additionally, the costs for the auction were considered too high.

Furthermore, Comp3 considers the disclosure of sensitive firm data for the due diligence and the missing possibility to exclude prospective buyers from the auction disadvantageous. Comp3 experienced the identification of prospective buyers, the complexity of the transaction, and the limited information about the buyers as less problematic. Price negotiations with prospective buyers were not considered problematic at all.

The most prominent advantages of the auction compared to bargaining negotiations was the acceleration of the transaction and the possibility to exploit even technologies with just low or moderate market values. The low total costs for the transaction was considered slightly advantageous. Furthermore, the fact that no knowledge transfer was necessary did not benefit Comp3. The standardized transaction process, the low external costs, the little need for internal resources, a higher quality of the technology valuation, the lower price expectations of the buyer, the standardized contracts and fees, as well as the possibility to reach higher sales prices were not of any relevance to Comp3.

The most prominent argument for Comp3 to choose the auction as an exploitation channel was the selection and marketing of the technology in other businesses, the identification of prospective buyers, the support for closing the contract, and the standardized legal framework provided by the auction firm. The marketing of the technology in other countries, the support throughout the information exchange between the seller and prospective buyers, and administration of the transaction through the auction firm was of less relevance to the company. The valuation of the technology through the auction firm and the reduced overall transaction costs were not relevant.

Comp3 appeared completely satisfied with the overall sales price and the duration of the transaction. However, Comp3 was not satisfied with the sales price compared to the development costs of the technology and the patent fees nor with the perceived reduced complexity of the auction compared to bargaining negotiations. All in all, Comp3 was not particularly satisfied with the transaction. Comp3 missed certain expertise and experience of the auction firm to handle such transactions. Thus, Comp3 recommended for future auctions to make use of expertise from a patent law firm that specializes in the exploitation of technologies and IP assets.

Summary
Comp3 completed a transaction for a technology that was never actively used internally. The technology comprised two patent families with an average age of about eight years for €25 thousand. The technology never had any operational and strategic relevance for the firm. Prior to the auction, Comp3 had tried to exploit the technology externally using internal resources. Comp3 expected the technology to have a moderate market potential in various businesses outside the firm's own boundaries as well as a moderate revenue potential.

The choice to exploit the technology externally was driven by the internal decision not to use the technology any further but to identify prospective buyers outside the firm's own boundaries. Twelve man days were necessary to complete the transaction and the total cost for the transaction was in the range of €10 thousand to €50 thousand, with the major cost drivers being the auction fees and internal costs (for example, salaries).

Major disadvantages of an auction were organizational difficulties and the high total costs. As major advantages, Comp3 mentioned the accelerated transaction process and the possibility to externally exploit technologies with low or moderate values. The auction firm was perceived to be particularly helpful for selecting the technology marketing activities and to identify prospective buyers. However, all in all, the firm was not satisfied with the transaction. Comp3 felt that the auction firm lacked substantial experience with such transactions and thus recommended the use of an experienced patent law firm to assist in future auctions. Moreover, the company was not satisfied with the ratio of costs compared to the sales price.

A Network Based Method for Chip Cards Transactions
Comp4 successfully sold a method for transactions with chip cards in networks. The related patents protect chip cards, middleware, and the software for the communication between different processes to ensure secure transmissions. The technology can be used for payments in open data networks, while the card can be used for specific transactions adjusted to the specific card type. The data exchange between the terminal and network can be encoded using various methods. The cards could serve multiple purposes, since the software can be adjusted.

The technology had a moderate strategic as well as operational relevance for Comp4 from 2002 to 2007. Prior to the auction, the technology had not reached a positive return on investment compared to the R&D spendings necessary to develop the technology (including patenting costs). The market and revenue potential was projected as moderate for different applications in various industries. The technology was developed in-house and Comp4 applied for a first patent on 6 February 2001. The technology is protected in the US, Australia, and Canada by one patent each and by two patents worldwide. Comp4 however sold only the US patent at the auction.

Internal Transaction Management

The decision to exploit the technology externally was driven mainly by the motive not to develop the technology any further. Furthermore, Comp4 aimed to become more active in the MfTI due to a strategic decision to exploit corporate assets more proactively in the future. Consequently, Comp4 wanted to show presence in the market. Furthermore, while the technology was not used internally, Comp4 was convinced of its significant application potential outside the firm's own boundaries.

The technology was selected by Comp4 through a detailed and systematic valuation exercise of the patent portfolio. Comp4's licensing committee was responsible for this process. The firm had approximately spent eleven man days until the transaction was completed: five were necessary to complete the planning of the transaction, three were spent on the negotiations, and three were spent on realizing the transaction. The internal costs of the transactions ranged from €10 thousand to €50 thousand. The main cost drivers were internal salaries.

Prior to the auction, Comp4 had tried to make use of the technology internally. However, Comp4 had not tried to exploit the technology externally prior to the auction, but had already auctioned other technologies. The buyer of the technology remained unknown to Comp4. Following the auction, the old and new owner did not agree on any further cooperation. Comp4 was not willing to deliver additional knowledge regarding the technology to the buyer, which was not requested by the buyer anyhow.

Transaction Success

The technology was sold for €85 thousand on the auction organized by OT on 6 April 2006. The transaction was initiated by an internal committee for licensing transactions. The most prominent problem of the transaction was the fact that certain prospective buyers could not be excluded from the auction by the seller. Furthermore, the transaction process was thought to be too complex and the total costs were considered too high. The disclosure of sensitive data, the limited knowledge on prospective buyers, and the expenses for the due diligence were considered problematic. However, Comp4 did not consider the coordination with the auction firm, the identification of prospective buyers, the price negotiations or the internal organization problematic.

Comp4 experienced the standardized transaction procedure, the high quality of the technology valuation, and the standardized contracts as particularly advantageous. Furthermore, the accelerated transaction, the high sales price, the possibility to sell low and moderate valued technologies, and the more realistic expectations regarding sales prices by buyers were additional advantages compared to bargaining negotiations.

Less advantageous aspects experienced by Comp4 include the overall transaction costs, the lower internal need for resources and expertise, and the higher transparency of the process. The standardized fee structure did not appear advantageous to Comp4. Furthermore, the fact that no additional knowledge transfer was requested by the buyer did not appear advantageous.

Comp4 experienced the role of the auction firm particularly helpful for the marketing of the technology in general and outside the firm's own boundaries, for the identification of prospective buyers, throughout the information exchange between the seller and prospective buyers, for the contract closing, and for the management of the transaction. Furthermore, the auction firm contributed to the reduction of the total transaction costs through the standardized legal framework. Comp4 considered the selection and valuation of the technology as well as the marketing thereof in other countries less important.

Overall, Comp4 claims to be completely satisfied with the transaction. The sales price was satisfactory. The sales price compared to the development and patenting cost was regarded as reasonable, as well as when comparing the price to the necessary resources for the transaction and the ease of the transaction. In the future, Comp4 will consider exploiting more technologies via technology auctions. According to Comp4, auctions are an appropriate way to exploit technologies externally.

Summary

Comp4 sold a technology for file transfers in networks for €85 thousand on the auction organized by OT on 06 April 2006. The technology was never actively used internally. Comp4's primary motive to sell the technology was a strategic decision not to develop the technology further and to become more active on the MfTI.

Comp4 perceived the fact that the seller cannot exclude prospective buyers from the auction as an eminent problem of technology auctions. Comp4 considered the high quality of the technology valuation, the standardized contracts, and standardized transaction structure particularly beneficial. The auction firm supported Comp4 particularly in the marketing of the technology, in identifying prospective buyers, and throughout the information exchange between the seller and the prospective buyers.

Overall, Comp4 experienced the transaction as very positive. Particularly, the sales price was very satisfactory. The statement that Comp4 will use auctions in future to sell technologies supports this conclusion.

9.2.2 Results from the Comparative Analysis

Successful transactions of the six technology auctions held between spring 2005 and autumn 2007 are investigated. Four technology transactions are presented that reached the highest or second highest sales prices at any of the six auctions. In the following they are compared in a cross-case analysis. I follow a similar structure as in the preceding descriptive analyses. As a basis for understanding the similarities and differences between the management of the different transactions, I first briefly compare the sellers, particularly with regard to their managerial competences, as well as the technologies. Second, I contrast the four cases with regard to the internal management of the transactions. Third, the firms' judgments of the advantages and disadvantages of the auction process are compared to those of bargaining negotiations and the role of the auction firm within the transactions is discussed, that is an assessment of how the firms perceived the success of the transactions via technology auctions.

As shown in Table 9, the external exploitation of technologies (ETE) had a high relevance for two firms (Comp2, Comp3) in 2004. For Comp1, the ETE was of moderate relevance in the daily business and for Comp4 the ETE had low relevance. The relevance of the ETE for the firm's business remained on a high level for Comp2 from 2007 to 2010. For Comp1, the ETE became less relevant and was expected to remain on a low level until 2010. For Comp3, the ETE became less relevant and was expected to remain on a moderate level. The ETE only became increasingly relevant for Comp4, increasing to a moderate / high level from 2004 to 2007 where it was expected to remain until 2010. Thus, three of the four firms expected the ETE to continue to be of moderate to high relevance in the future.[20]

Both SME had less resources dedicated to the ETE than both large firms and did not apply institutionalized tools for patent management (for example, use of commercial databases). Neither of them employed a dedicated patent manager. However, Comp2's business activities are focused on the implementation of patent strategies, thus it seems reasonable to assume that the firm employs certain patent management capabilities. Both large firms

(Comp3 and Comp4) run a patent management department and make use of patent management software and dedicated databases. Comp4 even has an IP committee in place that regularly reviews the firm's portfolio as well as an explicit patent strategy with certain elements particularly focused on the ETE. Thus, among the four firms, Comp4 employs the highest patent management competence. However, aside from their low ETE competence, both SME performed quite well with regard to the transaction volume.

Table 9.6 - Comparative properties of technologies

Firm Abbreviation	Comp1	Comp2	Comp3	Comp4
Seller Type	SME[a]			Large Firm
Transaction Volume (€)	50.000	649.044	25.000	85.178
Exploited Technology	Process for Manufacturing a Specific Surface	Technology for a Wireless File Based Multi Media Data Transfer	Miniaturized Analytical System	Method for Secure Transactions Using Chip Cards in a Network Environment
Application Areas	Process Technology, Nano-technology, Chemistry	Telecommunication (Wireless, Cellular, Optical)	Process Technology, Nano-technology, Chemistry	Telecommunication (Wireless, Cellular, Optical)
Average Age of Patents (Years)	6.1	5.5	7.0	8.1
Matter of Protection in Patent (Product / Process)	Process	Combination	Combination	Process
Development	Internally	Internally	Joint Venture	Internally
Strategic Relevance for Seller (2002/2007)	High / Low	High / High	Moderate / Low	Moderate / Moderate
Operational Relevance for Seller (2002/2007)	High / Low	Low / Low	Low / Low	Moderate / Moderate
Positive ROI[b] Reached Prior to the Auction	No	No	No	No

Notes:
[a] Following the EU definition for SMEs
[b] Return on investment of the sold technology

As shown in Table 9.6, the *sellers* of the four technologies represent a two-pair sample of two SME (Comp1 and Comp2) and two large firms (Comp3 and Comp4). All firms were active in different businesses. Comp1 was focused on bio and nanotechnology businesses. The Finnish-based Comp2 focuses on the implementation of patent strategies, Comp3 focuses on the chemical and pharmaceutical businesses, and Comp4 on the energy, electrical, and industrial businesses. Comp2 is the smallest firm with one employee and Comp4 is the largest with about 400 thousand employees.

Three of the four firms (except Comp3) sold technologies they had developed internally. The technology sold by Comp3 was developed in a joint venture. Two firms sold process technologies, while the other firms sold combinations of product and process technologies. None of the firms sold a dedicated product technology. Measuring the age of the technologies from the application date (on average across the patents in a bundle), none of the sold technologies can be regarded as either particularly old or young. However, the technologies sold by the large firms were on average 1.8 years older than those sold by the SME. The technologies had an average age of 6.67 years.

The technology sold by Comp3 consisted of the highest number of patents, but sold for the lowest price. The technology sold by the second largest firm (Comp4) sold for the second highest price, but consisted of only one patent. The technology that sold for the highest price was sold by an SME (the smallest firm among the four transactions) and consisted of three patents.

In 2002, the technologies sold by the two large firms (Comp3 and Comp4) had moderate strategic relevance. For Comp3, the relevance had recently decreased while it remained on a moderate level for Comp4. Among the SME, both technologies had a high strategic relevance in 2002. For Comp2, the relevance remained high until 2007 and for Comp1, it decreased to a low level. The technology sold by Comp1 had a high operative relevance for the firm in 2002, which decreased to a low level by 2007. The technologies of Comp2 and Comp3 were of low operative relevance and remained on that level, while, for Comp4, the technology was and remained on a moderate level. None of the technologies had ever reached a positive return on investments prior to their sale.

Table 9.7 - Perceived technology values

The economic potential of the sold technology outside own company's businesses with respect to...	Comp1	Comp2	Comp3	Comp4	Mean (m_c)
Transaction Volume [€]	50,000	694,044	25,000	85,178	202,31
a) ...the scope of different applications	3	4	2	2	2.75
b) ...application potential in a number of different businesses	4	4	2	2	3.00
c) ...the total potential sales volume	2	4	2	2	2.50
Mean (m_i)	3.0	4.0	2.0	2.0	2.75

Note: The five point Likert scale ranges from 0 (small/low) to 4 (large/high)

Comparing the transaction volumes to the economic potential of the technologies as perceived by the sellers (Table 9.7) reveals a somewhat inconsistent picture. Comp2 clearly indicated that its technology has a high economic potential with regard to its scope of different applications in a number of different fields and with regard to the total potential sales volume. In this regard, it is important to note that Comp2 was engaged in an infringement suit with regard to its technology with a large firm, which can be considered a good indication of the technology's economic potential. Comp4, which reached the second highest sales price, had however indicated an average, rather moderate, economic potential for the technology.

Comparing the average perception of the technologies' economic potential across the two firm groups reveals that both SME perceived their technologies to have a higher economic potential (Ø $m_{i(SME)}$=3.5) than the technologies sold by the large firms (Ø $m_{i(LF)}$=2.0). This observation is consistent with the higher average transaction volumes of the technologies sold by the SME (Ø $TV_{(SME)}$= €350 thousand) compared to those sold by the large firms (Ø $TV_{(LF)}$= €55 thousand).

Internal Transaction Management

As shown in Table 9.8, the two transactions conducted by the large firms were initiated by the departments responsible for the ETE activities or the head of the patent department (Comp3) or licensing department (Comp4). Comp1's CEO initiated the transaction while, at Comp2, an external party initiated the transaction. In both SME, the top level management (for example, CEO) was responsible for leading the transaction, while in both large firms, the responsibility for the transactions fell on dedicated departments (for example, the patent department or licensing council).

The primary motives to exploit the technologies were almost similar among the transactions. Three firms initiated the exploitation project driven by the motive that the firm had made the strategic decision not to develop the particular technology any further. However, with its business model focusing on the implementation of patenting strategies, Comp2 initiated the transaction with the primary motive to generate revenue streams from selling the technology. Furthermore, the firms indicated other motives that were important. The firms that wanted to enter the technology trade business, indicate presence on the MfTI, the opportunity to identify prospective buyers outside their own businesses, and the fact that the technology was not used internally as important motives. Thus, financial but also strategic motives drove the decision to choose a technology auction as external exploitation channel.

Prior to the auctions, all firms had attempted to exploit the technologies through other means. Comp2 had tried extensively to exploit the technology with its own resources, with the support of a patent law firm and even considered an own spin-off. The second SME (Comp1) had tried to exploit the technology externally with the help of another TMI, while both MNCs had tried to exploit the technologies using their own internal resources.

With regard to the monetary resources spent on completing the transactions, three of the firms had expenses in the range of between €10 thousand and €50 thousand. Only Comp2 had considerably higher costs for the transaction of more than €100 thousand. Comp2 explained these high costs in terms of the proportional commission fee ratio on the sales price as well as having to cover legal costs associated with an infringement suit prior to the auction. As a major cost driver of the transaction, three firms identified the proportional commission fees charged by the auction firms. The two large firms furthermore mentioned internal salaries and 'necessary time' as substantial cost drivers in the transaction project.

The resources required for completing the transaction varied between seven man days (Comp2) and 14 man days (Comp1). Three of the four firms needed more than ten man days to complete the transaction. With a little more than two working weeks, both large firms required almost similar resources to complete the transaction. On average, the firms spent 11 man days to complete the transaction.

Table 9.8 - Internal management of technology transactions

Firm Abbreviation	Comp1	Comp2	Comp3	Comp4
Seller Type	SME *		Large firm	
Initiator of the Project	CEO	External party	Head of the patent department	Licensing committee
Leading the Project	Director of Firm	CEO	Patent department	Licensing council
Primary Motive for the Project	Strategic Decision not to Develop the Technology further	Generating revenue Streams	Strategic Decision not to Develop the Technology further	Strategic Decision not to Develop the Technology further
Exploitation Prior to Auction	Internally, with Support of another TMI	Internally, with Patent Law Firm, as Spin-off	With Internal Resources	With Internal Resources
Transaction Costs	10,000 € - 50,000 €	> 100,000 €	10,000 € - 50,000 €	10,000 € - 50,000 €
Main Cost Drivers	Commission Fee to Auction Firm, Registration Fee	Commission Fees, Legal Costs for Law Suit, Interest Fees	Salaries, Commission Fees	Necessary Time
Necessary Resources (man-days)	14	7	12	11
... for Planing Stage	10 (71%)	5 (71%)	10 (83%)	5 (45%)
... for Negotiation Stage	1 (7%)	1 (14%)	0 (0%)	3 (27%)
... for Realization Stage	3 (21%)	1 (14%)	2 (17%)	3 (27%)
The Buyer	Unknown	Unknown	Unknown (probably 'Patent Troll')	Unknown
Additional Knowledge Transfer after the Auction	None	None (although the Firm was willing to)	None	None

Note: * Following the EU definition for SMEs.

A closer look at the distribution of this workload across the three process phases (applying the distinction proposed for Lichtenthaler's (2006a) ETE process model – see Chapter 3) revealed that all firms had spent the major share of resources on the planning phase. On average, the firms spent two thirds of the resources in this phase. On average, 20 per cent of the resources was spent on the final phase and the realization phase, while only a little more than ten per cent of the resources was spent on the negotiation phase. The distribution only differed – if slightly – in Comp4. The firm spent almost 50 per cent of the resources on the planning phase. The remaining 50 per cent was distributed almost equally between the other two phases. These results indicate that, in comparison to bargaining negotiations, firms save resources particularly with regard to the negotiation and realization phases. A possible explanation is that the auction firms conduct the negotiations (on the live auction event jointly for multiple technologies) and provide standard legal contracts for the realization of the transactions. The few resources required for the realization can be explained by the fact that none of the sellers performed additional knowledge transfers in the post-auction phase. In fact, none of the buyers had requested additional knowledge transfers. Except for Comp2, the other three firms were even not willing to do so and spend further resources on the technologies they had sold. All buyers of the four technologies remained anonymous after the transaction, thus are unknown to the sellers although Comp3 suspected that its buyer might be a patent troll.

Transaction Success

Within this second pre-study, I was particularly interested to understand which aspects of technology auctions sellers perceive as advantageous or disadvantageous in comparison to bargaining negotiations. Furthermore, I wanted to understand how firms perceive the role of the auction firm as TMI attempting to facilitate transactions.[21]

From Table 9.9, it appears that – on average – all firms agreed that auctions possess only few problems (overall m_i is 1.43). On average, Comp1, one of the two SME, was most skeptical with regard to patent auctions (m_i=2.1). The firm scored the highest value for three items (negotiating prices with prospective buyer; costs for the total transaction and disclosure of sensitive firm data). Comp2, having reached by far the highest sales price, perceived the fewest problems with the auction (m_i=0.3).

Table 9.9 - Perceived problems of transactions

Possible Problems of the Auctions	Comp1	Comp2	Comp3	Comp4	Mean (m_c)	M_1	M_2	M_3	M_x
a) Efforts to Prepare the Due Diligence	0	0	3	2	1,25	0	0	1	0
b) Coordination Problems with the Auction Firm	1	0	3	0	1,00	1	1	0	0
c) Difficulties in Identifying Prospective Buyer	1	0	1	0	0,50	1	1	1	**1**
d) Negotiating Prices with Prospective Buyer	4	0	0	0	1,00	1	1	0	0
e) Complexity of the Transaction	3	0	1	2	**1,50**	0	0	1	0
f) Costs for the Total Transaction	4	3	3	2	**3,00**	1	1	1	**1**
g) Disclosure of Sensitive Firm Data	4	0	2	2	**2,00**	0	0	1	0
h) Inability to Select / Exclude Certain Bidders (e.g. Competitors, "Trolls")	3	0	2	3	**2,00**	0	0	1	0
i) Internal Organizational Difficulties	1	0	4	0	1,25	0	1	0	0
j) Limited Information about Prospective Buyer	0	0	1	2	0,75	1	1	1	**1**
Mean (m_i)	2,10	0,30	2,00	1,30	1,43				

Note: Scale ranges from 0 (not problematic) to 4 (very problematic)

Applying the triple measure M_x, it appeared that the firms agreed on three items. As Table 9. shows, the sellers agreed that the overall costs for the transaction were problematic (primarily related to the relative nature of the commission fees). On the contrary, the firms agreed that identifying prospective buyers was not problematic. Surprisingly, this issue has often been cited in prior literature as a major problem when trading technologies (see Chapter 3). This problem seems to have been reduced through the auction transaction governance structure. Furthermore, the firms agreed that the limited information on the buyers was not problematic for concluding the transactions.

In addition to the items that appeared when applying the M_x measure (that is costs for the total transaction), three items scored above the overall mean value (m_i=1.43), indicating that those items are perceived – at least by some firms – as problematic. Three out of the four firms scored high values with regard to the disclosure of sensitive firm data and the inability to select / exclude certain bidders (for example, avoiding selling a technology to a competitor or patent troll). Furthermore, m_c was slightly higher for the item's complexity of the transaction than the overall mean value.

Table 9.10 - Perceived advantages of technology auction transactions

Possible Advantages of the Auctions	Comp1	Comp2	Comp3	Comp4	Mean (m_c)	M_1	M_2	M_3	M_x
a) Acceleration of the Transaction Process	0	4	2	2	**2,00**	0	1	1	0
b) Lower Total Costs of the Transaction Process	2	2	1	1	1,50	0	1	1	0
c) Lower External Costs (e.g. for Dedicated Patent Law Firm)	1	2	0	1	1,00	1	1	1	**1**
d) Lower Need for Internal Resources and Expertise	3	0	0	1	1,00	1	1	0	0
e) Standardized Transaction Procedure	3	3	0	3	**2,25**	0	1	1	0
f) Increased Quality of Technology Valuation	3	2	0	3	**2,00**	0	0	1	0
g) Increased Transparency in the Sales Process	4	4	0	1	**2,25**	0	0	1	0
h) Obtaining Higher Sales Prices	2	2	0	2	1,50	0	1	1	0
i) Possibility to Sell Low- and Medium-value Technologies	4	0	2	2	**2,00**	0	1	1	0
j) Reduction of Unreasonable Low Price Expectations of Buyers	2	2	0	2	1,50	0	1	1	0
k) Standardized Contracts	3	4	0	3	**2,50**	1	1	1	**1**
l) Standardized Lump-sum Payments	1	4	0	0	1,25	0	1	0	0
m) No Additional Know-how Transfer Requested by Buyer	1	2	1	0	1,00	1	1	1	**1**
Mean (m_i)	2,10	2,00	1,30	1,30	1,67				

Note: Scale ranges from 0 (not advantageous) to 4 (very advantageous)

As shown in Table 9.10, I derived 13 items from the literature as possible advantages of auctions in comparison to bargaining negotiations. Comparing the average m_i values, both SME perceived the auctions to be more advantageous than the large firms did. However, the overall mean (m_c=1.67) is not very high. Applying $M_{x,}$ three items appeared to be relevant. The firms agreed that two of the items are not particularly advantageous (lower external costs and no additional knowledge transfer requested) with values lower than the overall mean. All firms agreed that the standardized legal contracts are particularly advantageous.

However, comparing the m_c values for each item with the overall mean (m_i=1.67), various items appear to be advantageous. Three firms perceived the standardized transaction procedure as advantageous, giving the second highest score. Furthermore, both SME strongly agreed that the increased transparency by the sales process is advantageous. Moreover, three items scored above the overall mean. These include the acceleration of the transaction process, the increased quality of the technology valuation, as well as the possibility to sell low and medium-value technologies.

To better understand for which tasks the support of the auction firm was valued, I asked the sellers specifically about the role of the auction firm in the transaction. The results are presented in Table 9.11. From the 11 items tested, a quite homogeneous picture emerged. With regard to the role of the auction firm in the transaction process, eight items scored a value of 1 for M_x. With regard to the m_c values, the firms regarded the auction firms' role as highly beneficial (that is a score of above three) for four tasks. These include – in order of decreasing importance – the closing of the transaction, the provision of a standardized legal framework (including contracts), and, ranked equally, the support for information exchange between sellers and prospective buyers in the negotiation phase as well as the general process administration. While both SME valued the role of the auction firm moderately, Comp4 valued its role the most while, contrarily, Comp3 valued it the least. Furthermore, three of the seven items (where M_x has a value of one) ranked slightly above the overall mean (m_c). These include the selection of the particular technology, the marketing of the technology outside the firms' businesses, and the identification of prospective buyers (matchmaking). The firms further agreed that the role of the auction firm for the valuation of the technology / patent portfolio was not important. In addition, the firms disagreed on the role of the auction firm with regard to the marketing of technology in general, in other countries, and with regard to the reduction of total transaction costs.

Table 9.11 - Role of auction firm in the transaction

Importance of the Role of the Auction Firm	Comp1	Comp2	Comp3	Comp4	Mean (m_c)	M_1	M_2	M_3	M_x
a) Selecting the Particular Technology	3	4	2	1	**2.50**	1	1	1	**1**
b) Valuation of Technology / Patent Portfolio	1	1	0	1	0.75	1	1	1	**1**
c) Marketing the Technology:									
1. in general	1	4	1	4	**2.50**	1	0	1	0
2. in Other Countries	2	4	1	0	1.75	0	0	0	0
3. outside Own Company's Businesses	0	4	2	4	**2.50**	1	1	1	**1**
d) Identifying Potential Buyer (Matchmaking)	0	4	2	4	**2.50**	1	1	1	**1**
e) Support for Information Exchange between Seller and Buyer (Negotiation)	3	4	1	4	**3.00**	1	1	1	**1**
f) Transaction Closing	4	4	2	4	**3.50**	1	1	1	**1**
g) Provision of Standardized Legal Framework (incl. Standard Contracts)	4	4	2	3	**3.25**	1	1	1	**1**
h) General Process Administration	3	4	1	4	**3.00**	1	1	1	**1**
i) Reduction of Total Transaction Costs	1	0	0	3	1.00	1	1	0	0
Mean (m_i)	2.00	3.36	1.27	2.91	2.39				

Note: Scale ranges from 0 (no importance) to 4 (of decisive importance)

Finally, I assessed how firms perceived the success of the transactions. In general, it can be concluded that the firms perceived the auctions to be successful. As shown in Table 9.12, across all successful items, the overall mean score ($\varnothing m_i$) is 2.5. This finding is further confirmed by the fact that M_x has a value of one for the last of the five items, for which I asked the firms about their perception of the auction's success in general. With regard to the average m_i values of both groups, SME (Ø m_i=3.2) perceived the auctions as more successful than the large firms (Ø m_i=1.8). Furthermore, with regard to higher m_c values than the overall mean, the firms perceived the auctions as particularly successful with regard to the convenience of the transaction (for example, support of the auction firm) and the absolute achieved price.

Table 9.12 - Overall success of transactions

Satisfaction with the Transaction Regarding:	Comp1	Comp2	Comp3	Comp4	Mean (m_c)	M_1	M_2	M_3	M_x
Transaction Volume (€)	50,000	649,044	25,000	85,178					
a) The Absolute Achieved Price	1	4	2	3	**2.50**	1	0	1	0
b) The Sales Price in Comparison to Own Expenditures for the Technology (incl. Development, Administration, and Patent Filing Costs)	3	4	0	2	2.25	0	1	1	0
c) Resource Expenditures / Total Costs for the Transaction	2	4	0	2	2.00	0	1	1	0
d) Convenience of the Transaction (e.g. Support of the Auction Firm)	4	4	2	2	**3.00**	1	0	1	0
e) In general	2	4	1	4	**2.75**	1	1	1	**1**
Mean (m_i)	2.40	4.00	1.00	2.60	2.50				

Note: Scale ranges from 0 (no importance) to 4 (of decisive importance)

Furthermore, I found that the mean success measure across all items (m_i) relates perfectly with the ranking of the achieved sales prices. Achieving the highest sales price among the four firms, Comp2, perceived the auction as highly successful with regard to all five success items (m_i=4.0), while Comp4, achieving the second highest sales price, perceived the auction as a little less successful (m_i=2.6). Comp1, achieving the third highest sales price, ranked third with regard to its average perception of the auction success (m_i=2.4) and Comp3, achieving the lowest sales price, perceived the auction as the least successful (m_i=1.0).

Furthermore, the firms were asked to provide suggestions for improving the auction design. Only Comp3 and Comp2 provided suggestions. Comp3 suggested that it would be beneficial to the auction design if the auction firm acquire higher expertise with regard to patent valuation and would leave more operational freedom (flexibility) to the seller with respect to the auction process (for example, the contract design). Furthermore, Comp3 recommended employing additional expertise of specialized patent law firms. Comp2 suggested including information on the litigation status of the patents in the marketing material (for example, auction catalogue).

Conclusions

To conclude, the study investigated four successful technology transactions representing a two-pair sample, which includes two transactions where the sellers are SME and two where they are large firms. After all firms had attempted to exploit the technologies unsuccessfully through other means, they sold their technologies successfully through technology auctions. The transaction volumes range from €25 thousand to €650 thousand, with an average value of €202 thousand. The technologies were primarily sold due to strategic decisions of the firms not to develop them further.

With regard to transaction management, in both SME, the top level management (for example, CEO) was responsible for leading the transaction, while in both large firms, the responsibility for the transactions fell on dedicated departments (for example, the patent department or licensing council). Independent from their size, the firms spent similar resources to conduct the transaction, that is on average 11 man days, or less than €50 thousand. In contrast to bargaining negotiations, the firms saved resources, especially during the negotiation and realization phases of the transaction process. The commission fees to the auction firm were considered major cost drivers of the transactions. After the transaction, the buyers of the technologies remained unknown to the sellers, since all contracts and information exchange between the sellers and prospective buyers (for example, during the due diligence phase) were channeled via the auction firm. In addition to the transfer of the ownership rights, no additional knowledge transfer was offered, requested or agreed upon between the buyers and sellers.

In contrast to bargaining negotiations, this study's results indicate that firms perceive only few problems with the auction design. However, the firms perceived the overall cost of the transactions as too high (that is primarily the commission fees). The firms agreed that finding prospective buyers (which has commonly been cited as an obstacle in technology trade), appears to have been resolved by the technology auction design. A possible explanation is the outsourcing (disintegration) of this task to the auction firm that can reap economies of scale through joint marketing and standardization (for example, of the contracts). The sellers did not perceive the lack of information on the prospective buyers of their technologies as problematic. This may be due to the fact that the firms had sold technologies that did not have any strategic relevance for their businesses anymore. However, the firms indicated the disclosure of sensitive firm data and the inability to select / exclude certain prospective buyers in general as a disadvantage of the technology auction design.

The firms agreed that the standardized contracts are a major advantage of auctions. Furthermore, the standardized transaction procedure, the increased transparency, the acceleration of the transaction process, and the increased quality of the technology valuation are considered advantageous in comparison to bargaining negotiations. In general, the firms regard auctions as an appropriate exploitation channel to sell low and medium-value technologies.

Throughout the transaction, the sellers highly valued the auction firm's support. The firms indicated the importance of the auction firm's role in various tasks. Predominantly, the auction firms play a major role to close the transactions, to provide a standardized legal framework (including the contracts), and to support the information exchange between the prospective buyers and sellers in the negotiation phase of the transaction. Furthermore, the sellers acknowledged the auction firm's role in the general process administration, that is the auction firm as the process owner. This appreciation is further reflected by the fact that the firms acknowledged the convenience of the transaction as a particular success measure.

9.3 SYNTHESIS OF THE RESULTS

This section concludes the pre-studies that addressed the first specific research question. Moreover, I derive hypotheses that I tested in my quantitative study in order to address this study's second specific research question. This section synthesizes the results from both pre-studies on auction process structures and successful transactions based on, but also complemented by, insights from prior literature on the management of external technology transactions (Chapter 3), the auction literature (Chapter 5), and transaction governance structures (TGS) (Chapter 7). Lichtenthaler's (2006a) ETE process model and Kumar and Feldman's (1998) auction process model were used to generate an appropriate understanding of TGS for the investigated auctions. I then derive hypotheses that address the second research question. From transaction cost theory, Williamson's (1979) theoretical framework for the choice of governance structures served as a conceptual foundation.

9.3.1 The Transaction Governance Structure of Technology Auctions

The results from the first pre-study show that the TGS of technology auctions can be categorized into three phases in accordance with Lichtenthaler's (2006a) ETE process model, each being distinguished by a certain key event. The first phase is the pre-auction phase (planning), which is initiated through the auction announcement. During this phase, the auction firm invites sellers to register and to submit technologies that they propose for the next auction. The auction firm then collects these proposals and internally evaluates them. The publication of the auction catalogue containing the selected technologies initiates the second phase. During the due diligence period (negotiation), initial price negotiations take place that form the prospective buyers' willingness-to-pay (that is indifference price). However, the actual price negotiations (at the starting and closing of bids) take place during the live auction event. After the auction, in the post-auction phase (realization), the formal transfer of the payments, ownership rights (that is IPR), and eventually complementary (tacit) knowledge takes place as well as post-auction negotiations.

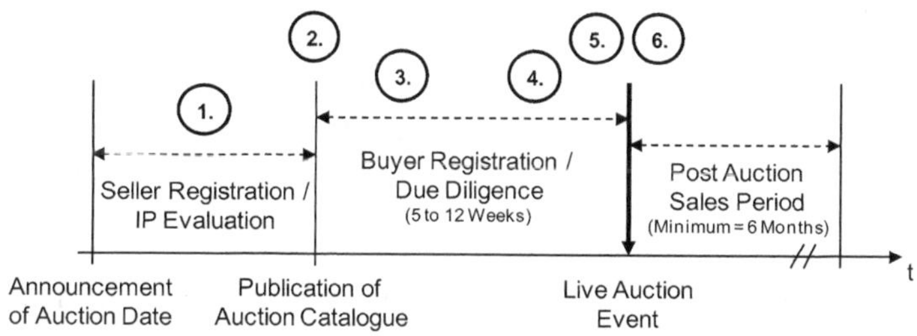

Figure 9.3 - Six standardization elements of technology auctions

The results of the second pre-study show that the TGS of technology auctions appears largely standardized. To a large extent, sellers and prospective buyers are restricted in their opportunity to adjust the rules governing the transaction, that is as commonly used in bargaining negotiations if firms govern the transaction themselves. Figure 9.3 illustrates the six identified elements in the investigated technology auctions that induce standardization into the TGS, which are discussed in chronological order along the auction process in the following section.

First, in the planning phase, the auction firms request *standardized participation fees from the sellers*. Claiming standardized participation fees thus reduces the complexity of the transactions. The auction firm does not need to negotiate fees with each participant for each transaction individually. For these fees, the auction firms conduct a first internal evaluation of submitted assets. If a technology has been positively evaluated and accepted for the auction, the auction firm commits itself to engage in search activities to identify possible buyers.

These search activities relate to the second aspect that induces standardization in the TGS. Having selected the technologies for an auction, the auction firm initiates *marketing activities jointly for all technologies*, primarily using auction catalogues. The sellers merely have to provide information about the technology, which the auction firms collect in order to prepare the auction catalogue. Within the catalogues, all technologies are presented in a similar, standardized way, mainly referring to the information contained in the patent documents. Little effort is made to highlight individual technology details or distinctive characteristics of certain technologies. By distributing the catalogue (that is a collection of technologies) to a wide range of prospective buyers across different industries, the auction firm takes care of the marketing efforts, which each seller would otherwise have had to conduct for each technology. The auction firms can thus realize economies of scale and reduce TC. The sellers themselves do not have to engage in marketing activities but rather outsource this task to the auction firms. Hence, as outlined in the definition of TMI (Chapter 4), the auction firm takes an economic risk. If a technology remains unsold, the auction firm must write off the related sunk costs for the efforts to prepare the transaction – although those costs are to some extent covered by the participation fees. However, the auction firms do not take ownership of technology assets; hence, they act as brokers but not as dealers who commonly take ownership of assets.

Third, when prospective buyers express their interest in a technology in the negotiation phase, they have to pay a *standardized buyer participation fee* in order to gain access to detailed information (for example, on the development status of a technology and the legal titles of related patents) and to conduct an in-depth evaluation (due diligence) of legal, technical, and economic issues related to the technology. Again, these fees are standardized for all prospective buyers, which reduces the complexity of the TGS by avoiding individual negotiations. For those fees, the auction firms cover the administrative cost of validating the buyers' identities, credit ratings, providing the necessary documents, assisting the buyers throughout the remaining auction process, and providing the necessary infrastructure for the due diligence (that is virtual due diligence rooms), once more realizing economies of scale that result in reduced TC.

Fourth, in order to evaluate technologies, the auction firms provide *virtual data rooms*, usually through third party service providers. As highlighted in Chapter 3, information asymmetry represents an immanent obstacle to technology transactions that prospective buyers need to overcome. In particular, the evaluation of technologies and related IP assets is often seen as difficult and resource consuming because legal, technical, and economic issues need to be checked intensively. This often requires intensive communication between a seller and the prospective buyers. Therefore, the due diligence process must be considered of critical importance in any transaction, as it is meant to adjust for information asymmetries. Although the auction firms had chosen professional service providers that offer a wide range of properties and flexibility in the due diligence process, their virtual data rooms obviously provide less flexibility than conducting a live and personal due diligence. Sellers and prospective buyers are restricted to exchange information or request and validate information that can be made available digitally (that is in explicit or codified form). For instance, prototypes or large drawings remain difficult to evaluate in virtual data rooms. Conducting interviews (for example, with lead inventors) is further hindered, because the communication between a seller and prospective buyers is channeled through the auction firm so that prospective buyers can remain anonymous. Moreover, the duration of the due diligence is standardized for all technologies. However, evaluating a complex technology (that is covered by a range of patent families) often requires more time and resources than evaluating a simple, discrete technology (for example, consisting of only one invention and / or being covered by just one patent). Thus, applying a standardized, virtual due diligence process for all similar technologies from the same service provider and for the same duration induces a further element of standardization into the TGS.[22]

Fifth, auction firms provide a *standardized legal framework* in the realization phase that includes standardized contracts that regulate, for instance, the transfer of the IP rights, payment schemes, and fees. In general, the prestudies show that sellers perceive such standardized legal frameworks as advantageous while they do not perceive it as problematic that buyers can remain anonymous. However, the standardized legal framework allows neither buyers nor sellers to individually adjust the contracts to the specific properties of a technology. For instance, it does not allow accounting for technological uncertainty by specifying a royalty scheme with individually negotiated royalty rates including further clauses (for example, grant back clauses), but only allows lump sum payments. However, sellers considered it particularly disadvantageous that the standardized legal framework does not allow for them to exclude certain prospective buyers, for instance the sellers' direct competitors to whom they would not usually sell or provide a license for a particular technology. Thus, although the standardized legal framework contributes to the reduction of TC, the sellers' responses indicate that the contracts might be rather over-standardized.

In contrast to the contracts used in the technology auctions, traditionally, contracts developed through bargaining negotiations (including pre-closing contracts such as a letter of intent and memorandum of understanding) allow for a high degree of flexibility. Oftentimes, they are developed on a case-by-case basis, sometimes with the support of specialized patent law firms; although sometimes large firms that employ in-house patent and licensing departments possess these capabilities internally. Thus, although highly flexible, these individually designed, idiosyncratic contracts are usually expensive and resource consuming. According to Bessy and Brousseau (1998: 456), in technology and knowledge transfers, contractual properties lead to relatively high TC. In contrast, the standardized legal framework developed by the auction firms and, at the same time, the standardized contracts that are applied to a range of technologies induce further standardization into TGS. This restricts sellers and buyers and prevents them from adjusting the TGS to their own needs and specific technology properties, thus again reducing TC.

Sixth, the auction firms further induce standardization to the TGS through a *fixed commission fee rate*, which is equal to a 15 per cent seller premium (disagio) and a 10 per cent buyer premium (agio) payable on the sales price. Once more, this standardization avoids individual, case-by-case negotiations about the fees payable to third parties (for example, patent law firms in bargaining negotiations). Although these fees increase the transparency of the transaction process, the auction firms do not differentiate the fees any further, some sellers complained about. The sellers considered the fees too high for high value technologies. Owing to the relative nature of these commission fees, the fees increase particularly for high value technologies, raising the TC more than proportionally above the administrative costs the auction firm has. Thus, the pre-studies show that, for high value technologies, the auctions appear less attractive for the sellers (and probably prospective buyers as well) as the commission fees reach a high level for which alternative bargaining negotiations then become attractive (that is potentially profitable). An improvement to the commission fee scheme might be to specify a maximum level or to define a decreasing fee scheme with increasing transaction values.

To conclude, the pre-studies have shown that the auction firms induce standardization into the TGS through six elements. Thereby, they are able to reduce the complexity of the TGS (see Figure 7.2). Subsequently, TC are reduced predominantly through economies of scale, bundled expertise, and by capitalizing on the auction firms' customer networks across a wide range of industries.[23] The pre-studies show that, in particular, the commission fees charged by the auction firms comprise the major share of the TC for the sellers and buyers, while the participation costs represent only a small share of them. Additionally, firms also have certain in-house costs (that is mainly salaries) particularly in the planning phase, in order to prepare the documents for the due diligence.

On the continuum of TGS (see Figure 7.1), those of technology auctions appear to be situated closely to the right extreme. Although they still have to be considered complex transactions, auctions of technologies almost represent 'spot market transactions'. Following the logic of the framework for the choice of transaction governance structures proposed by Williamson (1979) presented in Chapter 7, the highly standardized TGS of technology auctions can be expected to impose constraints on the technologies that can be auctioned. Consequently, the TGS of technology auctions appears preferable for certain technologies. As outlined in Chapter 6, technologies have different properties (for example, with regard to their technological complexity, uncertainty, and impact on subsequent technical developments). To better understand for which technologies the TGS of auctions appear particularly favorable, this issue was further studied in the following quantitative study.

In contrast to TGS of technology auctions, traditional bargaining negotiations are commonly conducted in private, bilateral settings and often negotiated directly between two internal patent / licensing departments of primarily large firms. For each technology transaction, idiosyncratic contracts, royalty schemes, and knowledge transfer agreements are developed on a case-by-case basis – often with the support of patent law firms. Thus, in contrast to technology auctions, the TGS of traditional technology transactions must be considered highly flexible. Hence, such TGS are situated rather close to the left extreme on the TGS continuum presented in Figure 7.1. One can further argue that, due to the low frequency of transactions, the highly specific nature of technologies, and the obstacles on the MfTI, idiosyncratic TGS has emerged as the most appropriate way to transact technologies. Evidently, these traditional transactions are feasible primarily for large firms who could afford in-house patent and licensing departments to develop and deal with highly specific TGS (for example, have the resources to design idiosyncratic contracts). This is in line with Ruzzier's (2009: 1) argument that 'a point repeatedly stressed by transaction cost economics is that the more specific the asset, the more likely it is that vertical integration will be optimal'. Particularly idiosyncratic transactions favor contracts that are individually designed to specific needs and requirements of the involved actors, for example, to incorporate tailored royalty schemes.

As it appears, the auction firms attempted to develop a novel transaction model enabling technologies transactions (that is heterogeneous and highly specific assets) at a lower TC. The extent to which this is possible is discussed in the following chapter. Within the remainder of this study, I take the previously mentioned argument further and investigate its expected consequences. If asset specificity and the frequency of transactions determine the choice of the appropriate TGS, would one not expect adjustments to the TGS to have implications for the types of technologies that can be transacted via this TGS (that is expressed in their properties)? Based on this reasoning, in the following, I develop and test hypotheses for technologies with their specific properties) for which auctions represent an appropriate transaction model.

9.3.2 Hypotheses for Auctioned Technologies

To address this study's second specific research question, I derive hypotheses based on the results of the pre-studies and the prior literature (presented in Chapter 3).[24] In the previous section, I conclude that the transaction governance structure (TGS) of the investigated technology auctions is highly standardized. Thus, on the TGS continuum (Figure 7.1), the TGS of technology auctions is considered to be situated close to the right extreme, that is almost representing spot market transactions. In contrast, bargaining negotiations, that are traditionally preferred for technology transactions (that is that allow high flexibility but are highly resource consuming) are rather situated towards the left extreme. This shift towards more standardized TGS should enable technologies to be traded more frequently for lower TC, but calls for the question whether such a TGS appears equally appropriate for all types of technologies. Following Williamson's (1979) argument presented in Chapter 7, I question whether this is the case. I argue that technologies are not equal and that they can be distinguished by certain properties (Chapter 6) along which one can determine whether technology auctions are a preferred transaction model. With regard to the discussions and tests of technology properties, I was constrained by the limited data sources which could be accessed to compile the dataset. In the following section, I develop hypotheses for properties that could be operationalized according to the available data.

Technological complexity

In Chapter 6, technological complexity was defined with reference to the trend towards increasingly complex technologies. Technologies can comprise a single component (that is one technical invention protected by a patent) or a set of complementary components (that is various inventions and several patents). While the first case refers to a discrete technology, the latter case refers to a complex technology. Hence, with an increasing number of components, a technology's complexity grows. Furthermore, in Chapter 6 distinctive technology properties are discussed that prevent an assessment of a technology's value (for example, patents' intangible nature, their technical and economic dimensions, and that they are unique by definition). When preparing negotiations or a transaction, prospective buyers commonly conduct in-depth due diligences to verify whether the technology covers all relevant IP assets, particularly patents for all technical components, to ensure freedom to operate. I subsequently discuss whether the TGS of technology auctions appear suitable for transactions of complex technologies.

Obviously, conducting a due diligence in order to reduce the investment risk is more resource consuming for complex technologies with a higher number of components (and related patents) than for discrete technologies. As is shown above, the auction firms have embedded measures to support the due diligence by the prospective buyers. Both the auction firms investigated in this study employed virtual online data rooms. Since online data rooms are nowadays commonly employed – even in large and certainly more complex mergers and acquisitions (Spedding, 2008) – they must also be regarded as a suitable measure for technology transactions (Lord and de Mergelina, 2004).

However, the TGS of technology auctions bear two major disadvantages. First, prospective buyers have so far only had a relatively limited amount of time to conduct the due diligence ranging from five to twelve weeks with an average of seven weeks. Whether this timeframe is considered sufficient in general is not clear.[25] However, if prospective buyers evaluate complex technologies, it seems reasonable to assume that this effort requires more resources than evaluating technologies with little complexity. Since each related patent needs to be evaluated, a due diligence becomes more expensive for each patent added to a technology. According to Aggarwal and Walden (2009: 27), 'the level of efforts required and the cost involved for this activity will be proportional to the number of searches. In other words, potentially patentable inventions will require searches even if the actual number of patents required may be less. Considering that the search is expensive, the search cost increases as the complexity of the bundle increases'. Seven weeks appears to be a fairly short timeframe for investigating the number of related aspects to evaluate a complex technology and its various components (for example, legal, technical, and economic dimensions of the related patent families). Furthermore, the period has to be accounted for some of the time spent on the internal decision making process within the management board on the prospective buyer's side. Decision makers involved in any transaction need to determine and agree on the 'ceiling price level' (see Chapter 3) or the indifference price (see Chapter 5) before participating in the live auction event.

Furthermore, to conduct a due diligence for highly complex technologies requires the exchange of more data and information between the seller and prospective buyers than for less complex technologies. For highly complex technologies, such data also involve a higher share of tacit knowledge components that are preferably exchanged through personal communication channels, for instance, through interviews with lead inventors. However, aside from the online data rooms, this case analysis reveals little indications that the auction firms have developed measures to facilitate such personal data and information exchange. Instead, it appeared that emphasis was put on the disconnection between sellers and prospective buyers in order to ensure that prospective buyers can remain anonymous. Furthermore, not only is the evaluation of a complex technology more difficult than that of a less complex (that is discrete) one, but the transfer of a complex technology requires more than simply transferring IP ownership rights in the post-auction phase. It also requires the transfer of tacit knowledge, for instance, by means of training engineers on the buyer side by lead inventors from the seller side. To specify the necessary transfer conditions in the contracts appears difficult within the highly standardized and inflexible legal framework of technology auctions. To conclude, the TGS of technology auctions appears less favorable for highly complex technologies that rather call for transactions with an idiosyncratic TGS, which allows for a high degree of flexibility. Hence, this argumentation leads to the following hypothesis:

> H_{1a}: A patented technology's technological complexity is <u>negatively</u> associated with its auction probability.

If, however, a complex technology finds a buyer who has conducted the due diligence in due time, the complex technology must have been worth making the effort. Applying the cost-based valuation approach, one can argue that it is commonly more resource consuming to develop complex technologies than to develop discrete technologies characterized by low complexity. Developing complex technologies commonly requires developing a number of components, their functions, and to define the interactions between the various components. Moreover, testing and validating the functionality becomes more resource consuming with an increasing complexity. In addition, in addition to the R&D investments to develop complex technologies, related patenting costs are consequently higher for complex technologies. For each invention added to a technology, patenting requires drafting of the application, but also payment of fees to the respective PTO. Hence, the patent maintenance costs (renewal fees) increase with a growing complexity of technologies to which more patents are related.

However, when applying the market-based approach for the valuation of technologies, these arguments do not hold. A technology's market value does not reflect the investments for the development of a technology, but rather its application potential for products, firms, and markets. This somewhat represents a prospective buyer's Willingness-to-pay. However, if a technology has at least found a buyer, there obviously needs to be some market value, even if the buyer wants to acquire the technology for passive purposes only (for example, to pull a competing technology out of the market). In such a situation, the seller will use its bargaining power to at least attempt to cover its R&D costs for developing the technology. Hence, sellers will charge more for a complex technology than for a less complex technology. From these arguments, the following hypothesis can be concluded:

> H_{1b}: If a patented technology is auctioned, its technological complexity is positively related to the sales price.

Technological uncertainty

In Chapter 6, I argue that technologies differ with regard to varying degrees of technological uncertainty. As uncertainty is among the main determinants proposed by Williamson (1979) and often discussed in the TC theory literature, the degree of technological uncertainty was expected to have an impact on the suitability of technologies for the highly standardized TGS of technology auctions.

The degree of uncertainty is of particular relevance during the due diligence, that is within the negotiation phase of a transaction, when prospective buyers evaluate a technology in order to establish a ceiling price level as a start for negotiations with the seller. In this instance, uncertainty can also be perceived as a risk that can cause a technology to fail, while failing has different meanings. For instance, a technology can provide the expected solution to a technical problem but not cost-effectively so that a final product cannot be sold on the market for a price below the users' maximum Willingness-to-pay or a technology may simply fail to solve a technical problem. When prospective buyers establish a ceiling price level, they obviously adjust it according to a technology's expected risk or its technological uncertainty. Quite commonly, prospective buyers do this by adjusting the discount rate in the net present value calculation, determining their Willingness-to-pay depending on the degree of expected uncertainty. Thus, an increasing uncertainty impacts their Willingness-to-pay negatively.

Uncertainty is often closely related to the age of technologies. For instance, with an increasing age, technologies become increasingly mature and thus characterized by lower uncertainty. If widely diffused and applied across a range of industries for various years, their related knowledge can be expected to have diffused within the expert community (commoditization). Thus, a transaction of a mature technology, with a high degree of common knowledge among experts, requires little tacit knowledge transfer.

Consequently, mature technologies are also less difficult to evaluate in the negotiation phase of a transaction than technologies with a high technological uncertainty, that is young technologies that have been developed recently. Young technologies, in contrast to mature technologies, still need to prove whether they can be used in any or various applications (for example, whether they will lead to cost reduction or higher quality in comparison to prior generation technologies). When transacting these technologies, the evaluation and due diligence becomes more complex and resource consuming. Highly uncertain technologies not only need to be evaluated with regard to the legal validity of the related IP rights, but also need to be carefully and comprehensively evaluated with regard to their technological and economic potential, that is their future potential to be applied in various applications across a range of industries.

One can further argue that conducting a due diligence becomes increasingly difficult the more distant a technology is from its application in marketable products. In an early lifecycle stage, young technologies often compete with alternative approaches to solve a technical problem. It is often nearly impossible to predict which of the competing technologies will successfully be established as a standard (dominant design) and when technology developments will have resulted in concrete products offered on the market. A recent example is the high-definition movie disc battle between the HD DVD (promoted by Toshiba) and Blue-ray Disc (promoted by Sony) as successors of the DVD. Early on, when the competing firms started their technology developments, it was hardly possible to predict who would win the 'race'. Nevertheless, a technology's potential to be established as a standard (dominant design) is not only influenced by its level of technological uncertainty, but other factors as well, such as the firm's licensing strategy, which proved to be important for instance in the VHS vs. Betamax case (Suarez, 2004).

Thus, if a prospective buyer considers acquiring a highly uncertain technology in an early development stage, the firm has to carefully evaluate the technology, accounting for the risk that the technology will fail or lose against another technology in an arms race (for example, through a risk and scenario analysis). Various complex questions arise, making such an exercise resource consuming and costly. Will future technologies build on the technology or improve upon it? Will the necessary complementary technologies be available in order to use the technology (for example, a video recorder is almost worth nothing if video tapes are no longer produced)?

This reasoning raises the question of whether the TGS of technology auctions appears equally suitable for trading technologies with high or low technological uncertainty. Similar arguments apply to this question as to the discussion of technological complexity. To evaluate a technology with a high uncertainty requires a more rigorous and intensive due diligence than evaluating a technology with a low degree of uncertainty. To evaluate a technology with a high degree of uncertainty, it might be necessary to involve third party experts to investigate a technology's market potential, or its likelihood to become a standard ('dominant design') in the future. Forecasting the future is never easy and commonly requires substantial amounts of time. An average of seven weeks available for the due diligence hardly allows for such an extensive procedure. An exception, however, would be if a technology has already been on a firm's shopping list. Furthermore, the limited flexibility of the virtual data rooms (that is restricted possibilities to exchange tacit knowledge with the seller, for example, through interviews with lead inventors) constrains prospective buyers to evaluate highly uncertain technologies. Aside of the due diligence procedure, buyers and sellers also want to account for the high uncertainty in contracts that can be adjusted and contain specific clauses and royalty schemes. Moreover, the standardized legal framework and the standardized contracts appear unsuitable for highly uncertain technologies.

Thus, for technologies with high uncertainty, these evaluation requirements can hardly be satisfied by the standardized TGS of technology auctions. Highly uncertain technologies rather call for bargaining negotiations characterized by an idiosyncratic TGS that allows full flexibility. Thus, I conclude that the TGS of technology auctions appears decreasingly favorable for technologies with an increasing uncertainty. Hence, I hypothesize that:

> H_{2a}: A patented technology's technological uncertainty is negatively associated with its auction probability.

However, if a technology was auctioned, its buyer had evaluated it as being worth buying. Thus, the buyer had thoroughly conducted a due diligence within the available timeframe, agreeing with the use of standardized contracts, and had come to the conclusion that the technology has future technical and economic potential. From this reasoning, the question emerges whether a buyer would pay more for a (young) technology with a high uncertainty than for a (mature) technology with a low uncertainty.

The younger a technology is, that is the more distant it is from an application, the higher the commonly associated economic risk is. Similar to the argumentation of technological complexity, a higher risk translates into a higher discount rate in the net present value calculation that determines the buyer's maximum Willingness-to-pay (ceiling price level). Thus, even for technologies with a high potential for various applications across a wide range of industries, the high economic potential will be negatively adjusted in the buyer's calculation. Consequently, in order to reach a similar ceiling price level, mature technologies (with lower degrees of uncertainty) need a considerably lower economic potential than highly uncertain technologies.

Another reasoning also predicts that mature technologies reach higher values than young technologies. For instance, Haupt et al. (2007) and Nordhaus (1967) show that the duration of patent protection upheld by the owner is a value indicator (at least as perceived by the owner). Otherwise, the owner would have no reason to pay the renewal fees to the PTO. According to OECD (2009: 140), patents are renewed only if the 'value of keeping the patent alive (based notably on the discounted expected stream of profits) is higher than the cost of renewing the patent'. This argument was an assumption for constructing the PatVal sample. Harhoff and Scherer (2000) only select patents that were held full term, because they assume that when an owner is willing to invest the renewal fees towards the maximum patent term (that is commonly 20 years), a patent must be of a certain value – at least to him. Accordingly, one can argue that the longer an owner keeps a patent valid, the higher the value appears, since renewal fees increase over time in most jurisdictions (OECD, 2009). Consequently, during the current period, the owner must weigh up the profitability of renewing the patent license for the following period (the protection option) against the costs of maintenance (Lanjouw and Schankerman, 1997).

With an increasing lifespan, technologies have an increasing likelihood to become subject to opposition procedures or infringements. A technology diffuses over its lifetime and more market actors become acquainted with it the older it gets ('commoditization'). The patent literature, for instance, Harhoff et al. (2003: 1343) finds that 'patents that are upheld against opposition and annulment procedures … are particularly valuable'. Such patents have been scrutinized within a court case and can thus be assumed to be particularly strong against invalidation claims and thus be a strong competitive 'weapon'. However, even if not litigated, technologies that have been used for a certain time can also be perceived as valuable because they have been protected by properly designed patents, which are rather difficult to invent around.

Hence, if technologies or their related patents have been continuously renewed – even against constantly increasing fees – and have either been successfully defended or not been offended, they can be considered to have a proven value. The risk accounted for in the net present value calculation of mature technologies also reduces the Willingness-to-pay to a lesser extent than for young, highly uncertain technologies. Consequently, young technologies would need to have an over-proportionally high economic potential for the results of their net present value calculation to become favorable over mature technologies. Hence, if mature technologies with a proven track record (for example, many applications as well as valid and strong patents) with a lower uncertainty are auctioned, they are likely to be auctioned for high values. Thus, I formulate the following hypothesis:

> H_{2b}: If a patented technology is auctioned, its technological uncertainty is negatively related to the sales price.

Technological Impact

In Chapter 1, I discussed the cumulative nature of technological progress, that is technologies developed today have an impact on subsequent (future) technological developments. These technologies differ with regard to their impact on subsequent technological developments. Hence, it has been widely argued that technologies with a higher impact will be more likely to find buyers and become transacted than technologies with little impact.

This aspect is further strengthened by the argument that if a certain firm that is working on a technology becomes aware of a similar technology, it might have an interest in acquiring it (when it is offered for sale) even if only to avoid that the related patents will be acquired and used by a competitor. Furthermore, competitors of the firm that own a technology might also have an incentive to acquire the technology or the related patents to build up bargaining power against the other firm. Thus, with an increasing range of uses for future technological developments, the number of prospective buyers who can participate in an auction increases. Hence, with an increasing number of bidders, the probability increases that at least one of these firms will acquire the technology. Hence, I conclude the following hypothesis:

> H_{3a}: A patented technology's technological impact on subsequent technological developments is positively associated with its auction probability.

As argued above, a technology's impact on subsequent technological developments indicates a use by others, that is if one or many future technologies build upon a technology. The higher the impact, the more uses a technology has. If a technology that is up for sale, or is offered at an auction, has many (potential) uses for future technological developments, more bidders will most likely be interested in acquiring it and hence participate in an auction. As reported in Chapter 6, in auctions, due to the increased competition, a higher number of bidders commonly leads to higher sales prices.

It has also been shown in various studies that investigate the value of patents that the impact of a patent is also positively related to a patent's value (see Chapter 6). Similarly, I argue that, if different patents are bundled together as sub-components of a technology, their joint impact is positively associated with the value of the technology. However, although not all patents may contribute equally to the technology's value, a technology's value might not be strictly proportional to the number of patents it consists of. As discussed in previous studies (for example, Granstrand (2000a)) different sub-components, but also patents, have different functions within technologies. While some broad patents might protect the generic core of the technology, others might refer to very specific (discrete) characteristics. Hence, I conclude that a technology's impact on subsequent technological developments is positively associated with a technology's value, which leads to the following hypothesis:

H_{3b}: If a patented technology is auctioned, its technological impact on subsequent technological developments is positively related to the sales price.

Technology quality

Chapter 6 shows that inventions developed by teams are commonly of a higher quality than those developed solely by individuals. Scholars have argued that this higher quality is a result of internal validation effects in teams, particularly because team members are often more critical to the work of the other team members. As this evidence is so convincing, one can expect that buyers perceive development teams as a quality indicator. A technology of higher quality would obviously have a higher probability to be sold and ultimately for higher prices.

However, one needs to question whether the auction TGS counters this effect. As discussed above, the due diligence procedure constrains the possibilities of prospective buyers when validating technologies. While prospective buyers would prefer, for example, to interview lead inventors for complex and uncertain technologies, this is more difficult during the negotiation phase of technology auctions. One could argue that this effect might be more severe if various inventors have contributed to the development of a technology and none of them could be interviewed. Such an effect could result in more difficulties when evaluating technologies developed by large teams disfavoring their probability to be auctioned. However, this must be considered a minor effect. While teams have a positive impact on a technology's quality, there is most likely little need to interview various inventors as necessary information could rather be captured by interviewing just the lead inventor. Thus, I can formulate the following hypothesis in line with previous literature:

> H_{4a}: A patented technology's technological quality is <u>positively</u> associated with its auction probability.

As higher quality is commonly positively related to value, the following hypothesis is formulated:

> H_{4b}: If a patented technology is auctioned, its technological quality is <u>positively</u> related to the sales price.

Value perception

Sellers commonly base their exploitation decision and the choice of the TGS for a technology on some kind of approximate value derived from one of the many valuation approaches (for example, cost-based approach). This value represents the seller's revenue expectations of a transaction. It appears reasonable to make two assumptions. First, sellers only engage in a transaction if the expected profit is positive. Thus, the expected revenues (incl. royalties) need to be higher than the TC. Second, one can further assume that, among different TGS, profit maximizing actors will prefer the one with the highest expected profits.[26] Having conducted a value assessment and deciding on a TGS among different alternatives (that is auctions vs. traditional bargaining negotiations), sellers have to consider two aspects in order to maximize their profits (see
Figure 9.4).

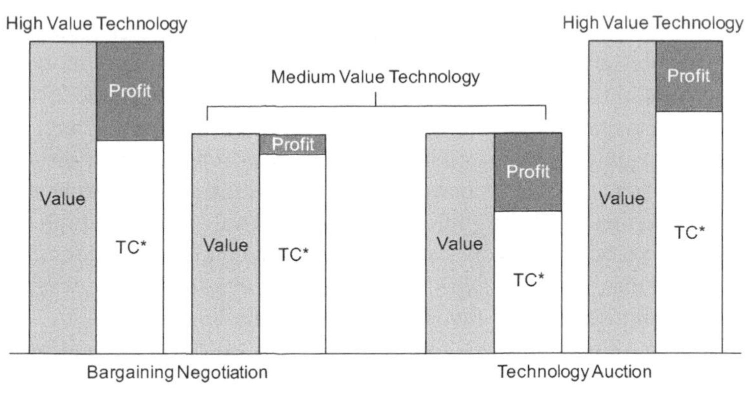

Figure 9.4 - Profit and cost comparison of bargaining negotiations / auctions

First, as sellers will attempt to *maximize their expected revenues*, they will prefer the TGS which enables them to exploit their maximum assessed value. Above I concluded that the TGS of technology auctions appears highly standardized. For instance, auctions employ standardized legal frameworks, including contracts and fees, and restrict due diligences to virtual data rooms. Thus, one can argue that a higher standardization limits sellers' possibilities to tailor transactions to favorable conditions, thereby constraining their possibilities to leverage higher revenues. Thus, a seller needs to consider a certain share of revenues expectations unattainable. In contrast, bargaining negotiations allow sellers to capture higher revenues because of the possibilities to tailor the TGS to the specific properties of any technology (for example, capitalize on information asymmetries). This leverage effect is obviously higher for high value technologies than for medium, or low value technologies. Thus, the high degree of standardization constrains sellers' possibilities to appropriate the optimal, maximum values from technologies, which they could capture from bargaining negotiations with idiosyncratically designed TGS.

Second, in Chapter 5, I discussed that auctions are generally preferred if the *number of bidders* is high. In that case, the competitive bidding will raise the price and the winner's curse effect will further lead to higher prices, at least leveraging on the common value element of a technology. However, for high value technologies, sellers have to distinguish two cases. With very *few prospective buyers*, the leverage effect can be considered marginal and sellers will prefer to conduct bargaining negotiations within an idiosyncratic TGS. Accordingly, if there are *multiple prospective buyers*, sellers would prefer an auction model. However, due to the high TC of auctions for high value technologies, (see below) sellers would again prefer idiosyncratic TGS and design an own competitive bidding situation in a more controlled private setting, that is a closed auction that would allow for more flexibility, for instance to exclude certain bidders (for example, direct competitors). To exclude certain bidders has also been reported as a disadvantage of the current TGS of technology auctions, which should also be considered more serious for high value technologies. Thus, sellers of high value technologies would most likely not prefer public technology auctions.

In order to maximize the expected profits (that is difference between revenues and costs), sellers will favor the TGS with the *lowest TC* (

Figure 9.4). In bargaining negotiations TC vary little whether a firm transacts a low, medium or high value technology. For instance, each technology needs to be marketed, time-specific contracts need to be negotiated, and additional knowledge transfers need to be regulated. If patent law firms are involved, their fees commonly depend on the workload (that is charged as man hours). Thus, depending on the perceived technology value, the flexibility of the TC for bargaining negotiations must be regarded as low.

In contrast, in Chapter 5, I conclude that technology auctions have lower TC than bargaining negotiations. This effect results from the higher degree of standardization (for example, standardized legal framework, including contracts with lump sum payments and economics of scale realized by the joint marketing of all technologies by the auctions firm and virtual data rooms). Owing to the reduction of TC, one can argue that sellers can even expect to exploit medium value technologies via auctions in a profitable manner that would not have been profitably transacted via bargaining negotiations (

Figure 9.4). Technology auctions thus lower the threshold value of technologies that can be transacted profitably. Thus, in an extreme case, owners who expect a technology to have a medium value would even be restricted to the use of technology auctions because bargaining negotiations would not allow for a profitable transaction or lead to higher profit expectations due to a lower TC.

However, although the TGS of auctions has a positive effect on lowering TC, resulting in an incentive to transact medium value technologies, this advantage decreases for higher value technologies. Owing to the commission fee structure (being a fixed share of the sales price), TC are considerably higher for high value technologies than for low value technologies. Thus, sellers can expect high TC for high value technologies similar to the TC of bargaining negotiations. In situations of similar TC, bargaining negotiations are preferred, because they do not constrain sellers and buyers compared to standardized auctions, but rather allow more flexibility for instance with regard to contract design, due diligence, and negotiations. Thus, sellers would favor idiosyncratic TGS over auctions, because they can maintain full control over the transaction and remain flexible. The flexibility of idiosyncratic TGS might further translate into possibilities to capitalize on information asymmetries, negotiate higher prices, and subsequently increase revenues and profits.

Thus, I conclude that bargaining negotiations allow for higher flexibility than auctions and thus allow sellers to further leverage revenue expectations. At the same time, bargaining negotiations have higher TC than auctions, however remain inelastic to high, medium or low value technologies. In contrast, in auctions, TC are generally lower than in bargaining negotiations but are substantially increased for high value technologies due to the commission fee structures. However, auctions allow for little adaptation to adjust TGS to individual technology properties. Therefore, the revenue leverage effect must be considered small. Thus, sellers of technologies with low or medium value expectations are more likely to prefer the auction model (as investigated in this study) over bargaining negotiations. Taking this argument to the extreme, due to a reduction of TC, technology auctions enable transactions of low and medium value technologies that cannot be transacted profitably using bargaining negotiations, due to negative profit expectations. Sellers of high value technologies however will less likely favor technology auctions, since they face higher profit expectations if they choose to design transactions with idiosyncratic governance structures (that is bargaining negotiations). My subsequent conclusion can be summarized into the following hypothesis:

> H_{5a}: The seller's value perception of a patented technology is negatively associated with its auction probability.

However, if a technology is successfully auctioned, one can assume that the buyer has successfully conducted the due diligence. In such a situation, the seller's expected value of the technology can be considered a positive indicator of the sales price. Although sellers' price expectations are known to be positively biased (see Chapter 6) in general, sellers are also known to have a reasonably realistic picture of technologies' market potential. Thus, the seller's value expectation is positively associated with its value, wherefore I hypothesize that:

> H_{5b}: If a patented technology is auctioned, the seller's value perception of a patented technology is positively related to the sales price.

NOTES

[1] For instance via a presentation that the CEO of OT, James Malackowski, gave on 10 February 2006 on the topic 'liquidity an auction environment can have on IP' according to Ocean Tomo (2007a).

[2] See Ocean Tomo (2007b).

[3] See Ocean Tomo (2007c) accessed 13 November 2007.

[4] In the following section, I refer primarily to the patents, which this study focuses on.

[5] A list containing the application areas offered for each of the six technology auctions from OT and IPA is provided in Annex II.

[6] Analyzing the terms and conditions of the five OT auctions, it was possible to identify adjustments to the auction procedures throughout the five OT auctions. A particular change in the procedures was the advanced choice of access channels for placing bids on the live auction, that is the bidders have the option to participate either in person, to connect via tele-

phone, to request double-blind bidding or opt for absentee bidding. Double blind bidding means that a representative of the OT is authorized to execute bids, that are requested by the bidder in written form, that is they follow the bidder's written instructions.

[7] According to Niioka (2006), a reason for the strategy change might be directly related to the reluctance of prospective buyers to bid for technologies with an undisclosed MRP.

[8] I found an indication that the auction was first announced in NZZ Online (2007) on 06 January 2007.

[9] IPB is a wholly owned subsidiary of Knowledge One Beteiligungs AG (K1B), a holding company that invests in knowledge-based firms (IP Bewertungs AG (2007)).

[10] A third revision of the catalogue was published about two months before the auction event on 23 March 2007. A supplement to the catalogue was published shortly before the auction.

[11] The descriptions of the lots vary slightly for IP assets other than patents. However, since patented technologies were by far the majority of offered assets, I do not report the small differences between how patented technologies and other assets were presented in the catalogue (for example, for trademarks, there were no technical drawings, but rather the trademark as such was presented).

[12] The reports for the first auction were prepared by TAEUS, an independent engineering firm based in Colorado Springs, US. TAEUS offered three types of reports prospective buyers could request for each IP asset, the T_{IP}Score, a free two page report summary, the slightly detailed T_{AM}Report for US$500, and the T_{LC}Report, a full report compiled using an internally developed process, called TAEUSworks for US$2 500.

[13] It is noteworthy that very few firms had performed technology auctions before IPA developed its auction process, thus it seems likely that IPA had developed their process taking into account learnings from OT.

[14] The definition of the terms and conditions is considered to take place only for the first auction with slight adjustments from auction to auction.

[15] Of course, prior to the first auctions, they needed to develop a general concept for the auction design.

[16] IPA offered a data room by Data Room Services GmbH & Co. KG. OT offered the online data room by IntraLinks, Inc.

[17] After the live auction event, Manfred Petri, the general agent from IPA, stated that 'six weeks simply wasn't enough time… [and that]… buyers need at least three months to do the complex legal checking to make informed bids' as reported by McClure (2008). While the due diligence process may have prevented the success of the Munich auction, six weeks should be adequate if sufficient notice is achieved and complete information is readily available to prospective buyers in a central library.

[18] The fees appear similar to those at art auctions that commonly charge a participation and commission fee. For instance, Sotheby's commonly charges a buyer's premium of 15 per cent over the final bid price and a standard (though negotiable) seller's commission of 20 per cent of the bid price (Lucking-Reiley, 2000).

[19] On a scale ranging from zero (of no importance) to four (of decisive importance), the firms' primary motives are considered those which scored 'four'.

[20] This picture is in contrast to the literature. According to, for example, Sheehan et al. (2004) many firms have experienced and will experience an increasing importance of the ETE for their businesses.

[21] The firms were asked to assess problems, advantages, and the role of the auction firm for different items that were derived from the literature and prior interviews. How the triple measure is calculated and should be interpreted (that is as a measure of agreement between the four case study firms) is presented in Chapter 8.

[22] Although virtual data rooms are standardized in the sense that personal data exchange is limited, according to Kummer and Sliskovic (2007), virtual data rooms also have some advantages over personal data rooms. These include (a) the possibility to allow parallel access to the data room by multiple buyers, which also eliminates the costs associated with setting

up multiple live data rooms and (b) the reduction of travel-related expenses by offering continuous and ubiquitous access to prospective buyers regardless of their location.

[23] Furthermore, additional knowledge was not transferred for any of the investigated transactions. The pure transfer of ownership rights further reduced the complexity and thus transaction costs. In contrast, the traditional models for the external exploitation of technology commonly assume that technology transfers require a transfer of complementary, tacit knowledge in addition to explicitly codified knowledge (for example, patent documents and blueprints) which can easily be transferred. Complementary knowledge could be prototypes (ranging from small tools to large scale production facilities) and tacit knowledge that can be sticky to, for example, lead inventors (and only transferred by training).

[24] For the definition of a hypothesis I follow Punch (2005: 38) who claims that a hypothesis is 'a predicted answer to a research question' with the hypothesis predicted a priori to the research. The hypotheses were defined according to the criteria for scientific and statistical hypotheses formulated by Albers (2007: 10) as well as Bortz and Döring (2006). Furthermore, I recognize that, according to Popper (1969), his hypotheses are falsifiable, that is hypotheses that are not falsifiable are non-scientific. This study primarily employed hypotheses that illustrate causal directional relations (if-then) in contrast to unidirectional hypotheses (Bortz and Döring (2006), and Burns (2000)).

[25] This issue is briefly discussed by McClure (2008).

[26] A seller would also take into account the risk for the transaction to fail as well as the likelyhood of success, however this aspect is beyond the scope of this research.

10. Analysis of Auctioned Technologies

10.1 OPERATIONALIZATION OF VARIABLES

Following Aggarwal and Walden (2009), above I define technologies as systems consisting of technical sub-components, that is inventions protected by patents (see Chapter 6). Consequently, for the operationalization, I primarily used measures developed in the patent literature[1] and particularly econometric studies of patent values, although this area of work is still largely in the research stage.[2] According to OECD (2009: 137), the 'evaluation and measurement are at a very crude stage compared to other areas where economic indicators are widely available (for example, exports and imports or R&D investments)'. In the following section, I refer extensively to the recent and comprehensive OECD (2009) report, which provides a summary of available and relevant measures.[3]

As a first step to develop the study's measures with all the variables in the dataset, a correlation analysis (see Annex III) was carried out to identify variables that are highly correlated (Backhaus et al., 2006). In a second step, following suggestions in the literature, I computed a number of alternative measures for some variables. In order to select the most suitable version to be used in the multivariate models, bivariate regressions were applied (Hair, 2006). Within the sensitivity analyses, however, some of the alternatively measured variables were also used.

Table 10.1 summarizes the 11 variables indicating the measurement scale for each variable (8 x metric, 8 x nominal), the value range (minimum and maximum values), and the source of the raw data (auction catalogue, patent database or post-auction press release). For three technology properties, I also used slightly alternative measures; therefore Table 10.1 includes 16 variables. From the 11 variables in the analytical models, two are used as dependent variables (sales status; sales price), five variables as independent variables, and three are included as control variables (covariates). All variables represent manifest variables. They are directly measured, thus are not operationalized through latent constructs (Bortz and Döring, 2006). As recommended by, for instance, Backhaus et al. (2006), since they were computed directly from attributes of the dataset, no factor analysis was conducted to validate them as otherwise.

Table 10.1 - Overview of variables

Measure	Label	Abbreviation	Source	Scale	Values (min - max) / Categories
Dependent Variables					
Sales Probability	Sales Status of Technology	SSLO	PR	Nominal	0=not sold;1=sold
Sales Price	Closing Bid in K EUR	BICL	PR	Metric	1.79 - 3650.87 KEUR
Independent Variables					
Technological Complexity	Number of Patents Comprising the Technology	NPLO	C	Metric	1 - 59
Technological Uncertainty	Average Remaining Protection Time of a Technology's Patents	ARPT	DB	Metric	0 years - 18.45 years
	Remaining Protection Time of the Youngest Patent of a Technology	Max_RPT	DB	Metric	0 years - 18.90 years
	Remaining Protection Time of the Oldest Patent of a Technology	Min_RPT	DB	Metric	0 years - 18.45 years
Technological Impact	Count of Sample Forward Citations	FORWC	C	Metric	0 - 87
	Availability of Sample Forward Citation	SFCA	C	Nominal	0=not available; 1=available
Technology Quality	Average Number of Inventors of Patents in a Technology	ANNI	DB	Metric	1 - 12
	Individual or Team Invention	Ind/Team	DB	Nominal	0=Individual Inventor; 1=Team Invention
Perceived Technology Value	Seller's Prediction of Expected Sales Price	VAEX	C	Metric	7.95 - 4075.5 KEUR
Control Variables					
Technology Source	Original Assignee Type	OATY	DB	Nominal	1=Individual Inventor, 2=University/ Research Insitute, 3=SME, 4=Large Firm
Auction	Auction Number in Dataset	AUNO	C	Nominal	1-6
Patent Jurisdiction	Country Designation of Patents in a Technology	PatCo	DB	Nominal	0=at least one non-US Patent in Technology; 1=US Patents only
Application Area	Technology Area	TARA#	C	Nominal	1 - 16

Note: C (auction catalogue); DB (patent database); PR (post-auction press release)

In the following section, I describe the measurement approaches used to compute the variables. First, I describe the variables' measurement approaches that were used as dependent variables for the analyses, followed by the independent variables and the covariates. Before moving on, however, it should be noted that I had to carefully assess whether the measures using patent data that are available from prior literature could be used in the study, particularly because I did not compute the variables from individual patents, but rather on a technology level (that is individual patent data were consolidated for patent bundles).

10.1.1 Dependent Variables

From the post-auction press releases that were published usually within a few days after the live auction event (see Table 8.1), I obtained two types of data that were coded into the two variables that were used as dependent variables in the regression models.

The first dependent variable is coded as a nominal-scaled (dichotomous) variable that measures a technology's sales status. While successfully transacted technologies were made explicit by mentioning sales prices, it became implicitly obvious which of the technologies listed in the auction catalogues remained unsold. The successfully auctioned technologies were coded as 1, while technologies that remained unsold were coded as 0.

In econometric patent value studies, values are commonly difficult to assess (see also Chapter 6). For instance, since most transactions are conducted confidentially prices remain undisclosed. However, in this study, technology prices could be directly measured. Moreover, sales prices of technologies could be derived from the post-auction press releases and were coded as the second dependent variable. I coded the variable according to the metric scale with values larger than zero, ranging from a minimum value of €1.790 thousand (that is equivalent to the lowest MRP of the six auctions) to a maximum of €3.65 million. The sales prices published in the post-auction press releases include the commission fees paid by the buyers.

With regard to the post-auction press releases, however, one particular issue needs mentioning in order to properly understand which of the technologies that were sold at the technology auctions were actually included in the post-auction press releases, and hence considered in the dataset. As illustrated in Figure 10.1, three types of transactions occur in relation to the technology auctions. Transactions are either closed during the live auction event, that is 'on the floor' (t_1) or within a post-auction period (t_2).[4] Transactions during this post-auction period can either occur shortly after the live auction event but prior to the publication of the post-auction press release (t_{2a}) or after the publication of the post-auction press release.

Figure 10.1 - Transactions closed at different occasions of the auction process

Obviously, all transactions closed during the live auction events (t_1) are included in the post-auction press release. From the transactions closed within the post-auction period (t_2), those that were closed prior to the publication date of the post-auction press release (t_{2a}) are included in the dataset.[5] Consequently, the transactions that closed after the publication date are not considered in the dataset.[6]

10.1.2 Independent Variables

From the auction catalogues and publicly available patent databases, I obtained data to compute variables that were used as independent variables in the regression models.

Technological complexity
Although the concept of technological complexity has been extensively employed in strategy and technology studies (Perrow (1984), Roberts (1990), Roberts and Gargano (1990), Demchak (1992), Tushman and Rosenkopf (1992), and Perrow (1994)), surprisingly, few empirical studies have employed rigorous measures of technological complexity or measured its impact on organizational performance (Singh, 1997). More recently, von Graevenitz et al. (2008: 12) have noted that 'despite the widely acknowledged notion of a technology's complexity, there is no direct measure of it nor is there an indirect construct related to complexity'. Owing to these difficulties, most studies of complexity contain relative measures, comparing the complexity of a product or technology to that of a closely related product or technology (Singh, 1997). However, among the few studies that have attempted to measure technological complexity, most are measured on levels that are not suitable for this study. For instance, measures for technological complexity on an industry or macro technology level are hardly useful to measure technological complexity on the micro technology level using patent data.

To my knowledge, regarding the use of patent data, technology complexity has been operationalized only to a limited extent.[7] Ahuja (2003) suggests measuring technological complexity using patent data on a firm level, which also appears applicable for individual patents. To measure the complexity of technologies developed by a firm, Ahuja (2003) uses the information on technological classes of the cited patents. For this operationalization, he refers to the argument that, if patents cite different technological classes, they can be regarded as having combined many relatively disparate technologies. Following this argument, Ahuja (2003) applies the Blau Index (1-Herfindahl) of patent classes cited by the firm in its patents of the previous six years, which I use as a complexity measure (Hall et al., 2001). Firms that consistently combine elements of distinct technological classes in their patents can thus be regarded as having more complex technologies. For instance, a patent that cites two technology classes may be considered less complex than a patent that combines knowledge from five different technology classes (see also Zander and Kogut (1995)).

Further measures were also suggested using patent data. For instance, Várdy (2010) suggested measuring technological complexity by means of the number of steps or tasks needed to produce some good or service. Lanjouw and Schankerman (1999) also measured firms' technological complexity using the average number of patent claims. Keep et al. (1994) measured complexity in terms of the total number of subclasses the average patent in a subclass is assigned to. However, as I had no access to citation and claim data, I could not apply the measures proposed above.

In a comprehensive study on technological complexity, Rycroft and Kash (1999) identify three different measures of technological complexity. The simplest measure relates to the number of components a technology has. The second measurement also takes into account the 'cybernetic contribution' from architectures that integrate components and subsystems through sophisticated feedback loops. The third measurement of complexity rests on the assumption that, although appreciation of the dynamic interaction between process and product innovation is essential, one must also consider the relationship between process and product technologies and the organizational systems that innovate them.

Thus, given the opportunity to extract a number of subsystems related to a technology from the auction catalogues, I was able to measure technological complexity with regard to its subsystems. From the three measures noted by Rycroft and Kash (1999), I thus applied the first measurement approach. However, I was not able to take into account any relationship (feedback) of the subsystems related to the architecture of the technology (Singh, 1997). This approach links directly to the study's technology definition that was proposed by Aggarwal and Walden (2009) (see Chapter 6). They argue that 'technologies are not created equal. Some are created to address more complex problems than others. Effective problem solving… involves breaking down the problem into smaller sub problems that are easier to comprehend and can be solved independently. Later the individual solutions can be integrated together as a solution to the complex problem. Since each invention in the technology can be independently protected as an IP by anyone in the industry… complexity is a measure of the size of the technology. A bigger technology requires more inventions'.

$$TechComp_i = \sum_{k=1}^{k} Patent_{ki}$$

Equation 3 - Measure for technological complexity

It is seldom possible to measure technological complexity taking the related components into account. For auctioned technologies, however, this data were available. Assuming that the number of patents reflects the number of a technology's subsystems, this data were compiled from the auction catalogues where sellers announced the number of complementary patent families that were bundled together and offered jointly for sale. For instance, Lot 17 of the OT auction conducted in the UK on 1 June 2007 comprises a technology entitled 'image processing sensory device for detecting vehicle operator drowsiness.' The technology consists of four complementary patent families, including a 'method and device for detecting drowsiness and preventing a driver of a motor vehicle from falling asleep', the 'image processing apparatus and method,' the 'method and apparatus for detection of drowsiness', and the 'methods and apparatus for identifying and localizing an area of relative movement in a scene'.

Following this reasoning and assuming that all necessary related inventions were included in a bundle to enable the buyer to fully utilize an offered technology, I measured the technological complexity as the count of the patents (as components) included in each technology offered for sale (see Equation 3). The variable of technological complexity is thus metric-scaled and can be any natural number equal to or above 1 (positive integer). It appears noteworthy that, for various technologies, the sellers offered only selected patents of a patent family (for example, only the German and US patents were for sale, but not the other patents covering Korea, Japan, etc.). I did not calculate each country-specific patent of the patent families that were included in the technology, but only counted each patent family as each patent family refers to one invention, that is component.

Technological uncertainty

Various different measures have been proposed and used for technological uncertainty in prior literature. For instance, Walker and Weber (1984) measure technological uncertainty as changes in specifications (the frequency of expected changes in specifications for the component) and technological improvements (the probability of future technological improvements of the component). Harrigan (1986) measures it as years to obsolescence. Robertson and Gatignon (1998) measure technological uncertainty according to six items relating to technology changes in a product category (technology is robust (reversed); lifecycles are short; technology is moving very fast; technology is on a constant plateau (reversed); technological pressure is intense; and technology moves rapidly from generation to generation). Díez-Vial (2007) measure technological uncertainty in terms of the degree of technological change – either in the specifications of the goods exchanged or in the skills and technologies that are necessary in each stage. In a questionnaire, firms were asked about the importance of constantly changing the equipment or updating the knowledge and qualities needed to efficiently carry out an activity. A variable was made up of three items, similar, to that of Poppo and Zenger (1998), and Walker and Weber (1984). However, this measure was not further specified. Davila (2000) relates the age of a technology to its uncertainty. Based on the argument of increasingly complex products, he argues that technologies with more components must be considered more novel. Thus, he uses the percentage of new parts as a proxy for uncertainty. Based on Simonin (1991) and Souder and Shrivastava (1985), as well as Steensma and Corley (2000) he measured technological uncertainty according to six items rated on a seven point Likert scale. The items captured the uncertainty associated with engineering design and the commercial success of the technology during a sourcing decision. Schilling and Steensma (2002) operationalize technological uncertainty as a construct with two dimensions. Commercial uncertainty relates to the uncertainty associated with product design and the market's acceptance of those products and processes. According to Clark and Wheelwright (1993), technical dynamism relates to the uncertainty of whether the underlying technology will be of ongoing value in the midst of a changing technological environment. Within a questionnaire, Perry et al. (2004) operationalized technological uncertainty on a seven point Likert scale in one question asking whether the technology for the end-product is uncertain, unpredictable or changing rapidly with respect to the investigated alliance. Applying a similar conceptualization as Robertson and Gatignon (1998), Arranz and Arroyabe (2008) measure technological uncertainty using a single item in a questionnaire using a five point Likert scale. To summarize, most of the previous measures applied measures with single or multiple-items within the questionnaires. However, the measures applied in the questionnaires could hardly be transferred to measure technological uncertainty using patent data.

Thus, referring at least to the measure used by Davila (2000) reflecting the common conceptualizations from the literature, I developed a measure that can be computed from the patent data. From the patent data, I computed the age of the technology. As argued in Chapter 6, while a young technology is commonly characterized by high uncertainty, the uncertainty decreases along the lifecycle. Mature technologies are fairly well known in the expert community (commoditization) and are thus characterized by lower uncertainty. A more detailed approach would also consider a patent's position within its related technology lifecycle. However, such a measure would require access to comprehensive data from the technology domain.[8] Consequently, this approach was dismissed. However, I considered the maximum protection time of a patent (being 20 years, with some exceptions) when computing the measure for technological uncertainty.

However, a problem was encountered while calculating the measure for technological uncertainty. As mentioned in Chapter 8, the data derived from the patent databases were not available on a technology level but rather for individual patents. Thus, I had to consolidate the individual patent data on a technology level.

$$TechUncer_i = \left(\frac{\sum_{k=1}^{k} RPT_{ki}\left(PRTE - \left(\frac{Auction\_Date_i - Application\_Date_i}{365}\right)\right)}{k} \right)$$

Equation 4 - Measure for average technological uncertainty

I developed a (first) measure as an average value of the related patents' remaining protection times (RPT). A patent's RPT is calculated as the difference between the maximum protection term and the age of the technology, which is calculated as the difference between the auction date and the patent's priority date (see Equation 4). For US patents, the priority dates are the 'issue date' issued before 1 June 1995 and the 'filing date' filed after 1 June 1995. Similar to those of non-US patents, the protection term (PRTE) is 17 years for US patents issued before 1 June 1995 and 20 years for all US patents filed after 1 June 1995.[9] The numerator is divided by 365 days to transfer the data unit from days into years. Similar to the conceptual notion mentioned above, the measure indicates that a patent with high RPT must be considered young, thus has a high technological uncertainty and vice versa.

In addition to the technology uncertainty measure based on the average RPT, I further employed two alternative measures. Instead of calculating the average RPT of the patents belonging to a technology, one could further argue that, as soon as the first patent of a technology expires (or is invalidated), the technical knowledge diffuses into the public domain. Hence, imitators will instantly learn about the technology resulting in a steep reduction of technological uncertainty. This argument rests upon the assumption that the earliest patent of a technology refers to the core of the technology on which subsequent incremental improvements along the innovation process were made, which were protected by later filed patents. However, one could argue that firms pursue evergreening strategies (see Granstrand (2003b)) attempting to uphold patent protection for as long as possible by filing subsequent patents. Competitors are only able to imitate the technology when the last patent expires.

Figure 10.2 illustrates how this argumentation can translate into two further measures. Within a technology, that is technology (PB_{xi}), the RPTs of the related patents vary on a continuum between RPT_{min} (indicating a patent with its protection about to expire soon) and RPT_{max} (indicating a recently applied patent with an RPT close to the full protection term of 20 years). With regard to the argumentation presented above, the first alternative measure (Max_RPT) is based on the RPT of the oldest patent belonging to a technology ($RPT(P_{x1})$). As soon as the first patent of a technology expires (or is invalidated), the technology can be imitated. The second alternative measure (Min_RPT) is based on the youngest patent with the shortest remaining patent protection time ($RPT(P_{xi})$).

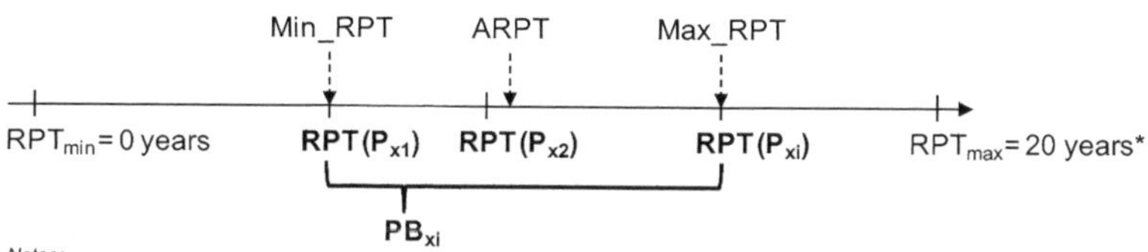

Notes:
PB (Patent bundle); RPT (Remaining protection time)
* For illustrative purposes, RPT_{max} is fixed to 20 years although in different patent systems it can be extended in particular cases

Figure 10.2 - Alternative measures for technological uncertainty

All three variables for technological uncertainty are metric-scaled, with their values always larger than zero and with a maximum of 20. To calculate the three variables, I obtained data on the priority, or the filing date for all patents in the dataset from various public patent databases (including USPTO, DEPATISnet, Google patents, epicene and WIPO, as well as the Hong Kong, Chinese, and Japanese PTOs' databases).

Technological Impact

Measuring the impact of a technology on subsequent technology developments is a difficult exercise using statistical indicators. One commonly suggested approach is to consult experts who analyze individual patents one by one in order to gather a somewhat reliable statement on the impact of a patent. This approach obviously becomes resource intensive for a large number of patents as indicated in this study's dataset. However, previous studies (for example, Trajtenberg (1990), Harhoff et al. (1999), Hall et al. (2005)) have found that the citations a patent receives from subsequently applied patents, labeled forward citations,[10] are actually correlated with a patent's impact on subsequent technological progress. Hall et al. (2005) state that patents that are cited more frequently can be assigned a higher impact. According to OECD (2009: 106), the 'number of citations a patent receives has been found to reflect, on average, the technological and commercial importance of a patent.' Furthermore, it has been consistently reported that patents that receive more citations than average are more likely to be renewed (Lanjouw et al., 1998) and opposed or litigated in tribunals (Harhoff and Reitzig (2004), Lanjouw and Schankerman (1997)).

However, critics still use forward citations as a measure of technological impact. In addition to the patent value, citations are also correlated with several aspects of the patent, for example, its legal robustness. Furthermore, citations are said to be a noisy measure of information flows (Singh (2005), Almeida and Kogut (1999)), particularly because many citations are added by the patent examiners and not the applicants or just to avoid infringements (for example, Alcácer and Gittelman (2004), and Harhoff et al. (2006)). Another problem of using forward citations is the fact that the number of forward citations is time dependent, that is they accumulate over time. As diffusion takes time, younger patents might not yet be well known and thus they may have a lower likelihood to be cited. According to OECD (2009), this phenomenon is known as the truncation effect where earlier patents have an intrinsically lower probability of being cited (Hall et al., 2001). However, despite these difficulties, according to Hall and Ziedonis (2001) and Troy and Werle (2008), a patent's forward citation can be regarded as a fairly good indicator of a patent's utilization, thus impacting subsequent inventions.

If forward citations of patents are perceived to indicate an invention's technological impact, then the accumulated number of the patents' forward citations relating to a technology can also be perceived as a measure of a technology's impact. To measure the technological impact, I used citation data that were presented in the auction catalogues for each technology.[11] However, the forward citations listed in the catalogues cannot be perceived as exhaustive, since they were labeled 'sample forward citations' in the catalogues. However, it appears reasonable to assume that this is also the reason why the auction firms listed selected forward citations in the catalogues, that is they constitute a marketing argument to attract prospective buyers' awareness. Thus, it can be assumed that the selection of forward citations was carefully planned. Auction firms commonly have access to large commercial patent databases, which they use to screen all patents for forward citations. Evidently, they selected forward citations that refer to primarily large, well known, and reputed firms (for example, Siemens AG, Daimler AG, Ford Corporation, and L'Air Liquide) to raise prospective buyers' awareness, while they might have omitted citations from small, relatively unknown firms. Thus, using these selected forward citations, my measure clearly does not represent the absolute number of forward citations as is commonly derived from patent databases. However, although the data might not represent an exhaustive list of forward citations, one could even argue that the pre-selection done by the auction firms added a quality dimension to the forward citation data – although maybe somewhat biased.

$$TechImpact_i = FORWC_i$$

Equation 5 : Measure of technological impact

Accordingly, I measured the technological impact of technologies on subsequent developments as the sum of the listed sample forward citations. To construct the measure for technology quality, I thus counted the number of forward citations ($FORWC_i$) for each technology as listed in the auction catalogues.

The resulting variable for technological impact ($TechImpact_i$) is metric-scaled with values equal to or larger than zero. Owing to the nature of forward citations, this study's measure of the technological impact can be expected to have a count data nature, thus appears to be skewed. Furthermore, I compiled another dichotomous measure to assess whether or not a technology has any citations. This measure is thus a nominal variable. Technologies that are indicated in the catalogue as having sample forward citations have a value of one and those that do not possess sample forward citations are indicated by the value zero.

Technology quality

As presented in Chapter 6, with a particular reference to Wuchty et al. (2007) and Schettino et al. (2008: 22), there is strong evidence that teams develop inventions of higher quality than individual inventors. A key argument holds that teams are more critical in validating and selecting the individual contributions of the team members. Furthermore, the number of inventors may proxy the cost of the research behind the invention, which is statistically related to the technical quality of the invention. Subsequently, one can conclude that development projects become increasingly expensive and research-intensive when more resources are involved (Guellec and de la Potterie, 2001).[12]

To measure a technology's quality, I thus constructed a measure similar to the one used by Gambardella et al. (2008) who measured the size of the group of collaborating inventors for patents. The number of inventors who have contributed to the development of an invention can be extracted from the patent application documents. Similar to a patent, a technology (being a bundle of various inventions protected by patents as defined by Aggarwal and Walden (2009) can be developed by an individual or a team. Following, for example, Schettino et al. (2008), a team of inventors is identified when at least two individuals are listed as inventors of a patent application.

Although I was able to derive each patent's number of inventors from the data derived from the patent databases, I had to consolidate this data for all patents belonging to a technology, similar to the process of measuring technological uncertainty.

Table 10.2 - Example of inventor team composition

Patent Number	Inventor Names
US000005967276A	LEICHLITER, AYNE K.; ALTON, ERLEN B.
US000006102178A	ALTON, ERLEN B.
US000006182808B1	ALTON, ERLEN B.; NAMK, MICHAEL; EYBERGEN, ILLIAM N.; SPRING, JAMES K.
US000006237735B1	ALTON, ERLEN B.; PRESTON, DAVID M.

However, this consolidation approach must be considered sensitive to two sources of error. First, when aggregating the number of inventors of each patent from a technology, I disregarded the identity of each inventor. This can cause an incorrect result if a technology consists of various patents that were each invented by different although individual inventors. In such cases, my algorithm would return the value 1 and thus indicate that the technology was invented by one inventor instead of a team. For technologies consisting of only one patent, the algorithm would however return correct results. Second, technologies could consist of multiple patents, each having more than one inventor, where the inventors of each patent differ. Subsequently, the following question arises: Was the whole bundle developed by a team jointly or was each patent developed independently by different persons?

To validate the data against these possible error sources, I conducted random cross-checks in the sample. No technology was found that represents the first error type, that is with more than one patent each developed by a different inventor. With regard to the second error type, a closer examination of the data revealed that such a grouping was not present in the dataset. In each of the investigated technologies, at least one inventor appeared in all or most of the patents belonging to that technology. A likely explanation for this is that one 'core inventor' developed the inventions relating to a technology jointly with other persons and at least knows all of them. Therefore, it appears correct to codify such a technology as a team invention. An example of such a case is illustrated in Table 10.2 for Lot64 of the first auction, where the inventor 'Erlen B. Alton' appears to have participated in the development of all four inventions that were bundled together by the seller and offered as one technology.

To develop a measure for technology quality, the number of inventors from each patent ($INNU_i$) belonging to a technology was calculated. In a second step, the accumulated number of inventors for all patents of the respective bundle was divided by the number of patents comprised in a technology (see Equation 6). If $TechQual_i$ is 1, I classified the technology as invented by an individual inventor. If $TechQual_i$ is larger than one, a technology can be classified as team invention. However, the measure used in this study does not account for the composition of collaborators to a team as suggested, for example, by Gruber et al. (2009).

$$TechQual_i = \frac{\sum_{k=1}^{k} INNU_{ki}}{k}$$

Equation 6 : Measure of technology quality

The measure is metric-scaled with values equal to or larger than one. As one can suspect that most inventor teams have low number of members, a count data nature of this measure appears likely (that is its distribution is expected to be skewed). Furthermore, I constructed one alternative measure differentiating only whether a technology was developed by one individual inventor or by a team similarly to the alternative measure of technological impact. This measure is dichotomous, nominal-scaled where a value one indicates a team invention that is of higher quality than a technology developed by a single inventor with the value zero.

Value Perception

Although firms can choose among different approaches (for example, cost, net present value, and option or market approaches) to assess the value of a technology measuring the sellers' value approximation is commonly difficult due to the multi-dimensional complexity of such an exercise as mentioned in Chapter 6.

By now, for some decades, patents and patent statistics have been under intensive scrutiny by economists to measure patent values (Harhoff et al., 2003).[13] According to OECD (2009) three main lines of work have been followed to estimate the private economic value of patents. These include surveys asking inventors or patent owners about their perception of the economic value of their patents, analyses of data from the patenting procedure (for example, grant or refusal of the application, citations, renewal, and geographical scope of protection) and estimates of values from financial data (for example, market value of firms and the value of initial public offerings (Hall et al., 2005). Currently, the 'PatVal' is the most prominent value approach as applied by Harhoff and Scherer (2000) and others who have traced back the inventors of a selected sample of patents filed at the German patent office[14] that were held full term applying a private value measure.[15] To assess the value perception, inventors were confronted with the following counterfactual question (Harhoff and Scherer, 2000: 560): 'If in 1980 you knew what you now know about the profit history of the invention abstracted here, what is the smallest amount for which you would have been willing to sell this patent to an independent third party, assuming that you had a bona fide offer to purchase and that the buyer would subsequently exercise its full patent rights?'.[16]

For this study, the auction catalogues provided an explicit, monetary measure of the sellers' value perception for the technologies.[17] For each technology, the auction catalogues specify a minimum expected sales value. Based on these values, I was able to directly measure the sellers' perceived technology values. Expected sales prices however were not specified in the IPA catalogue.[18] This approach appeared favorable particularly because where value expectation data were missing in the IPA auction, data were available indicating the opening bid for the auction. Since, for auction five, data were available for the value perception as well as for the opening bid, I could test and prove a correlation among those justifying the use of imputation techniques. Thus, for the perceived technology values of sellers, I compiled a metric-scaled measure with values larger than zero.

10.1.3 Control Variables

In addition to the effects captured by the independent variables, three effects are controlled for in this study's models. First, I control for the differences between the six auctions. As shown in Chapter 9, the six auctions included in the study's dataset were conducted within fairly similar and standardized transaction governance structures (TGS), for example, comparable legal terms and conditions. The auctions were organized by two different auction firms and took place in three different countries. OT organized five of the auctions (A_1, A_2, A_3, A_5, and A_6) while IPA organized one (A_4). Four auctions (A_1, A_2, A_3, and A_6) were held in the US, while two were held in Europe (A_4 in Germany and A_5 in the UK). As discussed in Section 5.3, IPA's auction design is largely based on OT's design with some slight adjustments (for example, introduced the possibility to auction licenses). The differences in the TGS between the auctions held in the UK and those held in the US were fairly small. As mentioned above, OT slightly adjusted the TGS of each of its following auctions (that is to account for a learning effect). Across all auctions the various key elements of the TGS have been similar, for instance all auctions applied the English bidding procedure. Despite the many similarities I wanted to control for effects resulting from the differences in the TGS and subsequently included the covariate 'auction number'. This variable was coded as six dichotomous dummy variables, while the dummy variable for the first auction served as a reference variable and was omitted in the models.

Second, I control for differences between the various nationalities of the patents that related to the technologies. As mentioned in Chapter 6, although similar to some extent, patent jurisdictions remain diverse in different countries. While the majority of patents in my dataset are US patents, a fraction of non-US patents was also auctioned. Thus, I differentiated between technologies that are protected only by US patents and those that include at least one non-US patent. Hence, I coded the dichotomous covariate 'patent nationality' with one indicating that the technology includes only US patents and zero indicating the presence of at least one non-US patent in the technology.

Third, I control for the different origins of the technologies as discussed in Chapter 6. Four dichotomous dummy variables were coded for the different technology source types. These include independent inventors, universities / research institutes, SME and large firms. The variable was coded according to the type of the patent assignees. Furthermore, a small fraction of technologies could not be assigned to any of these types. These few technologies were assigned to the group 'other'. As this group was considerably small, those technologies were not taken into account in the regression analyses. The reference variable for the dummy variables was the technology source type 'independent inventor'.[19]

Furthermore, a fourth covariate was to the models. However, they had to be omitted. As discussed in Chapter 6, patents' or technologies' values vary across different application areas. However, the application areas could not be considered in the models due to data limitations. Accounting for the 16 application areas through dummy variables would have resulted in the overfitting of models.

10.2 DESCRIPTIVE RESULTS

This chapter represents the descriptive results of the variables investigated. As outlined in Chapter 8, I present descriptive results for each variable primarily in two contingency tables (plotted against the auction numbers and technology source types) applying common descriptive measures such as count, mean, median, maximum, and minimum values. In a first step, the summary tables provide an overview of the dataset. To prevent biasing the results, I refer to the dataset of technologies (not for individual patents) restricted to those offered for sale using the English bidding procedure (see Chapter 5). Thus, I excluded the very few technologies that were offered as licenses or using Dutch bidding procedures, which mainly applied to A_4 by IPA. In reality, none of them were sold.

10.2.1 Summary Statistics

Table 10.3 provides an overview of the technologies that were offered and sold at the six auctions included in the dataset. In total, 393 technologies[20] were offered, which equals a mean of 66 technologies per auction. Of those technologies, 156 were sold while 237 remained unsold. A_1 and A_6 offered the highest number of technologies (77 technologies), accounting for the largest share of technologies (19.6 per cent) while A_5 only offered 47 technologies (12.0 per cent). The amount of technologies offered at A_3, A_4, and A_5 was below the mean.

Table 10.3 - Technologies offered and sold at the auctions

Auction Number													
1		2		3		4		5		6		Total	
Count	%	Count	%	Count	%	Count	%	Count	%	Count	%	Count	%
51	21.5	50	21.1	31	13.1	32	13.5	34	14.3	39	16.5	237	100.0
26	16.7	22	14.1	34	14.7	23	14.7	13	8.3	38	24.4	156	100.0
77	19.6	72	18.3	65	14.0	55	14.0	47	12.0	77	19.6	393	100.0

A_6 sold the most technologies (38 technologies), while only 13 technologies were sold at A_5. On average, 26 technologies were sold at each auction. The number of sold technologies was below the mean at A_2, A_4 and A_5. Taking a closer look at the sales ratio, the data reveal a mean sales ratio of 39.7 per cent. With respective sales ratios of 52.3 per cent and 27.7 per cent, it was the highest for A_3 and lowest for A_5. The sales ratios for A_3, A_4, and A_6 were above the mean.

Across the auctions, the data indicate a decreasing number of offered technologies. From the highest number of technologies offered at A_1, the number decreased steadily until A_5, with A_6 showing an increased number again. The sales ratio however shows a slight increasing trend from 33.8 per cent at A_1 to 49.4 per cent at A_6.

As shown in Table 10.4, 34.1 per cent of the 393 offered technologies were developed by large firms, 31.6 per cent by individual inventors, 27.0 per cent by SME, 6.9 per cent by universities / research institutes and a very small fraction (0.5 per cent) by others (for example, TMI have developed own technologies).

The numbers in Table 10.4 reveal that 74.1 per cent of the technologies developed by universities / research institutes were sold. This share is considerably higher than for technologies developed by independent inventors (45.2 per cent), SME (37.7 per cent), and large firms (28.4 per cent). Of the total number of sold technologies, those developed by independent inventors had the highest share (35.9 per cent), followed by SME (25.6 per cent), and large firms (24.4 per cent), while technologies developed by universities / research institutes accounted only for 12.8 per cent. Among the technologies that remained unsold, those developed by large firms had the highest share of 40.5 per cent, followed by those developed by independent inventors (28.7 per cent), and SME (27.8 per cent).

Table 10.4 - Sources of auctioned technologies

		Sales Status								
		Not Sold			Sold			Total		
		Count	Column %	Row %	Count	Count	Column %	Row %	Column %	Row %
Technology Source	Other	0	0.0	0.0	2	1.3	100.0	2	0.5	100.0
	Independent Inventor	68	28.7	54.8	56	35.9	45.2	124	31.6	100.0
	University / Research Lab	7	3.0	25.9	20	12.8	74.1	27	6.9	100.0
	SME	66	27.8	62.3	40	25.6	37.7	106	27.0	100.0
	Large Firm	96	40.5	71.6	38	24.4	28.4	134	34.1	100.0
	Total	237	100.0	60.3	156	100.0	39.7	393	100.0	100.0

Table 10.5 illustrates the distribution of offered, unsold, and sold technologies across *16 application areas* (see Chapter 8 and the coherence table in Annex II). Ranked according to the total number of offered technologies, Table 10.5 shows that about one third of all technologies were offered in two application areas. Altogether, 79 technologies (20.1 per cent) were offered for applications in 'Telecommunications (wireless, cellular, optical and other)' and 60 technologies (15.3 per cent) were offered for applications in '(Micro) Electronics / Computer Systems & Technologies'. These top two application areas, together with seven application areas in which 20 to 30 technologies were offered, accounted for 80 per cent of all offered technologies. The lowest numbers of technologies were offered for application in 'Energy / Utilities / Power Technologies / Oil & Gas', and 'Aviation & Aerospace'. In total, 11 technologies were offered as application areas that could not be classified into any of the application areas.

Looking at the sold technologies, Table 10.5 further shows that those technologies offered in the top two application areas were also the best sellers in absolute terms. A total of 32 technologies were sold for applications in 'Telecommunications (wireless, cellular, optical & other)' and 31 for '(Micro)Electronics / Computer Systems & Technologies'. Furthermore, only two areas sold more than 10 technologies: 19 technologies were sold for applications in the area of 'Internet / Web Services / E-Commerce / Internet TV' and 14 technologies for applications in the area of 'Digital & Home media / Entertainment'. Technologies sold in these four areas jointly account for 61.5 per cent of all the sold technologies. The remaining 38.5

per cent of the sold technologies are scattered across the other 12 application areas.

Table 10.5 - Distribution of auctioned technologies across application areas

		Sales Status				
		Not Sold		Sold		Total
		Count	%	Count	%	Count
Technology Area (n=16)	Telecommunications (Wireless, Cellular, Optical & Other)	47	59.5	32	40.5	79
	(Micro)Electronics / Computer Systems & Technologies	29	48.3	31	51.7	60
	Internet /Web Services / E-Commerce / Internet TV	11	36.7	19	63.3	30
	Chemical Engineering / Applied Sciences / Advanced Materials / Nanotechnology	22	73.3	8	26.7	30
	Medical & Life Sci / Devices / Telemedicine / Bio-chip / Bio-tech	20	71.4	8	28.6	28
	Location-based Systems / Applications / Logistics	15	65.2	8	24.8	23
	Digital & Home Media / Entertainment	9	39.1	14	60.9	23
	Automotive Systems & Devices / Industrial Machinery	14	63.6	8	36.4	22
	Electrical / Mechanical Engineering / Process Automation	17	85.0	3	15.0	20
	Display / Multimedia / User Interface Technologies	13	68.4	6	31.6	19
	Consumer Products / Electronics	13	81.3	3	18.8	16
	Green Energy / Technology / Industrial Systems	8	72.7	3	27.3	11
	Business Methods / Financial Services	6	60.0	4	40.0	10
	Energy / Utilities / Power Technologies / Oil & Gas	4	50.0	4	50.0	8
	Aviation & Aerospace	0	0.0	3	100.0	3
	Miscellaneous	9	81.8	2	18.2	11
	Total	237	60.3	156	39.7	393

Note: Ranked by number of offered technologies

With regard to the sales ratios, the top ranking application area (that is 'Telecommunications (wireless, cellular, optical & other)') only ranks moderately with 40.5 per cent. Five application areas have sales ratios equal to or higher than 50 per cent. It is important to note that all three of the technologies that were offered for applications in 'Aviation & Aerospace' were sold (that is 100 per cent sales ratio). Furthermore, among those areas with a sales ratio equal to or above 50 per cent were three with more than ten technologies sold in absolute numbers. These include 'Internet / Web Services / E-Commerce / Internet TV' (63.3 per cent), 'Digital & Home media / Entertainment' (60.9 per cent) and '(Micro)Electronics / Computer Systems & Technologies' (51.7 per cent).

Aside from the sales status, this study's dataset allows further insights into the sales prices. Table 10.6 shows how the technology sales distribute across the 16 application areas ranked by average sales prices of the application area. In total, at the six auctions, technologies were auctioned for €30.28 million, with mean sales of €5.05 million per auction. The mean sales price across all technologies is €194.15 thousand, while the corresponding median value is €77.23 thousand; thus 60.2 per cent lower. Technology prices range from a minimum of €1.79 thousand to a maximum of

€3.65 million. The highest price was paid for a technology with applications in 'Internet / Web Services / E-Commerce / Internet TV.'

Table 10.6 - Sales prices across application areas

		Sales Price [T€]					
		Mean	Median	Minimum	Maximum	Count	Sum
Technology Area (n=16)	Digital & Home Media / Entertainment	513,36	356,92	28,34	2,227.01	23	7,187.08
	Internet /Web Services / E-Commerce / Internet TV	416,24	178,16	7,72	3,650.87	30	7,908.57
	Display / Multimedia / User Interface Technologies	392,90	251,01	8,75	1,255.25	19	2,357.39
	Location-based Systems / Applications / Logistics	355,99	79,86	8,10	2,105.53	23	2,847.9
	Business Methods / Financial Services	138,74	108,54	57,92	279,95	10	554,95
	Telecommunications (Wireless, Cellular, Optical & Other)	138,38	88,28	8,75	649,04	79	4,428.06
	Consumer Products / Electronics	127,81	182,21	8,75	192,46	16	383,42
	Aviation & Aerospace	123,48	69,99	17,00	283,44	3	370,43
	(Micro)Electronics / Computer Systems & Technologies	103,22	64,79	5,00	851,78	60	3,199.78
	Green Energy / Technology / Industrial Systems	86,81	16,00	5,00	239,42	11	260,42
	Medical & Life Sci / Devices / Telemedicine / Bio-chip / Bio-tech	58,05	18,50	16,00	297,44	28	464,43
	Chemical Engineering / Applied Sciences / Advanced Materials / Nanotechnology	16,09	9,49	1,79	50,00	30	128,70
	Miscellaneous	14,08	14,08	12,15	16,00	11	28,15
	Automotive Systems & Devices / Industrial Machinery	12,80	8,97	7,17	40,49	22	102,40
	Electrical / Mechanical Engineering / Process Automation	12,33	11,00	11,00	15,00	20	37,00
	Energy / Utilities / Power Technologies / Oil & Gas	7,12	8,86	1,79	8,97	8	28,48
	Total	194,15	77,23	1,79	3,650.87	393	30,287.16

Note: Ranked by mean sales price

In four application areas, the mean sales price was above the total mean sales price. Table 10.6 reveals the highest mean sales price of €513.36 thousand for technologies with applications in the 'Digital & Home media/Entertainment' area. Total sales of this area amounted to €7.19 million, thus 23.7 per cent of the total sales. Technologies with applications in 'Internet / Web Services / E-Commerce / Internet TV' rank second with regard to the mean sales price. On average, €416.24 thousand is paid for a technology in this area. Total sales of this area however rank first with €7.90 million, or 26.1 per cent. Only two further application areas have mean sales prices above the total mean of €194.15 thousand. Technologies in the area of 'Display / Multimedia / User Interface Technologies' sell for an average of €392.90 thousand and technologies in the area of 'Location-based Systems / Applications / Logistics' for €355.99 thousand. However, the total sales of each of these two areas account for less than ten per cent of the total sales. Rather the two areas 'Telecommunications (wireless, cellular, optical & other)' (14.6 per cent) and '(Micro)Electronics / Computer Systems & Technologies' (10.6 per cent) each account for more than 10 per cent of the total sales.

Thus, from Table 10.5 and Table 10.6, I can conclude that two areas are among those with high sales ratios and high mean sales prices. These include 'Internet / Web Services / E-Commerce / Internet TV' with a sales ratio of 63.3 per cent (1st rank) and a mean sales price of €416.24 thousand (2nd rank) and 'Digital & Home media / Entertainment' with a sales ratio of

60.9 per cent (2nd rank) and a mean sales price of €513.36 thousand (1st rank). Together the technologies sold in these two areas account for 21.2 per cent of all sold technologies.

Furthermore, two other application areas play an important role. In total, 32 technologies were sold, with a sales ratio of 40.5 per cent in the area of 'Telecommunications (wireless, cellular, optical & other)' with a mean sales price of €138.38 thousand (below the total average), with 14.6 per cent still accounting for a large share of total sales. Finally, 31 technologies, with a sales ratio of 51.7 per cent, were sold in the area of '(Micro)Electronics / Computer Systems & Technologies' with a mean sales prices of €103 22 thousand, accounting for €3.20 million and thus a share of 10.6 per cent of total sales. With a mean sales ratio of 54.1 per cent, these four areas account for 61.5 per cent of all sold technologies and 75.0 per cent of total sales.

Table 10.7 - Overview of sales prices across the technology auctions

		Auction Number						
		1	2	3	4	5	6	Total
Sales Price [T€]	Mean	94.87	219.30	247.47	17.78	463.75	214.32	194.15
	Median	8.97	148.73	99.21	16.00	141.98	115.85	77.23
	Minimum	1.79	8.75	8.10	5.00	21.09	7.72	1.79
	Maximum	1,255.25	787.35	2,227.01	50.00	3,650.87	1,351.54	3,650.87
	Count	26	22	34	23	13	38	156
	Sum	2,466.61	4,824.69	8,414.04	409.00	6,028.81	8,144.01	30,287.16

Table 10.7 further reveals how the sold technologies and sales prices are distributed across the six auctions. While, on average, 26 technologies were sold at each auction, the technologies are not equally distributed across the auctions. A minimum of 13 technologies was sold at A_5, while the highest number of technologies was sold at A_6 (38 technologies). The sales per auction varied from a lowest result of €409.00 thousand at A_4 to the highest results at A_3 with about 20 times more sales (€8.41 million) than at A_4.

Table 10.8 - Sources of technology and technology sales prices

		Technology Source					
		Other	Independent Inventor	University / Research Lab	SME	Large Firm	Total
Sales Price [T€]	Mean	404.91	171.59	26.65	355.48	134.64	194.15
	Median	404.91	68.99	17.50	115.85	77.23	77.23
	Minimum	404.91	5.00	8.97	8.75	1.79	1.79
	Maximum	404.91	1,351.54	77.23	3,650.87	851.78	3,650.87
	Count	2	56	20	40	38	156
	Sum	809.82	9,608.93	533.06	14,219	5,116.35	30,287.16

As displayed in Table 10.7, the 23 technologies sold at A_4 reached the lowest mean sales price of €17.78 thousand. In contrast, the 13 technologies sold at A_5 achieved the highest mean sales price of €463.75 thousand. However, the technologies sold at A_2 achieved the highest median sales price of €148.73 thousand while, at A_1 the median value is considerably lower amounting to only €8.97 thousand. With €3.65 million, the highest absolute sales price was achieved for a technology at A_5[21], while a technology at A_1 was sold for the lowest price of €1.79 thousand. Table 10.7 does not reveal any consistent trend although, across the first three US auctions, the total sales and mean sales prices increased steadily.

Table 10.8 presents descriptive statistics for sales prices across different technology sources. Aside from the category 'others', technologies that were developed by SME achieved the highest mean sales price of €355.48 thousand as well as the median value of €115.85 thousand. In contrast, technologies developed by universities/ research institutes had the lowest mean sales price of €26.65 thousand and a median value of €17.50 thousand. The highest sales price across the six auctions (€3.65 million) was paid for a technology developed by an SME. The lowest sales price, for which a technology was sold, was €1.79 thousand. This technology was developed in the laboratory of a large firm.

Table 10.9 - Comparison of expected value and sales prices

		Auction Number						
		1	2	3	4	5	6	Total
Expected Value [T€]	Mean	714.54	376.27	526.46	-	406.93	519.96	525.56
	Median	407.55	238.59	202.46	-	220.00	315.95	280.84
	Maximum	4,075.50	1,590.60	3,681.00	-	2,756.00	3,475.40	4,075.50
	Minimum	81.51	7.95	22.09	-	50.00	35.11	7.95
Sales Price [T€]	Mean	94.87	219.30	247.47	17.78	463.755	214.32	194.15
	Median	8.97	148.73	99.21	16.00	141.98	115.85	77.23
	Maximum	1,255.25	787.35	2,227.01	50.00	3,650.87	1,351.54	3,650.87
	Minimum	1.79	8.75	8.10	5.00	21.09	7.72	1.79

As noted in Chapter 6, sellers are known to be positively biased when estimating values for their own technologies. Table 10.9 indicates that this effect seems to be present in this study's dataset. This table compares the expected values predicted by the sellers with the actual sales prices of the technologies across five auctions (for A_4, no expected values were available[22]). Across four auctions (except A_5), the mean and median expected values were above the sales prices. For A_5, the median expected value was higher than the sales price, but not the mean value. Across all auctions, the total mean expected value was 2.71 times that of the total sales price. The total median expected value was 3.63 times that of the total median sales price.

10.2.2 Technology Properties

In addition to the summary statistics provided above, the following section provides more detailed descriptive statistics for the five technology properties that serve as independent variables in the study's regression models (see operationalization in Section 8.3). Table 10.10 presents univariate statistics for these measures for both sold and unsold technologies.

Table 10.10 reveals that across all 393 technologies the technological complexity ranges from a minimum of one to a maximum of 59 with a mean of 3.72 and a median value of two. The technological uncertainty ranges from a minimum 0.73 to 18.45 with a mean value of 11.61 and a median value of 12.32. The technological impact ranges from zero to 87 with a mean of 6.76 and a median value of 3.00. The technological quality ranges from 1.00 to 12.00 with a mean of 1.99 and a median value of 1.95.[23] The expected values range from €7.95 thousand to €4.08 million with a mean of €527.05 thousand and a median value of €280.84 thousand.

Table 10.10 - Descriptive statistics for technology properties

			Technological Complexity	Technological Uncertainty	Technological Impact	Technology Quality	Expected Value [T€]
Sales Status	Not Sold	Mean	4.52	11.63	6.46	1.96	537.62
		Median	2.00	12.33	2.00	1.71	294.48
		Maximum	59.00	18.45	87.00	12.00	4,075.5
		Minimum	1.00	0.83	0.00	1.00	7.95
	Sold	Mean	2.50	11.58	7.21	2.05	509.26
		Median	1.00	12.17	3.00	2.00	280.84
		Maximum	18.00	17.34	60.00	6.00	4,075.5
		Minimum	1.00	0.73	0.00	1.00	22.09
	Total	Mean	3.72	11.61	6.76	1.99	527.05
		Median	2.00	12.32	3.00	1.95	280.84
		Maximum	59.00	18.45	87.00	12.00	4,075.5
		Minimum	1.00	0.73	0.00	1.00	7.95

Across the 237 unsold technologies, the technological complexity ranges from 1.00 to 59.00 with a mean of 4.52 and a median value of 2.00. The technological uncertainty ranges from 0.83 to 18.45 with a mean of 11.63 and a median of 12.33. The technological impact ranges from zero to 87.00 with a mean of 6.46 and a median value of 2.00. The technology quality ranges from 1.00 to 12.00 with a mean of 1.96 and a median value of 1.71.

The expected values range from €7.95 thousand to €4.08 million with a mean of €537.62 thousand and a median value of €294.48 thousand.

Among the 156 sold technologies, the technological complexity ranges from 1.00 to 18.00 with a mean of 2.50 and a median value of 1.00. The technological uncertainty ranges from 0.73 to 17.34 with a mean of 11.58 and a median value of 12.17. While the technological impact ranges from 0 to 60.00 with a mean of 7.21 and a median value of 3.00, the technology quality ranges from 1.00 to 6.00 with a mean of 2.05 and a median value of 2.00. The expected values range from €22.09 thousand to €4.08 million with a mean of €509.26 thousand and a median value of €280.84 thousand.

Comparing the mean and median values of the unsold and sold technologies, the following picture emerges. The mean and median values of the unsold technologies' technological complexity are higher by 66.7 per cent and 100.0 per cent, respectively, than those of the sold technologies. The mean and median values of the technological uncertainty are almost similar compared to the unsold and the sold technologies, with the mean and median values of the unsold technologies being slightly greater. The mean and median values of the technological impact of the unsold technologies are lower by 10.0 per cent and 33.3 per cent, respectively, than those of the sold technologies. Both the mean and median values of technology quality are lower by 4.4 per cent and 14.5 per cent, respectively, than those of the sold technologies. The mean and median values of the expected value of the unsold technologies are higher by 5.6 per cent and 4.9 per cent, respectively than those of the sold technologies. These descriptive results are in line with the hypotheses derived in Section 5.3.

Technological Complexity

Table 10.11 and Table 10.12 display detailed descriptive statistics for technological complexity (that is measured by the number of related patent families; see Section 6.1) for the technologies offered at the six auctions. In total, the mean technological complexity across all technologies is 3.72.
The total median value is 2.00, which is 46.2 per cent lower than the total mean value. The mean value of the sold technologies is 2.49 and the median value is 1.00, thus, respectively, 33.1 per cent and 50.0 per cent lower than the mean. The mean value of the unsold technologies is 4.52, which is 21.5 per cent higher than the total mean value, with the median value remaining at 2.00. In the group of sold technologies, the corresponding mean and most median values are always higher than of those of the group of unsold technologies.

The technology with the highest technological complexity, with a value of 59, was offered at A_1, but remained unsold. Among the sold technologies, the technology with the highest technological complexity had a value of 18 and was also offered at A_1. At A_1, both sold and unsold technologies had the highest mean values (5.53) of the six auctions. However, A_5 showed the highest median values. Technologies offered at A_6 had the lowest mean values for technological complexity for both sold and unsold technologies.

Table 10.11 - Technological complexity across auctions

			Auction Number						
			1	2	3	4	5	6	Total
Technological Complexity	Not Sold	Mean	6.43	4.40	2.84	5.31	5.47	2.05	4.52
		Median	3.00	2.00	1.00	3.00	4.00	1.00	2.00
		Count	51	50	31	32	34	39	237
		Minimum	1.00	1.00	1.00	1.00	1.00	1.00	1.00
		Maximum	59.00	29.00	17.00	21.00	25.00	14.00	59.00
	Sold	Mean	3.77	1.77	2.68	2.43	3.00	1.74	2.49
		Median	1.50	1.00	1.00	2.00	3.00	1.00	1.00
		Count	26	22	34	23	13	38	156
		Minimum	1.00	1.00	1.00	1.00	1.00	1.00	1.00
		Maximum	18.00	4.00	17.00	12.00	9.00	10.00	18.00
	Total	Mean	5.53	3.60	2.75	4.11	4.79	1.90	3.72
		Median	2.00	2.00	1.00	2.00	3.00	1.00	2.00
		Count	77	72	65	55	47	77	393
		Minimum	1.00	1.00	1.00	1.00	1.00	1.00	1.00
		Maximum	59.00	29.00	17.00	21.00	25.00	14.00	59.00

Looking at the distribution of technological complexity from technologies stemming from different technology sources, Table 10.12 reveals that the technologies developed by large firms had the highest total mean technological complexity of 6.28. Aside from the category, 'other', which includes only very few technologies, the technologies developed by SME ranked the second highest with a mean value of 2.92, which is considerably lower (53.5 per cent) than the technological complexity of those technologies developed by large firms. Ranking third, the technologies developed by universities / research institutes had a mean technological complexity of 2.26 followed by technologies developed by independent inventors with a mean value of 1.94.

This ranking appears similar across all sold and unsold technologies. The median values across all categories are always lower than the corresponding mean values. The technology with the highest technological complexity of 59.00, was developed by a large firm but remained unsold. The highest technological complexity value for a sold technology was 18.00. This technology was also developed by a large firm.

Table 10.12 - Technological complexity for different sources

			Technology Source					
			Other	Independent Inventor	University / Research Lab	SME	Large Firm	Total
Technological Complexity	Not Sold	Mean	-	2.22	3.00	3.20	7.18	4.52
		Median	-	2.00	2.00	2.00	4.00	2.00
		Count	0	68	7	66	96	237
		Minimum	-	1.00	1.00	1.00	1.00	1.00
		Maximum	-	15.00	12.00	17.00	59.00	59.00
	Sold	Mean	4.50	1.59	2.00	2.48	4.00	2.49
		Median	4.50	1.00	2.00	1.00	2.00	1.00
		Count	2	56	20	40	38	156
		Minimum	4.00	1.00	1.00	1.00	1.00	1.00
		Maximum	5.00	10.00	4.00	17.00	18.00	18.00
	Total	Mean	4.50	1.94	2.26	2.92	6.28	3.72
		Median	4.50	1.00	2.00	1.50	3.00	2.00
		Count	2	124	27	106	134	393
		Minimum	4.00	1.00	1.00	1.00	1.00	1.00
		Maximum	5.00	15.00	12.00	17.00	59.00	59.00

Technological Uncertainty

More details on the technologies' technological uncertainty (that is measured according to the average remaining lifetime of the corresponding patents; see Section 6.1) are presented in the following two tables.

Table 10.13 reveals a total mean technological uncertainty of 11.61 across all technologies. The highest technological uncertainty of 18.45 relates to a technology that was offered at A_4, but remained unsold. The lowest technological uncertainty of 0.73 relates to a technology that was (surprisingly) sold at A_2. Except for A_6, the mean technological uncertainty at the auctions was lower for the sold technologies than the unsold technologies. However, this pattern does not hold true for median values. In A_1 and A_4, the median values for technological uncertainty were lower for the unsold technologies. Overall, Table 10.13 displays little variation of technological uncertainty across all auctions in absolute numbers. However, among the sold (unsold) technologies, the mean technological uncertainty varied by 18.3 per cent (16.2 per cent) between the lowest mean value at A_1 and the highest mean value at A_4.

Taking into account that the measure of technological uncertainty can range from zero to 20, the results for the mean values of technological uncertainty show that, on average, the sold and unsold technologies are, with a little variation, in an area of moderate technological uncertainty. However, some technologies show extreme values.

Table 10.13 - Technological uncertainty across auctions

			Auction Number						
			1	2	3	4	5	6	Total
Technological Uncertainty	Not Sold	Mean	10.73	11.87	12.28	12.80	11.87	10.82	11.63
		Median	11.14	13.12	13.78	13.10	12.32	11.86	12.33
		Count	51	50	31	32	34	39	237
		Minimum	2.14	2.53	3.49	5.70	3.97	.83	.83
		Maximum	18.02	17.42	18.33	18.45	15.53	17.29	18.45
	Sold	Mean	10.19	11.51	12.13	12.48	11.04	11.66	11.58
		Median	11.93	12.45	12.54	13.13	11.67	11.73	12.17
		Count	25	22	34	23	13	38	155
		Minimum	2.45	.73	4.95	4.52	2.36	7.03	.73
		Maximum	16.58	17.34	16.38	16.81	14.93	16.07	17.34
	Total	Mean	10.55	11.76	12.20	12.66	11.64	11.24	11.61
		Median	11.19	12.88	12.79	13.13	12.21	11.75	12.32
		Count	76	72	65	55	47	77	392
		Minimum	2.14	.73	3.49	4.52	2.36	.83	.73
		Maximum	18.02	17.42	18.33	18.45	15.53	17.29	18.45

An addition, Table 10.14 displays the distribution of technological uncertainty for the technologies stemming from different technology sources. Except for technologies developed by SME, the mean technological uncertainty was higher for technologies that remained unsold. Among the sold technologies, those developed by universities / research institutes show the highest mean technological uncertainty of 12.75, followed by technologies developed by SME (12.65), by independent inventors (11.45), and by large firms with the lowest mean technological uncertainty of 10.03. Among the unsold technologies the technologies developed by universities / research institutes show the highest mean technological uncertainty of 14.72, followed by the technologies developed by independent inventors (12.02), by SME (11.63), and by large firms with the lowest mean technological uncertainty of 11.13. A similar pattern holds true for the median values. The mean technological uncertainty varies between the lowest and highest values for sold (unsold) technologies by 21.3 per cent (24.4 per cent).

The technologies with the lowest and highest technological uncertainty of 0.73, and, respectively, 18.45, were developed by independent inventors. The technology with the absolute highest technological uncertainty remained unsold, while the technology with the lowest technological uncertainty was sold. However, another technology with a similarly low technological uncertainty of 0.83, which was developed by an SME, remained unsold.

Table 10.14 - Technological uncertainty for different sources

			Technology Source					
			Other	Independent Inventor	University / Research Lab	SME	Large Firm	Total
Technological Uncertainty	Not Sold	Mean	-	12.02	14.72	11.63	11.13	11.63
		Median	-	12.62	15.90	12.56	11.81	12.33
		Count	0	68	7	66	96	237
		Minimum	-	1.94	10.06	.83	2.14	.83
		Maximum	-	18.45	17015	18.33	16.16	18.45
	Sold	Mean	10.50	11.45	12.75	12.65	10.03	11.58
		Median	10.50	12.34	13.39	13.19	10.69	12.17
		Count	2	56	20	40	37	155
		Minimum	10.12	.73	7.08	2.36	2.45	.73
		Maximum	10.88	16.38	16.81	17.34	15.89	17.34
	Total	Mean	10.50	11.76	13.26	12.02	10.82	11.61
		Median	10.50	12.55	13.48	12.72	11.47	12.32
		Count	2	124	27	106	133	392
		Minimum	10.12	.73	7.08	.83	2.14	.73
		Maximum	10.88	18.45	17.15	18.33	16.16	18.45

Technological Impact

As discussed in Section 6.1, a technology's impact on subsequent technological progress can be measured by its forward citations. Table 10.15 and Table 10.16 present descriptive results with regard to the technological impact of the technologies included in this study's dataset. Table 10.15 shows that the total mean value of the technological impact is 6.77.

The total mean value of the sold technologies' technological impact is 11.2 per cent higher than those that remained unsold. Across all auctions, the mean technological impact of the sold technologies is higher than the mean value of the unsold technologies. Technologies sold (unsold) at A_5 (A_6) had the highest mean technological impact of 12.15 (10.92) compared to the lowest mean value of 1.26 (1.22) at A_4. The technology with the highest technological impact of 87 was offered at A_6, but remained unsold. Various technologies in both groups have a technological impact of zero.

The technologies offered and sold by IPA at A_4 have by far the lowest values for technological impact, that is, by far lower than the average mean value of all OT auctions. Furthermore, it appears that technologies offered and sold at A_5 and A_6 have a particularly high technological impact with a mean value of almost double those of A_1 to A_3. Moreover, it appears that, despite the moderate mean values, technologies offered and sold at A_1 show comparably small differences between their mean and median values.

Table 10.16 displays the distribution of the technological impact across technologies stemming from different technology sources. All sold technologies have higher mean values for technological impact across the different types of technology sources than the unsold technologies.

Table 10.15 - Technological impact across auctions

			Auction Number						
			1	2	3	4	5	6	Total
Technological Impact	Not Sold	Mean	5.60	5.14	5.74	1.22	10.09	10.92	6.46
		Median	4.50	1.00	1.00	.00	5.00	4.00	2.00
		Count	51	50	31	32	34	39	237
		Minimum	.00	.00	.00	.00	.00	.00	.00
		Maximum	17.00	28.00	31.00	10.00	66.00	87.00	87.00
	Sold	Mean	5.62	5.32	6.41	1.26	12.15	12.08	7.22
		Median	6.00	2.00	3.00	.00	8.00	7.00	3.00
		Count	26	22	34	23	13	38	156
		Minimum	.00	.00	.00	00	.00	.00	.00
		Maximum	13.00	27.00	36.00	7.00	41.00	60.00	60.00
	Total	Mean	5.61	5.19	6.09	1.24	10.66	11.49	6.77
		Median	5.00	1.00	2.00	.00	5.00	7.00	3.00
		Count	77	72	65	55	47	77	393
		Minimum	.00	.00	.00	.00	.00	.00	.00
		Maximum	17.00	28.00	36.00	10.00	66.00	87.00	87.00

Among the sold technologies, those developed by independent inventors have the highest mean technological impact of 8.11, followed by those technologies developed by SME (8.00) and large firms (7.50). While the technological impacts of technologies stemming from these three sources appear in a similar range, the mean technological impact of technologies developed by universities / research institutes is considerably lower with a value of 2.30. A possible reason for this is that university inventions (see Chapter 6) are commonly more distant from the market than corporate inventions. Thus, university inventions might rather have a larger time lag before they receive citations (see also Hall et al. (2001)). Among the unsold technologies, those developed by large firms have the highest mean technological impact of 7.22, followed by technologies developed by independent inventors (6.51) and SME (5.85). Again, the mean technological impact of the technologies developed by universities / research institutes is considerably lower with 1.71. With regard to the median values, across both groups, the ranking follows a similar order, although this ranking is different compared to the ranking based on mean values. In both groups, technologies developed by large firms have the highest median technological impact, followed by those developed by SME, independent inventors and universities / research institutes.

Table 10.16 - Technological impact for different sources

			Technology Source					
			Other	Independent Inventor	University / Research Lab	SME	Large Firm	Total
Technological Impact	Not Sold	Mean	-	6.51	1.71	5.85	7.22	6.46
		Median	-	1.00	.00	2.00	4.00	2.00
		Count	0	68	7	66	96	237
		Minimum	-	.00	.00	.00	.00	.00
		Maximum	-	87.00	10.00	31.00	66.00	87.00
	Sold	Mean	11.00	8.1	2.30	8.00	7.50	7.22
		Median	11.00	1.00	1.00	3.00	5.00	3.00
		Count	2	56	20	40	38	156
		Minimum	10.00	.00	.00	.00	.00	.00
		Maximum	12.00	60.00	8.00	41.00	37.00	60.00
	Total	Mean	11.00	7.23	2.15	6.66	7.30	6.77
		Median	11.00	1.00	.00	2.00	4.00	3.00
		Count	2	124	27	106	134	393
		Minimum	10.00	.00	.00	.00	.00	.00
		Maximum	12.00	87.00	10.00	41.00	66.00	87.00

The technology with the highest mean technological impact of 87 was developed by an independent inventor, although it remained unsold. In the group of sold technologies, the technology with the highest mean technological impact of 60 was also developed by an independent inventor. Technologies with zero technological impact stem from all technology sources.

Technology Quality

Table 10.17 shows that the mean quality (for operationalization by the number of inventors; see Section 6.1) of the sold technologies is higher than for those that remained unsold. Looking at the individual auctions, the table shows that, for A_1 and A_2, the technologies' mean quality is higher in the group of unsold technologies, while, from A_3 onwards the mean quality is higher in the group of sold technologies. Except in A_6, this pattern is very similar for the median values. For this auction, the median values are similar for both groups. Among the sold (unsold) technologies, the mean value for quality is the highest at A_4 (A_2) and the lowest at A_2 (A_3). A technology at A_2 showed the highest quality of 12.00, however, remained unsold. Among the sold technologies, the highest quality is 6.00 (A_6). Various technologies across all auctions have the minimum quality of one.

Table 10.18 displays the distribution of the quality of technologies stemming from different technology sources. In the group of technologies that were sold, technologies developed by universities / research institutes have the highest quality of 2.87, followed by technologies developed by large firms (2.49), SME (2.07), and individual inventors (1.41). Among the technologies that remained unsold, the technologies developed by large firms have the highest quality (2.50), followed by technologies developed by SME (1.93), universities / research institutes (1.73), and independent inventors (1.23).

Table 10.17 - Technology quality across auctions

			Auction Number						
			1	2	3	4	5	6	Total
Technology Quality	Not Sold	Mean	2.15	2.45	1.54	1.96	1.70	1.63	1.96
		Median	2.00	2.00	1.20	1.87	1.59	1.00	1.71
		Count	51	50	31	32	34	39	237
		Minimum	1.00	1.00	1.00	1.00	1.00	1.00	1.00
		Maximum	4.34	12.00	3.00	8.36	3.00	6.00	12.00
	Sold	Mean	1.97	1.68	1.81	2.87	2.10	2.00	2.04
		Median	2.00	1.75	1.35	3.00	1.67	1.00	2.00
		Count	26	22	34	23	13	38	156
		Minimum	1.00	1.00	1.00	1.00	1.00	1.00	1.00
		Maximum	4.00	4.00	5.00	5.58	4.40	6.00	6.00
	Total	Mean	2.09	2.21	1.68	2.34	1.81	1.81	1.99
		Median	2.00	2.00	1.20	2.00	1.67	1.00	2.00
		Count	77	72	65	55	47	77	393
		Minimum	1.00	1.00	1.00	1.00	1.00	1.00	1.00
		Maximum	4.34	12.00	5.00	8.36	4.40	6.00	12.00

Table 10.18 - Technology quality for different sources

			Technology Source					
			Other	Independent Inventor	University / Research Lab	SME	Large Firm	Total
Technology Quality	Not Sold	Mean	-	1.23	1.73	1.93	2.50	1.96
		Median	-	1.00	2.00	1.92	2.21	1.71
		Count	0	68	7	66	96	237
		Minimum	-	1.00	1.00	1.00	1.00	1.00
		Maximum	-	3.00	3.00	5.00	12.00	12.00
	Sold	Mean	2.60	1.41	2.87	2.07	2.49	2.04
		Median	2.60	1.00	3.00	2.00	2.00	2.00
		Count	2	56	20	40	38	156
		Minimum	2.20	1.00	1.00	1.00	1.00	1.00
		Maximum	3.00	4.00	5.00	5.00	6.00	6.00
	Total	Mean	2.60	1.31	2.57	1.98	2.50	1.99
		Median	2.60	1.00	2.33	2.00	2.16	2.00
		Count	2	124	27	106	134	393
		Minimum	2.20	1.00	1.00	1.00	1.00	1.00
		Maximum	3.00	4.00	5.00	5.00	12.00	12.00

The technology with the highest quality (with a value of 12.00) that remained unsold at A_2 was developed by a large firm. Nevertheless, the technology with the highest quality (with a value of 6.00) that was sold was also developed by a large firm. It is interesting to note that the median quality of technologies developed by universities / research institutes is higher than the corresponding mean values of both groups.

Technology Value Perception

Prior to any transactions, sellers commonly conduct a technology value assessment. The following tables present the sellers' value expectations of the technologies that are included in this study's dataset. Table 10.19 displays sellers' value perceptions of the individual auctions, although these data were not available for the IPA auction (A_4).

Table 10.19 - Value perception of sellers across auctions

			Auction Number						
			1	2	3	4	5	6	Total
Expected Value [T€]	Not Sold	Mean	797.52	397.74	366.12	-	445.59	551.04	537.62
		Median	407.55	218.71	276.08	-	202.50	351.05	294.48
		Count	51	50	31	32	34	39	237
		Minimum	81.51	7.95	51.53	-	50.00	35.11	7.95
		Maximum	4,075.50	1,590.60	1,104.30	-	2,756.00	3,475.40	4,075.50
	Sold	Mean	551.76	315.89	707.71	-	294.45	491.47	505.44
		Median	407.55	308.18	147.24	-	238.00	263.29	268.92
		Count	26	22	34	23	13	38	156
		Minimum	81.51	39.77	22.09	-	62.00	52.66	22.09
		Maximum	4,075.50	994.13	3,681.00	-	675.00	2,808.40	4,075.50
	Total	Mean	714.54	376.27	526.46	-	406.93	519.96	525.56
		Median	407.55	238.59	202.46	-	220.00	315.95	280.84
		Count	77	72	65	55	47	77	393
		Minimum	81.51	7.95	22.09	-	50.00	35.11	7.95
		Maximum	4,075.50	1,590.60	3,681.00	-	2,756.00	3,475.40	4,075.50

The table shows that the total mean value of the unsold technologies (€537.62 thousand) is above the mean value of the sold technologies of €505.44 thousand. With regard to the individual auctions, the same is true for all auctions, except A_3. At A_3, the mean expected value is considerably higher (by 93.3 per cent) for the sold technologies compared to the unsold technologies. Similarly, the total median expected value is also higher for the unsold technologies than the sold technologies. However, across the individual auctions, the picture is mixed. In both groups, the median expected value is considerably lower (almost 50 per cent) than the mean value.

At A_1 sellers expected a highest maximum value of €4.08 million although this was for different technologies, some of which remained unsold. The lowest expected value was €7.95 thousand for a technology that remained unsold at A_2. The mean expected values in the group of technologies that were sold (remained unsold) varied by 240.3 per cent (217.8 per cent) from the highest mean value of €707.71 thousand at A_3 (€797.52 thousand at A_1) to the lowest value of €294.45 thousand at A_5 (€366.12 thousand at A_3).

Table 10.20 - Value perceptions of sellers for different sources

			Technology Source					
			Other	Independent Inventor	University / Research Lab	SME	Large Firm	Total
Expected Value [T€]	Not Sold	Mean	-	505.06	736.20	484.69	590.15	537.62
		Median	-	318.12	736.20	280.84	247.80	294.48
		Count	0	68	7	66	96	237
		Minimum	-	15.91	736.20	7.95	27.84	7.95
		Maximum	-	3,475.40	736.20	1,870.00	4,075.50	4,075.50
	Sold	Mean	736.20	502.54	496.74	597.41	404.51	505.44
		Median	736.20	228.19	175.53	306.07	247.50	268.92
		Count	2	56	20	40	38	156
		Minimum	736.20	22.09	73.62	36.81	73.62	22.09
		Maximum	736.20	2,808.40	1,630.20	3,681.00	4,075.50	4,075.50
	Total	Mean	736.20	503.99	530.94	527.75	535.74	525.56
		Median	736.20	318.12	175.53	280.84	247.80	280.84
		Count	2	124	27	106	134	393
		Minimum	736.20	15.91	73.62	7.95	27.84	7.95
		Maximum	736.20	3,475.40	1,630.20	3,681.00	4,075.50	4,075.50

In addition, Table 10.20 provides further insights into the expected values of technologies stemming from different sources. In total, from the highest mean expected value of large firms (€535.74 thousand) to the lowest mean expected value of independent inventors (€503.99 thousand), the total mean expected values of the technologies stemming from the different sources vary only by 6.3 per cent. However, within the two groups, the variation is larger. Among the sold (unsold) technologies, the mean expected values vary by 47.7 per cent (52.9 per cent) from the highest value of €597.41 thousand (€736.20 thousand) to the lowest value of €404.51 thousand (€484.69 thousand), aside from the category 'others'.

The highest mean values of €4.08 million were expected for technologies developed by large firms, of which some were sold while others remained unsold. Among the technologies that were sold, the highest mean value of €597.41 thousand was expected for those developed by SME, while, among the group of technologies that remained unsold, the highest value of

€736.20 thousand was expected for those developed by a university / research institute.

10.2.3 Summary of Major Descriptive Results

At the six auctions, a total of 393 technologies were offered from which 156 (39.7 per cent) were successfully auctioned while 237 (60.3 per cent) remained unsold. Almost one third of the technologies that were offered were developed by large firms, while the rest were developed by independent inventors, SME, and a smaller fraction by universities / research institutes. However, those technologies developed by universities / research institutes had the highest sales ratio, followed by those developed by independent inventors, SME, and then large firms. At the six auctions, a total of 156 technologies were auctioned for €30.29 million, with a mean of €5.05 million sales per auction. The mean sales price of the technologies was €194.15 thousand, while the median sales price of €77.23 thousand was considerably lower (by 60.2 per cent).

In general, comparing actual realized sales prices with the sellers' expected sales prices revealed that the actual realized sales prices were considerably lower than the values expected from the sellers. While I found a mean sales price of €194.15 thousand the sellers' expected mean sales price was €525.56 thousand. These results indicate that sellers overvalue the prices they can expect for their technologies on the MfTI. The data presented in this study however might contribute somewhat to correct this situation.

The prices paid for the technologies in this study's dataset range from a minimum of €1.79 thousand to a maximum of €3.65 million. The technology that sold for the highest price was for applications in 'Internet / Web Services / E-Commerce / Internet TV'. A large share of the technologies was offered and sold in four different application areas. With an average sales ratio of 54.1 per cent these four application areas together account for 61.5 per cent of all sold technologies and 75.0 per cent of total sales. These four areas include technologies sold for applications in 'Internet / Web Services / E-Commerce / Internet TV' with a sales ratio of 63.3 per cent and a mean sales price of €416.24 thousand. Technologies sold for applications in 'Digital & Home media / Entertainment' have a sales ratio of 60.9 per cent and a mean sales price of €513.36 thousand. Technologies sold for applications in 'Telecommunications (wireless, cellular, optical & other)' have a sales ratio of 40.5 per cent and a mean sales price of €138.38 thousand. Technologies sold for applications in '(Micro)Electronics / Computer Systems & Technologies' have a sales ratio of 40.5 per cent and a mean sales price of €103.22 thousand .

Table 10.21 - Distributions of technology properties and sales prices

Descriptive Statistics

	N	Minimum	Maximum	Mean	Skewness		Kurtosis	
	Statistic	Statistic	Statistic	Statistic	Statistic	Std. Error	Statistic	Std. Error
Technological Complexity	392	1.00	59.00	3.7245	4.582	.123	32.949	.246
Technological Uncertainty	392	.73	18.45	11.6081	-.815	.123	.353	.246
Technological Impact	389	.00	87.00	6.7584	3.104	.124	13.845	.247
Technology Quality	392	1.00	12.00	1.9910	2.592	.123	12.689	.246
Expected Value [T€]	298	7.95	4,075.50	527.0541	2.959	.141	9.957	.281
Sales Price [T€]	155	1.79	3,650.87	195.3316	5.328	.195	35.241	.387
Valid N (listwise)	111							

With regard to the five technology properties, the descriptive results revealed a mean technological complexity of 3.72 (sold: 2.49; unsold: 4.52), a mean technological uncertainty of 11.61 (sold: 11.58; unsold: 11.63), a total mean technological impact of 6.76 (sold: 7.21; unsold: 6.46), a total technology quality of 1.99 (sold: 2.05; unsold: 1.96), and a mean expected value of €527.05 thousand (sold: €509.26 thousand; unsold: €537.62 thousand). Thus, the sold technologies have lower mean values for technological complexity, technological uncertainty, and the expected sales prices, while they have higher mean and median values for technological impact and technology quality. Comparing these two groups in general, with regard to the sales probability, this pattern seems to reflect the hypotheses developed in Section 9.3.

However, this conclusion can hardly be drawn solely from the descriptive statistics, since the differences between the mean and median values indicate that the values of the five technology properties as well as of the sales prices are not normally distributed, but are rather skewed. Table 10.21 further reveals relatively high values for skewness and kurtosis. Moreover, a Kolmogorov-Smirnov test proved that the technology properties and sales prices are not normally distributed. This skewness however is not unexpected as prior studies (for example, Winkelmann (2008)) have indicated the count data nature of many patent indicators on which this study's operationalization is largely based. In this study's variables technological uncertainty appears right skewed. In contrast, the other technology properties and sales prices appear to be left skewed.

Multivariate statistical analyses are necessary and will be pursued in the following section to test the hypotheses. For the regression analyses, the non-normal distributions have been accounted for and corrected through transformations of the variables. In the following section, I present results of inferential statistical analyses, which are used to further test which of the technology properties have an impact on the sales probability or the sales price of technologies. In contrast to the descriptive statistics provided in this chapter, the inferential statistics provided in the next chapter use multivariate statistical procedures – more specifically, multivariate, logistical, and log-linear OLS regression analyses.

10.3 REGRESSION RESULTS

Following the results of the descriptive analyses the results of the regression analyses are presented in this section. These results address the study's second research question (Chapter 2), that is 'Which patented technologies should be preferably auctioned?'

I present inferential statistical results of investigations of causal relationships between a set of independent variables (that is the five technology characteristics) and two dependent variables (that is sales decision, sales price) in order to test the hypotheses which were developed from the theory and the two pre-studies presented in Section 9.3. The hypotheses were translated into a conceptual model with five technology properties as depicted in Figure 10.3, which was inspired by the conceptual framework proposed by Punch (2005).

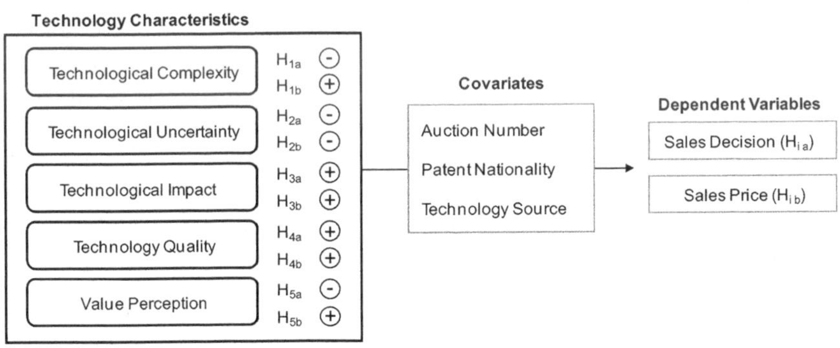

Figure 10.3 - Conceptual model for regression analyses

The conceptual model indicates the individual directional hypotheses with the directions indicated in the circles: '+' indicates a positive relation and '-' indicates a negative relation. The indexes of the hypotheses indicate to which of the two dependent variables the hypotheses are related. The index 'a' indicates that the hypothesis is related to the dependent variable, 'sales probability', which is a dichotomous variable. The corresponding analytical model is a multivariate logistical regression. The index 'b' indicates that the hypothesis is related to the dependent variable 'sales price', which is a metric-scaled variable. The corresponding analytical model is a multivariate log-linear OLS regression model. The choice of these two regression types was motivated in Chapter 2. As outlined in Chapter 8, whether my data fulfilled the necessary assumptions of both models was tested thoroughly according to Hair (2006) and Gujarati (2003). As illustrated in Figure 10.3, in addition to the five technology properties, three control variables (covariates) are employed following Giuri et al. (2005), as well as Lamoreaux and Sokoloff (1998).

To develop the multivariate analytical models, the variables were operationalized according to the measures described in Section 6.1. For some variables, alternative measures were computed that are very similar. For instance, technological uncertainty could have been operationalized based on the average, minimum or maximum remaining protection time calculated from the related patents. Bivariate regression analyses were used to choose the preferred operationalization. If variables could be operationalized with different scales (for example, a metric-scaled continuous variable or a selected number of dichotomous dummy variables), I attempted to use the measures with the highest explanatory power, that is measurement scales with the highest level of detail. For the independent and dependent variables, bivariate one-way Pearson correlation coefficients are displayed in Annex III.

To test these hypotheses, I interpret the direction of the relationships between each independent and the dependent variable indicated by the prefix of the coefficients (positive / negative relation) as well as their significance levels. Furthermore, taking several independent variables into account in one model, the results of the regression analyses indicate accumulated effects and can thus determine a measure for the total share of effects explained by the model (goodness-of-fit). According to Gujarati (2003), this measure indicates how well the sample regression line fits the data. If all measures would lie on the regression line, they would have a 'perfect' fit, although this is rarely the case. Therefore, the aim is rather to keep the residuals around the regression line as small as possible. For the logistic regression models, the four goodness-of-fit measures are used that are presented in Chapter 8.[24]

First, I present results from logistic regression analyses where the dependent variable is the 'sales decision'. This variable differentiates two cases, that is whether technologies were sold or not. Thus, with this logistic regression analyses, I investigate whether the independent variables influence the probability that a technology will be sold. For these models, all the offered technologies, except those developed by the group 'other', were taken into account. For the presentation of the logistic regression results, the best practices as identified by Hoetker (2007) and Peng et al. (2002) were considered. Second, I present the results of log-linear OLS regression analyses, where the dependent variable is the sales price. For these models, only the technologies that were sold at the six auctions were taken into account, except those developed by 'others'. For both models, results of various sensitivity analyses are also presented that were computed for selected subsamples (for example, only taking the four US auctions into account), which also use alternatively measured variables. Finally, I summarize the major regression results. Cases that could have caused bias to the results were not included in the dataset. For instance, only technologies were considered that were offered for sale (not for example, for license) and technologies that were auctioned using the English auction type (that is Dutch auctions were excluded).[25]

10.3.1 Determinants of Sales Probability

For the logistic regression model, the dichotomous dependent variable 'sales decision' was coded following the convention for binary logistic regressions. Thus, I coded the dependent class of greatest interest as one ('the event occurring') and the other class as zero ('the event not occurring') and coded its expected correlates as +1 to assure positive correlation (Garson, 2010). If not coded meaningfully, logistic coefficients would be difficult to interpret. Following the presentation of the results for the 'full' untransformed and transformed models, I present results of various sensitivity analyses.

Table 10.22 displays the regression results of two multivariate logistic models. The 'untransformed model' (left) displays the results of the variables, which were all entered in their original form. The 'Lg10-transformed model' (right) displays the results of a model for which the five independent, metric-scaled variables were entered as log10-transformed variables to correct for non-normal distributions.

Both models have reasonably good Nagelkerke R^2 values. With 0.173, or 0.198 for the transformed model[26], the values are just below the 0.2 to 0.4 range, which is commonly considered to indicate 'very good' results (see Section 8.2). As the -2 log likelihood measure has no absolute maximum value, its values are generally difficult to judge in absolute terms. However, a lower value for the transformed model indicates a better quality of that model. With 66.2 per cent correct classifications for the untransformed and

69.6 per cent for the transformed models, those values can be considered of moderate quality. Furthermore, the Hosmer and Lemeshow values are non-significant, and thus indicate well-fitting models. Thus, as all four measures have reasonably good values, I can conclude that both models show a good fit.

Table 10.22 - Multivariate logistic regression models

Models [a]	Full Logit Models for all Auctions					
	Untransformed Model			Lg10-transformed Model [b]		
	Coef.	Sig.	SE	Coef.	Sig.	SE
(Constant)	-1.897	0.054 [f]	0.985	0.090	0.090 [f]	1.315
Control Variables						
Auction_2 [c]	-0.408	0.283	0.380	-0.118	0.769	0.400
Auction_3	0.459	0.225	0.378	0.793	0.050 [g]	0.405
Auction_4	1.197	0.199	0.932	1.506	0.121	0.971
Auction_5	-0.046	0.919	0.454	0.253	0.593	0.473
Auction_6	0.081	0.831	0.379	0.312	0.423	0.390
Uni / research lab [d]	1.471	0.010 [h]	0.574	1.276	0.029 [g]	0.584
SME	-0.424	0.156	0.298	-0.435	0.162	0.312
Large_firm	-0.633	0.055 [f]	0.331	-0.902	0.009 [h]	0.344
Patent_nationality	1.829	0.035 [g]	0.867	1.733	0.053 [f]	0.897
Hypothesized Variables						
Technological_complexity	-0.108	0.008 [h]	0.041	-1.278	0.002 [h]	0.419
Technological_uncertainty [e]	-0.015	0.621	0.030	0.100	0.845	0.514
Technological_impact	0.014	0.282	0.013	0.526	0.073 [f]	0.293
Technology_quality	0.165	0.089 [f]	0.097	1.293	0.022 [g]	0.565
Value_perception	0.000	0.569	0.000	-0.032	0.910	0.280
Nagelkerke R^2	0.173			0.198		
-2 Log Likelihood	471.65			438.50		
Classification Correct	66.20			69.60		
Hosmer and Lemeshow	0.166			0.829		

Notes:
[a] For all models n=369; dependent variable is "Sales decision" (reference category is 0=not-sold)
[b] Five hypothesized variables were entered as Lg10 transformed variables
[c] Base variable is the first auction
[d] Base variable is "independent inventors"
[e] Measured as minimum remaining protection time
[f] Indicates 10% significance
[g] Indicates 5% significance
[h] Indicates 1% significance

Comparing both models, it appears that the transformed model must be considered to have a slightly better goodness-of-fit. With a value of 0.198 for Nagelkerke's R^2, the transformed model has a 14.5 per cent higher value than the untransformed model with a Nagelkerke R^2 value of 0.173. The -2 log-likelihood value of the transformed model of 438.50 is 7.0 per cent lower than that of the untransformed model with a value of 471.65. The share of correct classifications is also 5.1 per cent higher than that of the transformed model compared to the untransformed model. The Hosmer and Lemeshow value is almost five times higher than the value of the untransformed model. In both models displayed in Table 10.22, the coefficients of four of the five hypothesized variables have prefixes, confirming the directions predicted by the hypotheses. Both models correctly predict the negative relation of technological complexity and sales probability on a one per cent significance level. Both models also correctly predict the positive relation of technology quality and sales probability. While the transformed model predicts this relationship to have a significance level of five per cent, the 'original' model predicts that this relationship will have a significance level of ten per cent. Furthermore, both models correctly predict the positive relationship of technological impact and sales probability. The transformed model predicts this relationship to have a significance level of ten per cent. Furthermore, the transformed model correctly predicts the negative relationship of value perception and sales probability, although this relationship appears insignificant. Although the untransformed model predicts another direction (positive prefix), the prediction of the transformed model has a higher weight, since it has a better overall goodness-of-fit. However, the transformed model does not predict the hypothesized direction of the relationship between value perception and sales probability nor does it indicate any significant effect.

Table 10.23 - Nagelkerke R^2 change statistics for logistic regression

		Change Statistics			
		Untransformed Model		Lg10-transformed Model [b]	
Models [a]	Added Variable	Nagelkerke R^2	Change of Nagelkerke R^2	Nagelkerke R^2	Change of Nagelkerke R^2
1	Auction_2, Auction_3, Auction_4, Auction_5, Auction_6	0.046	0.046	0.053	0.053
2	Technology Source (Uni / Research Lab; SME; Large Firm)	0.112	0.066	0.123	0.070
3	Patent_nationality	0.133	0.021	0.145	0.022
4	Technological_complexity	0.156	0.023	0.170	0.025
5	Technological_uncertainty [c]	0.159	0.003	0.171	0.001
6	Technological_impact	0.163	0.004	0.181	0.010
7	Technology_quality	0.172	0.009	0.198	0.017
8	Value_perception	0.173	0.001	0.198	0.000

Notes:
[a] For all models n=369; dependent variable is the sales status "Sales decision" (reference category is 0=not-sold)
[b] Five hypothesized variables w ere entered as Lg10 transformed variables
[c] Measured as minimum remaining protection time

Thus, I can conclude that both models confirm the predictions of three of the five hypothesized relationships (H_{1a}, H_{3a}, and H_{4a}). Furthermore, the transformed model, which indicates better goodness-of-fit measures than the untransformed model, also predicts the right direction of H_{5a} (value perception); although this effect is not significant.

Table 10.23 displays change statistics for Nagelkerke R^2, one commonly used 'goodness of fit' measure, if individual variables are systematically entered into the untransformed or the transformed model. Table 10.23 reveals that each variable entered into the model increases the value of Nagelkerke's R^2, thus has a positive contribution to the explanatory power of the models. One exception is the variable, 'value perception,' in the transformed model, which has a zero impact. Aside from the control variables, in both models, technological complexity has the largest effect, followed by technology quality, technological impact, technological uncertainty, and value perception.

Furthermore, in both models, the same three covariates show significant effects with equal prefixes. Moreover, technologies developed by universities / research institutes seem to have a positive and significant impact on the sales probability. This effect is significant at the one per cent level in the untransformed model and significant at the five per cent level in the transformed model. Technologies developed by large firms, however, show a negative, significant effect in both models. This effect is significant at the ten per cent level in the untransformed model and significant at the one per cent level in the transformed model. The nationality of the related patent also has a positive, significant impact in both models. Thus, technologies with only US patents have a higher probability of being sold. The effect is significant at the five per cent level in the untransformed model and at the ten per cent level in the transformed model. Furthermore, in the transformed model, the third auction has a positive impact on the sales probability being significant at the five per cent level. Otherwise, none of the models predict any significant impact of the other auctions.

The transformed logistic model was further subjected to sensitivity analyses as described in Section 8.2. In order to test the robustness of the model, I computed the regression results for various sub-samples. Sub-samples were filtered for (a) 25 per cent, 50 per cent, and 57 per cent quartiles of the value ranges (tails) of the sales price and (b) compiled for different sets of auctions (that is US / EU, first / subsequent auctions in a country).

a) Robustness across quartiles of value ranges

Table 10.24 and Table 10.25 present the regression results of the sensitivity analyses of the logistic model using the log10-transformed variables across eight different value ranges of the technologies' sales prices. Table 10.24 presents the results of four sub-samples ($Model_1$ - $Model_4$) of technologies that were both sold for high prices, that is low price technologies were excluded below a certain cut-off point (right tail of the price distribution). Table 10.25 presents results from three sub-samples ($Model_6$ - $Model_8$), where high price technologies were excluded for sales prices higher than a certain cut-off point (left tail of the price distribution). Table 10.25 also presents the regression results of the mid-range valued technologies, that is sold for prices ranging in the mid 50 per cent quartile ($Model_5$).

In general, the results displayed in Table 10.24 confirm the results presented in Table 10.22. In all models, those three hypothesized variables (technological complexity, technological impact, and technology quality) were found to be significant. Moreover, the coefficients of the three variables indicate similar directions. Similar to the full model, technological complexity was significant and negative at the one per cent level for $Model_1$, $Model_2$, and $Model_3$. In $Model_4$, the coefficient was only significant at the five per cent level. While technological impact is positive and significant at the ten per cent level in the full model, it is even significant at the one per cent level in $Model_2$, $Model_3$, and $Model_4$. However, regarding technology quality, the picture is mixed. While it is positive and significant in the original model at the five per cent level and at the one per cent level in $Model_1$ and $Model_2$, it is only significant at the ten per cent level in $Model_3$ and $Model_4$. Furthermore, in contrast to the original model, in $Model_4$, the value perception is also highly significant at the one per cent level with a high positive coefficient indicating a strong positive impact of this variable on the sales probability of highly valuable technologies.

It is important to note that the coefficients for the three significant variables are also the highest in $Model_4$, which indicates the strongest impact of the sales probability for high value technologies.

With regard to the covariates, all models confirm the positive significant effect of A_3. However, the positive significant effect of universities / research institutes is only robust in $Model_1$ and $Model_2$, but disappears for high value technologies with sales prices above €77 thousand. Likewise, the pattern represents the negative significant impact of large firms. As it turns out, the positive significant effect of patent nationality is not robust for technologies that sold for more than €5 thousand.

In addition, $Model_2$, $Model_3$, and $Model_4$ show further significant effects not observed in the full model. In those three models, A_2, A_5, and A_6 show positive significant effects. In $Model_2$, for all technologies that sold for more than €15 thousand, A_4 also indicates a positive significant effect. Thus, in $Model_2$, all auctions have a positive significant impact, with A_4 having the strongest impact followed by A_3, A_6, A_5, and A_2. In $Model_3$ and $Model_4$, the strong positive effect of A_4 disappears and the coefficient changes the prefix to a negative value.

Thus, from the sensitivity analysis across the right tail of the sales price, I can conclude that the significant effects of the three hypothesized variables (technological complexity, technological impact, and technology quality) are robust across all models. In addition, for high value technologies, the value perception also has a positive, significant impact.

Table 10.24 - Logistic regression models across value ranges - right tail

Models [a]	Log Transformed Model			$Model_1$ BICL >5			$Model_2$ Above 25% Quartile (>15)			$Model_3$ Above 50% Quartile (>77)			$Model_4$ Above 75% Quartile (>190)		
	Coef.	Sig.	SE	Coef.	Sig.	SE	Coef.	Sig.	SE	Coef.	Sig.	SE	Coef.	Sig.	SE
(Constant)	-2.232	0.090 [e]	1.315	-2.544	0.057 [f]	1.336	-4.758	0.002 [g]	-4.758	-6.280	0.000 [g]	1.741	-11.118	0.000 [g]	2.412
Control Variables															
Auction_2 [b]	-0.118	0.769	0.400	0.013	0.974	0.408	1.400	0.023 [f]	0.614	1.366	0.029 [f]	0.625	2.038	0.020 [f]	0.877
Auction_3	0.793	0.050 [f]	0.405	0.905	0.028 [f]	0.412	2.406	0.000 [g]	0.617	2.215	0.000 [g]	0.627	2.077	0.028 [f]	0.944
Auction_4	1.506	0.121	0.971	1.499	0.130	0.989	2.570	0.031 [f]	1.189	-17.190	0.998	6765.541	-16.639	0.998	6682.9
Auction_5	0.253	0.593	0.473	0.372	0.437	0.479	1.914	0.004 [g]	0.661	1.742	0.011 [f]	0.681	1.731	0.090 [e]	1.022
Auction_6	0.312	0.423	0.390	0.435	0.274	0.397	1.938	0.001 [g]	0.601	1.716	0.005 [g]	0.614	1.834	0.041 [f]	0.897
Univ. / research institute [c]	1.276	0.029 [f]	0.584	1.476	0.015 [f]	0.606	1.490	0.029 [f]	0.680	1.252	0.338	1.306	-17.132	0.999	12402.7
SME	-0.435	0.162	0.312	-0.396	0.206	0.313	-0.516	0.135	0.345	-0.150	0.693	0.379	-0.476	0.368	0.529
Large firm	-0.902	0.009 [g]	0.344	-0.906	0.009 [g]	0.349	-0.772	0.046 [f]	0.387	-0.366	0.399	0.435	-0.397	0.518	0.615
Patent_nationality	1.733	0.053 [e]	0.897	1.821	0.045 [f]	0.910	1.571	0.102	0.961	1.546	0.174	1.137	0.368	0.770	1.257
Hypothesized Variables															
Technological_complexity	-1.278	0.002 [g]	0.419	-1.209	0.004 [g]	0.422	-1.317	0.005 [g]	0.464	-1.510	0.004 [g]	0.518	-1.719	0.011 [f]	0.679
Technological_uncertainty [d]	0.100	0.845	0.514	0.149	0.777	0.524	0.436	0.443	0.568	0.602	0.338	0.628	0.084	0.904	0.698
Technological_impact	0.526	0.073 [e]	0.293	0.526	0.078 [e]	0.299	0.879	0.007 [g]	0.327	1.105	0.002 [g]	0.365	1.299	0.009 [g]	0.498
Technology_quality	1.293	0.022 [f]	0.565	1.232	0.031 [f]	0.571	1.657	0.008 [g]	0.620	1.163	0.090 [e]	0.685	1.768	0.067 [e]	0.965
Value_perception	-0.032	0.910	0.280	-0.016	0.955	0.282	0.111	0.724	0.314	0.555	0.108	0.345	2.576	0.000 [g]	0.588
Observations (n)		369			365			336			302			260	
Share Sold Patents (m/n)		40.11%			39.45%			34.23%			26.82%			15.00%	
Nagelkerke R^2		0.198			0.206			0.271			0.303			0.403	
-2 Log Likelihood		438.502			429.462			358.352			280.649			151.957	
Classification Correct		69.6			70.4			71.7			75.5			86.5	
Hosmer and Lemeshow		0.829			0.379			0.373			0.899			0.778	

Notes:

[a] Dependent variable is always "sales decision" (reference category is 0=not-sold)
[b] Base variable is "first auction"
[c] Base variable is "independent inventors"
[d] Measured as minimum remaining protection time
[e] Indicates 10% significance
[f] Indicates 5% significance
[g] Indicates 1% significance

However, mixed results were found for the covariates. Aside from the positive significant effect of A_3, the effects of technology source types disappear for high value technologies, while other effects from the different auctions appear in some of the models.

It is important to note that, from $Model_1$ towards $Model_4$, three of the goodness-of-fit measures continuously improved. The Hosmer and Lemeshow measure however fluctuates, but always shows highly insignificant values, as it should. According to the goodness-of-fit measures, $Model_4$ best predicts the sales probability.

Table 10.25 presents the results of the four further models to test the robustness of the full model. One model ($Model_5$) predicts the sales probability of technologies with moderate sales prices excluding the extreme low and high price technologies, that is in the mid 50 per cent quartile range with sales prices higher than €15 thousand but lower than €190 thousand. Three further models ($Model_6$, $Model_7$, and $Model_8$) are computed for the left tail of the sales price distribution, that is sub-samples in the lower 75 per cent / 50 per cent / 25 per cent quartiles systematically excluding the high value technologies.

Systematically excluding the high value technologies, Table 10.25 shows that the coefficients for technological complexity are negative in all models. Similar to the full model, the effect is significant in $Model_6$ and $Model_8$, but rather at the ten per cent level. $Model_5$ and $Model_7$ do not show a significant impact of technological complexity. The positive effect of technological impact found in the full model is also present in $Model_5$. Compared to the full model, the significance level decreased from five per cent to ten per cent. The positive significant effect of technology quality was found for all models, except $Model_8$. However, these levels vary. Similar to the full model, $Model_5$ and $Model_7$ show significance levels of five per cent. In $Model_6$, the effect is only significant at the ten per cent level. The prefix changes in $Model_7$ and $Model_8$, causing a robust effect in terms of technology quality. Across all models, the prefix is similarly positive and it only becomes insignificant in $Model_8$. The effect of value perception is negative in all models and significant in all models except $Model_8$. The negative prefix is in line with the hypothesized direction but in the opposite direction of the full model's prefix. Except technological complexity, none of the hypothesized variables show a significant effect in $Model_8$.

With regard to the covariates, the following picture emerges from Table 10.25. Only the negative significant impact of large firms appears robust across all models. The positive effect of patent nationality disappears in $Model_7$, although the prefix remains similarly positive across all models. The positive effect of A_3 disappears in $Model_6$ and $Model_7$. Furthermore,

$Model_5$ and $Model_6$ show positive significant effects for A_4. $Model_3$ and $Model_7$ show positive and negative effects for A_5 and A_6. Thus, when systematically excluding the high value technologies from Table 10.25, one can conclude that only the effect of technology quality is stable, although not for the technologies with sales prices lower than €15 thousand. Concerning the covariates, only the negative impact of large firms on the sales probability was found to be stable, although in three of the four models, the positive effect of the patent nationality also remained stable.

To conclude, the results of the sensitivity analyses presented in Table 10.24 and Table 10.25 show that the logistic model better predicts the sales probability of technologies with high sales prices. $Model_4$ has the highest Nagelkerke R^2 of 0.403 with eight significant effects. Furthermore, the model seems to fairly predict the sales probability of technologies with moderate sales prices higher than €15 thousand ($Model_2$) and lower than €190 thousand ($Model_5$). Both models showed 10 significant effects. The model however seems less robust for technologies with low sales prices (that is below €15 thousand). $Model_8$ shows only one significant effect among the hypothesized variables.

b) Robustness across auctions

In further sensitivity analyses, I investigated how the regression model behaves when computed for four auction sub-sets. Owing to too small case numbers and thus overfitting problems, the model could hardly be computed for individual auctions. Table 10.26 reveals the results from sensitivity analyses for two pairs of sub-samples. $Model_9$ and $Model_{10}$ only represent results from analyses of the US auctions as well as the European auctions in Germany and the UK. $Model_{11}$ and $Model_{12}$ present the results of first auctions in each country (US, Germany, and the UK) compared to the subsequent auctions. In my dataset, these results are only available for the US.

Regarding the full logistic model, the models reveal similar effects to those revealed in Table 10.22. Among the hypothesized variables, the positive effect of technology quality is the most robust. Across all models, the coefficient is strongly positive. Similar to the full model, the effect of technology quality is significant at the five per cent level for all models except $Model_{11}$. In this model, the effect is significant at the ten per cent level. In addition, similar to the full model, the effect of technology complexity has a negative prefix in all models. Similar to the full model, the effect is significant at the one per cent level in $Model_{11}$, but only significant at the five per cent level in $Model_9$ and $Model_{10}$.

Table 10.25 - Logistic regression models across value ranges - left tail

Models [a]	Model$_4$ Above 75% Quartile (>190)			Model$_5$ Middle 50% Quartile (>15 & <190)			Model$_6$ Lower 75% Quartile (<190)			Model$_7$ Lower 50% Quartile (<77)			Model$_8$ Lower 25% Quartile (<15)		
	Coef.	Sig.	SE	Coef.	Sig.	SE	Coef.	Sig.	SE	Coef.	Sig.	SE	Coef.	Sig.	SE
(Constant)	-11.118	0.000 [g]	2.412	-4.011	0.093 *	2.384	-0.606	0.738	1.810	1.721	0.409	2.082	0.479	0.847	2.475
Control Variables															
Auction_2 [b]	2.038	0.020 [f]	0.877	1.055	0.221	0.863	-0.946	0.047 [f]	0.477	-1.756	0.003 [g]	0.582	-2.320	0.001 [g]	0.704
Auction_3	2.077	0.028 [f]	0.944	2.832	0.001 [g]	0.834	0.455	0.300	0.439	-0.620	0.264	0.556	-1.507	0.028 [f]	0.687
Auction_4	-16.639	0.998	6682.9	4.682	0.007 [g]	1.728	2.348	0.074 [e]	1.313	1.654	0.265	1.484	1.533	0.361	1.678
Auction_5	1.731	0.090 [e]	1.022	2.281	0.010 [g]	0.886	-0.186	0.733	0.544	-1.729	0.046 [f]	0.867	-19.842	0.997	6143.0
Auction_6	1.834	0.041 [f]	0.897	2.473	0.003 [g]	0.827	0.026	0.953	0.432	-1.052	0.054 [e]	0.546	-2.801	0.001 [g]	0.858
Univ. / research institute [c]	-17.132	0.999	12402.7	1.403	0.045 [f]	0.701	1.369	0.022 [f]	0.596	1.115	0.081 [e]	0.639	0.126	0.898	0.987
SME	-0.476	0.368	0.529	-0.542	0.204	0.426	-0.450	0.213	0.361	-1.130	0.022 [f]	0.493	-0.571	0.375	0.644
Large firm	-0.397	0.518	0.615	-0.925	0.056 [e]	0.484	-1.057	0.009 [g]	0.404	-1.690	0.001 [g]	0.529	-1.722	0.015 [f]	0.710
Patent_nationality	0.368	0.770	1.257	2.656	0.075 [e]	1.493	2.567	0.045 [f]	1.282	2.242	0.132	1.487	3.402	0.056 [e]	1.781
Hypothesized Variables															
Technological_complexity	-1.719	0.011 [f]	0.679	-0.857	0.161	0.611	-0.901	0.083 [e]	0.520	-0.846	0.205	0.667	-1.550	0.089 [e]	0.912
Technological_uncertainty [d]	0.084	0.904	0.698	1.549	0.146	1.066	0.368	0.626	0.755	-0.855	0.299	0.823	-1.697	0.105	1.047
Technological_impact	1.299	0.009 [g]	0.498	0.831	0.049 [f]	0.423	0.185	0.604	0.356	-0.615	0.171	0.449	-0.827	0.163	0.593
Technology_quality	1.768	0.067 [e]	0.965	1.680	0.026 [f]	0.753	1.150	0.080 [e]	0.656	1.575	0.068 [f]	0.863	0.993	0.421	1.233
Value_perception	2.576	0.000 [g]	0.588	-1.535	0.001 [g]	0.451	-1.143	0.001 [g]	0.355	-1.164	0.007 [g]	0.429	-0.621	0.232	0.519
Observations (n)	260			297			330			288			252		
Share Sold Patents (m/n)	15.00%			25.59%			33.03%			23.26%			12.30%		
Nagelkerke R^2	0.403			0.355			0.271			0.346			0.309		
-2 Log Likelihood	151.957			255.858			347.304			237.434			143.347		
Classification Correct	86.5			79.1			75.5			83.0			38.5		
Hosmer and Lemeshow	0.778			0.44			0.487			0.086			0.909		

Notes:

[a] Dependent variable is always "sales decision" (reference category is 0=not-sold); for all models BICL is larger than €15 thousand
[b] Base variable is "first auction"
[c] Base variable is "independent inventors"
[d] Measured as minimum remaining protection time
[e] Indicates 10% significance
[f] Indicates 5% significance
[g] Indicates 1% significance

However, the positive effect of technological impact appears to be less robust. Although this effect has a positive prefix in all models and shows an even higher significance level of one per cent compared to the full model in $Model_9$ and $Model_{12}$, the effect is not significant in $Model_{10}$ and $Model_{11}$.

With regard to the covariates, Table 10.26 reveals the following picture:[27] The positive effect of universities / research institutes remains similarly significant at the five per cent level in $Model_{10}$ and $Model_{11}$. In $Model_9$ and $Model_{12}$, the effect remains in the same direction (positive), but is not significant. The strong negative effect of large firms disappears in $Model_{10}$ and $Model_{11}$ and shows up only in $Model_9$ and $Model_{12}$. While even in $Model_{10}$ and $Model_{11}$ the prefix changes its direction, the effect remains negative with a significance level of one per cent in $Model_{12}$ and five per cent in $Model_9$. The positive impact of A_3 disappears in the two models where it can be measured ($Model_9$ and $Model_{12}$).

Thus, I can conclude from Table 10.26 that technology quality has a robust positive effect on the sales probability across all auction sub-samples, regardless of whether the auction took place in the US or Europe, or whether it was the first or a subsequent auction. The effect of technological complexity remains relatively robust and negative for all sub-samples, although it is not significant for the subsequent auctions.

The effect of technological impact is robust only for the US and subsequent auctions, but is not significant for the EU and first auctions, although always positive. With regard to the covariates, I can conclude from Table 10.26 that the technologies developed by large firms have a negative impact on the sales probability in the US only, while the effect could not be shown for those auctions that took place in Europe. For these auctions, the effect appears rather positive, but not significant. In contrast, the positive effect of technologies developed by universities / research institutes could only be proven for EU auctions and not for those in the US, although in those auctions at least the direction of the prefix remains similar. The positive effect of A_3 does not seem robust. It is positive for those models, in which it was measured, but not significant.

Table 10.26 - Logistic regression models for auction sub-samples

Models [a]	Log Transformed Model			Model$_9$ US Auctions (A_1, A_2, A_3, A_6)			Model$_{10}$ EU Auctions (A_4, A_5)			Model$_{11}$ First Auctions (A_1, A_4, A_5)			Model$_{12}$ Subsequent Auctions (A_2, A_3, A_6)		
	Coef.	Sig.	SE	Coef.	Sig.	SE	Coef.	Sig.	SE	Coef.	Sig.	SE	Coef.	Sig.	SE
(Constant)	-2.232	0.090 [e]	1.315	-3.305	0.015	1.353	-1.541	0.451	2.046	-3.806	0.052	1.962	-1.611	0.217	1.306
Control Variables															
Auction_2 [b]	-0.118	0.769	0.400	1.393	0.024	0.618							-0.483	0.240	0.411
Auction_3	0.793	0.050 [f]	0.405	2.288	0.000	0.619							0.431	0.270	0.391
Auction_4	1.506	0.121	0.971				-1.062	0.215	0.857	1.267	0.167	0.917			
Auction_5	0.253	0.593	0.473							2.163	0.003	0.736			
Auction_6	0.312	0.423	0.390	1.820	0.003	0.604									
Univ. / research institute [c]	1.276	0.029 [f]	0.584	1.000	0.416	1.230	2.000	0.032 [f]	0.934	1.920	0.030 [f]	0.884	0.979	0.427	1.231
SME	-0.435	0.162	0.312	-0.456	0.234	0.384	-0.294	0.739	0.882	-0.357	0.647	0.779	-0.523	0.184	0.394
Large firm	-0.902	0.009 [g]	0.344	-0.951	0.028 [f]	0.434	0.459	0.614	0.908	0.258	0.738	0.773	-1.036	0.024 [g]	0.458
Patent_nationality	1.733	0.053 [e]	0.897												
Hypothesized Variables															
Technological_complexity	-1.278	0.002 [g]	0.419	-1.258	0.025 [f]	0.561	-1.899	0.040 [f]	0.925	-2.381	0.005 [g]	0.839	-0.914	0.114	0.578
Technological_uncertainty [d]	0.100	0.845	0.514	0.792	0.288	0.745	-0.066	0.940	0.879	-0.382	0.626	0.782	1.079	0.182	0.808
Technological_impact	0.526	0.073 [e]	0.293	1.049	0.006 [g]	0.381	0.236	0.752	0.745	0.461	0.486	0.661	1.038	0.008 [g]	0.394
Technology_quality	1.293	0.022 [f]	0.565	1.168	0.100 [e]	0.711	3.148	0.020 [f]	1.356	2.499	0.040 [f]	1.216	1.322	0.070 [f]	0.730
Value_perception	-0.032	0.910	0.280	0.049	0.885	0.340	0.237	0.787	0.875	0.518	0.431	0.658	-0.030	0.932	0.358
Observations (n)		369			242			94			140			196	
Share Sold Patents (m/n)		40.11%			55.13%			44.62%			30.84%			71.93%	
Nagelkerke R^2		0.198			0.255			0.356			0.364			0.168	
-2 Log Likelihood		438.502			265.294			88.823			114.095			240.353	
Classification Correct		69.6			70.2			79.8			83.6			66.3	
Hosmer and Lemeshow		0.829			0.64			0.623			0.035			0.675	

Notes:

[a] Dependent variable is always "sales decision" (reference category is 0=not-sold); for all models BICL is larger than €15 thousand
[b] Base variable is "first auction"
[c] Base variable is "independent inventors"
[d] Measured as minimum remaining protection time
[e] Indicates 10% significance
[f] Indicates 5% significance
[g] Indicates 1% significance

10.3.2 Determinants of Sales Price

Following the analyses of determinants for the sales probability, I present the results of multivariate log-linear OLS regressions for which I tested the impact of the five technology properties on the sales price (see Chapter 12 for tests of assumptions). The sales price is a metric-scaled variable ranging from a minimum of €1.79 thousand to a maximum of €3.65 million (see for example, Table 10.7). Owing to skewed distributions of most of the independent variables as well as the dependent variable, the metric-scaled independent and dependent variables were transformed. Various transformations were tested (for example, square root, x^2, x^3, and inverse), but the best results were found for the log10-transformed model, to which I refer in the following section. Following the presentation of the results for the 'full' untransformed and log10-transformed models, I present results from various sensitivity analyses. Obviously, the sample for the multivariate OLS regressions represent a sub-sample from the dataset as it refers to sold technologies only (n=156).

Table 10.27 displays the regression results of the multivariate OLS models. The untransformed model (left) displays the results of all the variables that were entered in their original form. The Log10-transformed model (right) displays the results where the five independent, metric-scaled variables and the dependent variable were entered as log10-transformed variables.

The untransformed model has a higher adjusted R^2 value than the transformed model. However from Chapter 8, we know that the untransformed model is subject to heteroscedasticity, which biases the properties of OLS estimators. These are no longer efficient, that is the t and F tests can be misleading although their significance values are not affected by heteroscedasticity (Gujarati, 2003). However, comparing the untransformed and transformed models, Table 10.27 displays rather consistent results. With an average of 0.775, the adjusted R^2 value of both models must be considered of high explanatory power.

First of all, the results of both models displayed in Table 10.27 show a high number of one per cent significance effects among the hypothesized variables as well as the covariates. Across all models, three of the hypothesized variables are highly significant at the one per cent level. The sellers' value perception has the strongest, positive effect on the sales price.[28] Furthermore, both technological complexity and technological impact show highly significant effects on the sales price at the one per cent level. Technological complexity (technological impact) has a negative (positive) impact on the sales price across all models.

Table 10.27 - OLS regression models

Models [a]	Full OLS Models for all Auctions					
	Untransformed Model [b]			Lg10-transformed Model [c]		
	Stand. Coef.	Sig.	SE	Stand. Coef.	Sig.	SE
(Constant)		0.000 [i]	90.455		0.001 [i]	0.300
Control Variables						
Auction_2 [d]	0.235	0.000 [i]	44.362	0.553	0.000 [i]	0.102
Auction_3	0.173	0.000 [i]	46.256	0.473	0.000 [i]	0.107
Auction_4	0.251	0.001 [i]	86.163	0.232	0.027 [h]	0.201
Auction_5	0.317	0.000 [i]	53.476	0.499	0.000 [i]	0.123
Auction_6	0.052	0.241	46.457	0.474	0.000 [i]	0.106
Univ. / research institutes [e]	-0.036	0.340	62.607	-0.057	0.308	0.148
SME	0.155	0.000 [i]	36.363	0.115	0.045 [h]	0.086
Large_firm	0.133	0.003 [i]	38.877	0.091	0.164	0.092
Patent_nationality	0.167	0.011 [h]	73.390	0.158	0.096 [g]	0.172
Hypothesized Variables						
Technological_complexity	-0.640	0.000 [i]	3.270	-0.279	0.000 [i]	0.101
Technological_uncertainty [f]	-0.048	0.201	3.434	-0.049	0.359	0.126
Technological_impact	0.278	0.000 [i]	1.472	0.289	0.000 [i]	0.076
Technology_quality	0.048	0.172	11.545	0.016	0.758	0.153
Value_perception	0.956	0.000 [i]	0.023	0.535	0.000 [i]	0.073
Adjusted R^2		0.848			0.701	

Notes:
[a] For all models n=156
[b] Dependent variable is the Log10-transformed sales price
[c] Five hypothesized variables w ere entered as Log10 transformed variables
[d] Base variable is the first auction
[e] Base variable is "independent inventors"
[f] Measured as minimum remaining protection time
[g] Indicates 10% significance
[h] Indicates 5% significance
[i] Indicates 1% significance

With regard to the covariates, A_2, A_3, A_4, A_5, and technologies developed by SME as well as patent nationality show a significant effect in both models. A further covariate (large firm) is significant at the one per cent level in the untransformed model and at A_6 also in the transformed model. Consistent in both models, technologies developed by universities / research institutes have a negative, although insignificant effect. Although technologies developed by universities / research institutes do not show a significant impact in any of the models, the negative coefficients in both models indicate that their technologies would sell for lower prices than those developed by independent inventors. Among the auction number dummy variables, the ranking of the effect size varies between the untransformed and transformed models. In the untransformed model, A_5 has the strongest impact compared to A_1, followed by A_4, A_2, and A_3. In the transformed model, A_2 has the highest impact compared to A_1, followed by A_5, A_3, A_6, and A_4. Among the different sources of technology, the ranking is consistent across both models. In both models, technologies developed by SME have the highest positive impact compared to individual inventors, followed by large firms and universities / research institutes. Nevertheless, only SME have a significant effect in both models.

Table 10.28 - Change statistics for OLS regression models

Model	Added Variable	Change Statistics			
		Untransformed Model [a]		Log-transformed Model [b]	
		Adjusted R^2	Change of Adjusted R^2	Adjusted R^2	Change of Adjusted R^2
1	Auction_2, Auction_3, Auction_4, Auction_5, Auction_6	0.046	0.046	0.355	0.355
2	Technology Source (Uni / research lab; SME; large firm)	0.057	0.011	0.364	0.009
3	Patent_nationality	0.057	0.000	0.366	0.002
4	Technological_complexity	0.051	-0.006	0.368	0.002
5	Technological_uncertainty [c]	0.061	0.010	0.389	0.021
6	Technological_impact	0.183	0.122	0.461	0.072
7	Technology_quality	0.178	-0.005	0.457	-0.004
8	Value_perception	0.848	0.670	0.701	0.244

Notes:
[a] Dependent variable is the sales price
[b] Five hypothesized variables were entered as Log10 transformed variables; dependent variable is the Log10-transformed sales price
[c] Measured as minimum remaining protection time

Table 10.28 displays *change statistics* if individual variables are systematically entered into the untransformed or transformed model. Table 10.28 reveals that most variables contribute to an increasing R^2. Among the hypothesized variables, the largest contribution comprises the seller's value perception. This variable contributes 67.0 per cent to the untransformed model's overall adjusted R^2 of 0.848. In the transformed model, it contributes 24.4 per cent. Ranked second, technological impact contributed 12.2 per cent to the untransformed model and 7.2 per cent to the transformed model. Among the hypothesized variables, in both models technological uncertainty makes a further, although small, positive contribution. While technological complexity and technology quality show no contributions in the transformed model, they show a negative contribution in the untransformed model (which may indicate that the model is unreliable). Among the *covariates*, the auction dummies have the highest contribution of 4.6 per cent (35.5 per cent) in the untransformed (transformed) model, followed by technology source type dummies with a contribution of 1.1 per cent (0.9 per cent) in the untransformed (transformed) model. In both models, the patent nationality makes (almost) no contribution to the models' goodness-of-fit of the models.

In both models presented in Table 10.27, the coefficients of four of the five hypothesized variables have prefixes in the directions predicted in the hypotheses. Both models correctly predicted the positive effects of technological impact and value perception on the sales price at a one per cent significance level. Comparing the size of both effects, Table 10.27 shows that the value perception effect is stronger in both models than the effect of technological impact. Furthermore, in both models, the prefixes of technological uncertainty (negative) and of technology quality (positive) are similar to the hypothesized directions. Furthermore, both models show a highly significant effect of technological complexity on the one per cent level. However, the direction of this effect is opposite to the hypothesized direction.

Thus, I can conclude that both models clearly confirm the predictions of two hypotheses (H_{3b} and H_{5b}). Furthermore, both models consistently show the effect directions as predicted by H_{2b} and H_{4b}. However, these effects appear to be insignificant. Furthermore, both models also show a highly significant effect of technological complexity, however, with an opposite prefix than predicted by H_{1b}. In general, both models show high goodness-of-fit measures with an average adjusted R^2 of 0.775.

The effect of H_{2b} becomes clearer if technological uncertainty is further tested. Therefore, the sample is split into two equal sub-samples that represent cohorts of young or old technologies, each with 50% of the number of cases (that is a cut-off point with an average remaining protection time of 12.17 years).[29]

Table 10.29 reveals the log-linear OLS regression results of the six models. For each cohort, three models were computed where technological uncertainty was operationalized differently according to the average, minimum, and maximum remaining protection time of the related patents (see Section 6.1). For both cohorts, the table shows strong significant effects of technological uncertainty on the one per cent level in four of the six models. However, the models indicate that the relationship between technological uncertainty and sales prices might not be linear but rather u-shaped. In the cohort of young technologies, technological uncertainty has a negative, significant impact on the sales price. In the cohort of old technologies, technological uncertainty, however, has a positive, significant effect on the sales price.[30]

Similar to the full models displayed in Table 10.27, all the models in Table 10.29 confirm the significant positive effect of value perception. Moreover, all the models show a negative effect of technological complexity, although this effect is only significant in four models. Similarly, the models reveal that the positive effect of technological impact is also always positive, but only significant for old technologies. Furthermore, the models displayed in Table 10.29 reveal another interesting effect. In five of the six models, the effect of technology quality is significant. In the models for old technologies, the effect is positive and significant at the one per cent level as predicted by H_{4b}. However, in the three models for the sub-sample of young technologies, the effect is negative, significant at a ten per cent level in $Model_1$, five per cent level in $Model_2$, and slightly above the 10% level in $Model_3$. Thus, similar to technological uncertainty, the models indicate that the relationship on technology quality and the sales price might be u-shaped rather than linear.

In Table 10.29, similar effects of the covariates are found for the auctions' and patents' nationalities compared to those revealed in the models displayed in Table 10.27. However, old technologies developed by universities / research institutes also show a negative significant effect. In two of the three models, the impact of technologies developed by SME is slightly above the ten per cent significance level, thus it is less robust. For young technologies, however, the effect of patent nationality is not significant in any of the three models.

Similar to the logistic regression model, the transformed log-linear OLS regression model was further subjected to sensitivity analyses as described in Section 4.2. In order to test the robustness of the model, I computed the regression results of the sub-samples. Similar to the sensitivity analysis conducted in the previous chapter, sub-samples were filtered according to (a) 25 per cent, 50 per cent, and 75 per cent quartiles of the sales price's value ranges and (b) compiled for sub-sets of auctions (that is US / EU; first / subsequent auctions in a country).

a) Robustness across quartiles of value ranges

Across eight different value ranges of the technologies' sales prices, Table 10.30 and Table 10.31 present multivariate OLS regression results using the log10-transformed dependent and independent variables. Table 10.30 presents the results of four sub-samples ($Model_1$ - $Model_4$) of technologies that were sold at high prices, that is low price technologies below certain cut-off points were excluded (right tail of the sales price distribution). Table 10.31 presents results from three sub-samples ($Model_6$ – $Model_8$) where high price technologies were excluded if their sales prices were higher than certain cut-off points (left tail of the sales price distribution). Table 10.31 also presents the regression results of the mid-range valued technologies, that is sold for prices ranging in the mid 50 per cent quartile ($Model_5$).

Table 10.29 - Log-linear OLS regression models for sub-samples of young / old technologies

Models [a]	Log Transformed Model All Auctions			Model$_1$ Top 50% ARPT (young techn.)			Model$_2$ Top 50% ARPT (young techn.)			Model$_3$ Top 50% ARPT (young techn.)			Model$_4$ Low 50% ARPT (old techn.)			Model$_5$ Low 50% ARPT (old techn.)			Model$_6$ Low 50% ARPT (old techn.)		
	Stand. Coef.	Sig.	SE	Stand. Coef.	Sig.	SE	Stand. Coef.	Sig.	SE	Stand. Coef.	Sig.	SE	Stand. Coef.	Sig.	SE	Stand. Coef.	Sig.	SE	Stand. Coef.	Sig.	SE
(Constant)		0,001 [i]	0,300		0,109	1,213		0,001 [i]	0,684		0,757	1,210		0,000 [i]	0,306		0,000 [i]	0,381		0,000 [i]	0,356
Control Variables																					
Auction_2 [b]	0,553	0,000 [i]	0,102	0,764	0,000 [i]	0,126	0,778	0,000 [i]	0,114	0,776	0,000 [i]	0,129	0,370	0,000 [i]	0,107	0,394	0,000 [i]	0,140	0,392	0,000 [i]	0,116
Auction_3	0,473	0,000 [i]	0,107	0,815	0,000 [i]	0,126	0,827	0,000 [i]	0,114	0,821	0,000 [i]	0,129	0,131	0,020 [h]	0,115	0,189	0,009 [i]	0,148	0,134	0,030 [h]	0,127
Auction_4	0,232	0,027 [h]	0,201	0,149	0,333	0,241	0,157	0,255	0,216	0,183	0,251	0,250	0,256	0,005 [i]	0,209	0,289	0,013 [h]	0,271	0,287	0,004 [i]	0,227
Auction_5	0,499	0,000 [i]	0,123	0,550	0,000 [i]	0,149	0,553	0,000 [i]	0,134	0,565	0,000 [i]	0,153	0,387	0,000 [i]	0,129	0,455	0,000 [i]	0,165	0,421	0,000 [i]	0,139
Auction_6	0,474	0,000 [i]	0,106	0,653	0,000 [i]	0,131	0,659	0,000 [i]	0,118	0,670	0,000 [i]	0,135	0,227	0,001 [i]	0,111	0,301	0,000 [i]	0,142	0,258	0,000 [i]	0,120
Univ. / research institute [c]	-0,057	0,308	0,148	0,014	0,845	0,155	0,025	0,715	0,140	0,003	0,974	0,160	-0,173	0,002 [i]	0,181	-0,159	0,022 [h]	0,235	-0,174	0,003 [i]	0,197
SME	0,115	0,045 [h]	0,086	0,161	0,041 [h]	0,098	0,173	0,016 [h]	0,089	0,170	0,037 [h]	0,102	0,109	0,038 [h]	0,092	0,093	0,171	0,120	0,087	0,126	0,100
Large_firm	0,091	0,164	0,092	0,055	0,523	0,115	0,053	0,500	0,104	0,073	0,413	0,119	-0,036	0,568	0,095	-0,009	0,918	0,123	-0,022	0,751	0,103
Patent_nationality	0,158	0,096 [g]	0,172	-0,096	0,468	0,199	-0,091	0,448	0,181	-0,085	0,533	0,206	0,361	0,000 [i]	0,177	0,359	0,001 [i]	0,230	0,370	0,000 [i]	0,193
Hypothesized Variables																					
Technological_complexity	-0,279	0,000 [i]	0,101	-0,014	0,874	0,139	-0,142	0,091 [g]	0,129	-0,011	0,917	0,164	-0,378	0,000 [i]	0,091	-0,371	0,000 [i]	0,126	-0,485	0,000 [i]	0,106
Technological_uncertainty [d]				-0,140	0,050 [h]	0,989							0,264	0,000 [i]	0,159						
Technological_uncertainty [e]	-0,049	0,359	0,126				-0,309	0,000 [i]	0,532							0,042	0,514	0,134			
Technological_uncertainty [f]										-0,050	0,520	0,975							0,230	0,000 [i]	0,200
Technological_impact	0,289	0,000 [i]	0,076	0,055	0,513	0,120	0,024	0,735	0,104	0,107	0,198	0,119	0,346	0,000 [i]	0,077	0,280	0,000 [i]	0,100	0,333	0,000 [i]	0,084
Technology_quality	0,016	0,758	0,153	-0,120	0,085 [g]	0,171	-0,134	0,036 [h]	0,156	-0,115	0,107	0,176	0,192	0,000 [i]	0,167	0,202	0,003 [i]	0,218	0,188	0,001 [i]	0,183
Value_perception	0,535	0,000 [i]	0,073	0,408	0,000 [i]	0,087	0,410	0,000 [i]	0,079	0,400	0,000 [i]	0,089	0,651	0,000 [i]	0,074	0,631	0,000 [i]	0,096	0,656	0,000	0,081
N		156			76			76			76			76			76			76	
Adjusted R^2		0,701			0,726			0,779			0,711			0,871			0,796			0,846	

Notes:

[a] Dependent variable is the Log10 transformed sales price
[b] Base variable for is "first auction"
[c] Base variable is "independent inventors"
[d] Measured as average remaining protection time
[e] Measured as minimum remaining protection time
[f] Measured as maximum remaining protection time
[g] Indicates 10% significance
[h] Indicates 5% significance
[i] Indicates 1% significance

In general, the results displayed in Table 10.30 confirm the results of the full transformed log-linear model presented in Table 10.27. In all models, technological complexity and value perception have robust effects. Similar to the full model, technological complexity is always negatively significant at the one per cent level in $Model_1$, although it is only significant at the five per cent level in $Model_2$, $Model_3$, and $Model_4$. The significant positive effect of value perception at the one per cent level is found in all four models. The effect of technological impact is also found to be positive in all four models, being significant at the five per cent level in $Model_1$ and at the ten per cent level in $Model_3$. In $Model_2$, the significance level is slightly above ten per cent. Furthermore, the prefixes of technological uncertainty (negative) and technology quality (positive) are robust and similar to the full model in all four models, although the effects are not significant.

With regard to the covariates, only $Model_1$ shows significant effects of the auctions similar to the full model. In $Model_1$, A_3 has the strongest significant positive impact, followed by A_6, A_2, and A_5. In contrast to the full model, A_4 has no significant effect. Among the other models, A_4 only has a negative significant effect in $Model_2$, thus a different prefix than in the full model. The positive significant effect of technologies developed by SME found in the full model is also present in $Model_1$ and $Model_2$, but disappears in the other models. Patent nationality is only significant in $Model_3$, but with a different prefix than in the full model. The four models show an average adjusted R^2 of 0.597, that is 14.9 per cent lower than that of the full model.

Thus, from the sensitivity analysis of the sales prices' right tail, I can conclude that the effects of the hypothesized variables that were found in the full model were robust in the four models based on sub-samples of the right tail of the sales price distribution.

While the effects of technological complexity and value perception must be considered strongly robust, the effect of technological impact can be considered robust. However, mixed results were found with regard to covariates. While most of the covariates had an impact in the full model, they do not show effects in the other models, except $Model_1$. Thus, their effects disappear for transactions of high value technologies. $Model_2$ has the highest adjust R^2 value of all models, with only four significant effects.

Table 10.30 - Log-linear OLS regression models across value ranges - right tail

Models [a]	Log Transformed Model			Model$_1$			Model$_2$			Model$_3$			Model$_4$		
	All Auctions			(BICL>5)			Above 25% Quartile (BICL>15)			Above 50% Quartile (BICL>77)			Above 75% Quartile (BICL>190)		
	Stand. Coef.	Sig.	SE	Stand. Coef.	Sig.	SE	Stand. Coef.	Sig.	SE	Stand. Coef.	Sig.	SE	Stand. Coef.	Sig.	SE
(Constant)		0.001 [g]	0.300		0.879	0.400		0.010 [f]	0.299		0.000 [g]	0.353		0.004 [f]	0.543
Control Variables															
Auction 2 [b]	0.553	0.000 [g]	0.102	0.464	0.000 [g]	0.130	-0.004	0.969	0.144	0.035	0.797	0.135	-0.152	0.618	0.212
Auction_3	0.473	0.000 [g]	0.107	0.529	0.000 [g]	0.120	-0.121	0.315	0.143	-0.102	0.527	0.136	-0.147	0.674	0.263
Auction_4	0.232	0.027 [f]	0.201	-0.056	0.709	0.281	-0.719	0.000 [g]	0.247						
Auction_5	0.499	0.000 [g]	0.123	0.411	0.000 [g]	0.149	-0.063	0.488	0.150	-0.038	0.763	0.141	-0.223	0.383	0.231
Auction_6	0.474	0.000 [g]	0.106	0.511	0.000 [g]	0.119	-0.188	0.127	0.138	-0.217	0.191	0.131	-0.260	0.453	0.218
Univ. / research institute [c]	-0.057	0.308	0.148	-0.017	0.832	0.149	0.000	0.995	0.107	-0.075	0.292	0.175			
SME	0.115	0.045 [f]	0.086	0.129	0.049 [f]	0.095	0.149	0.008 [f]	0.064	0.033	0.701	0.068	0.218	0.186	0.105
Large_firm	0.091	0.164	0.092	0.094	0.196	0.109	0.094	0.104	0.074	0.016	0.864	0.083	0.186	0.363	0.146
Patent_nationality	0.158	0.096 [e]	0.172	-0.177	0.205	0.266	-0.111	0.379	0.190	-0.126	0.097 [e]	0.260	-0.217	0.216	0.332
Hypothesized Variables															
Technological_complexity	-0.279	0.000 [g]	0.101	-0.257	0.000 [g]	0.132	-0.164	0.004 [f]	0.094	-0.252	0.004 [f]	0.099	-0.482	0.017 [f]	0.176
Technological_uncertainty [d]	-0.049	0.359	0.126	-0.075	0.253	0.172	-0.073	0.192	0.115	-0.075	0.378	0.116	-0.111	0.514	0.142
Technological_impact	0.289	0.000 [g]	0.076	0.222	0.003 [f]	0.089	0.093	0.125	0.059	0.170	0.068 [e]	0.065	0.209	0.268	0.106
Technology_quality	0.016	0.758	0.153	0.024	0.699	0.172	0.037	0.490	0.116	0.036	0.637	0.123	0.029	0.847	0.207
Value_perception	0.535	0.000 [g]	0.073	0.433	0.000 [g]	0.091	0.620	0.000 [g]	0.061	0.770	0.000 [g]	0.064	0.589	0.001 [g]	0.136
N	156			152			117			82			39		
Adjusted R^2	0.701			0.596			0.771			0.638			0.382		

Notes:

- [a] Dependent variable is the Log10 transformed sales price
- [b] Base variable for is the first auction
- [c] Base variable is "independent inventors"
- [d] Measured as minimum remaining protection time
- [e] Indicates 10% significance
- [f] Indicates 5% significance
- [g] Indicates 1% significance

Table 10.31 shows four further models to test the robustness of the full model for the left tail of the sales price distribution. One model ($Model_5$) is computed so that it predicts the sales price of technologies with moderate values excluding the extreme low and high price technologies, that is in the mid 50 per cent quartile range, which is higher than €15 thousand but lower than €190 thousand. Three further models ($Model_6$, $Model_7$, and $Model_8$) are computed for the left tail of the sales price distribution, that is sub-samples in the lower 75 per cent / 50 per cent / 25 per cent quartiles, systematically excluding the high value technologies.

While the effects that were found in the full model are confirmed in the models, they appear less robust in general. The effect of technological complexity is found only significant in $Model_7$ where the prefix is different from the full model. In $Model_7$, the effect of value perception is found to be significant, but also with a different prefix than in the full model. In $Model_6$ (and almost in $Model_7$), technological impact has a positive significant impact similar to its effect in the full model.

With regard to the *covariates*, in $Model_6$ and $Model_7$, all auctions (except for A_4 in $Model_7$) show significant positive effects. In $Model_8$, the effects disappear except for A_4 and A_6. In $Model_6$ and $Model_7$, technologies developed by universities / research institute have a significant positive effect. In $Model_5$ and $Model_6$, SME also have a positive effect. In $Model_5$ ($Model_8$), large firms have a positive (negative) significant effect. Moreover, in $Model_8$, the patent nationality has a positive significant effect similar to the full model. These three models show an average adjusted R^2 of 0.364, that is 48.1 per cent lower than that of the full model.

For technologies with mid-range values (that is in the 50 per cent quartile), $Model_5$ shows similar prefixes for the hypothesized variables (except for technology quality) as in the full model. However, only the effect of value perception appears to be significant at the one per cent level. With regard to the covariates, A_4 has a negative significant effect at the five per cent level. Among the different technology sources, SME rank the highest with a positive significant effect, followed by large firms.

Table 10.31 - Log-linear OLS regression models across value ranges - left tail

Models [a]	Log Transformed Model All Auctions			$Model_5$ Middle 50% Quartile (15<BICL<190)			$Model_6$ Lower 75% Quartile (BICL<190)			$Model_7$ Lower 50% Quartile (BICL<77)			$Model_8$ Lower 25% Quartile (BICL<15)		
	Stand. Coef.	**Sig.**	**SE**	**Stand. Coef.**	**Sig.**	**SE**	**Stand. Coef.**	**Sig.**	**SE**	**Stand. Coef.**	**Sig.**	**SE**	**Stand. Coef.**	**Sig.**	**SE**
(Constant)		0.001 [g]	0.300		0.004 [f]	0.349		0.443	0.440		0.028 [f]	0.416		0.695	0.491
Control Variables															
Auction 2 [b]	0.553	0.000 [g]	0.102	0.052	0.723	0.162	0.408	0.000 [g]	0.140	0.343	0.007 [g]	0.136	-0.162	0.445	0.123
Auction_3	0.473	0.000 [g]	0.107	0.055	0.789	0.150	0.682	0.000 [g]	0.116	0.416	0.004 [g]	0.133	-0.185	0.468	0.147
Auction_4	0.232	0.027 [f]	0.201	-0.703	0.031 [f]	0.263	0.203	0.369	0.273	0.547	0.087 [e]	0.224	1.252	0.011 [f]	0.237
Auction_5	0.499	0.000 [g]	0.123	0.037	0.808	0.165	0.541	0.000 [g]	0.155	0.306	0.010 [g]	0.226			
Auction_6	0.474	0.000 [g]	0.106	-0.070	0.732	0.152	0.684	0.000 [g]	0.114	0.656	0.000 [g]	0.123	0.465	0.039 [f]	0.138
Univ. / research institute [c]	-0.057	0.308	0.148	0.062	0.573	0.090	0.178	0.072 [e]	0.127	0.290	0.064 [e]	0.119	0.078	0.780	0.223
SME	0.115	0.045 [f]	0.086	0.315	0.000 [g]	0.066	0.263	0.002 [f]	0.096	0.177	0.163	0.119	0.118	0.622	0.136
Large_firm	0.091	0.164	0.092	0.214	0.018 [f]	0.074	0.108	0.238	0.111	-0.107	0.450	0.124	-0.790	0.019 [f]	0.138
Patent_nationality	0.158	0.096 [e]	0.172	0.077	0.755	0.207	-0.020	0.924	0.267	0.169	0.571	0.217	1.329	0.010 [g]	0.259
Hypothesized Variables															
Technological_complexity	-0.279	0.000 [g]	0.101	-0.074	0.402	0.102	-0.007	0.935	0.159	0.250	0.061 [e]	0.172	0.256	0.355	0.230
Technological_uncertainty [d]	-0.049	0.359	0.126	-0.040	0.634	0.200	0.142	0.095 [e]	0.246	0.178	0.154	0.217	-0.068	0.769	0.204
Technological_impact	0.289	0.000 [g]	0.076	0.036	0.683	0.062	0.250	0.005 [g]	0.093	0.213	0.103	0.107	0.072	0.809	0.132
Technology_quality	0.016	0.758	0.153	-0.006	0.946	0.116	-0.034	0.693	0.176	-0.042	0.763	0.200	-0.587	0.066	0.261
Value_perception	0.535	0.000 [g]	0.073	0.351	0.000 [g]	0.086	-0.009	0.910	0.122	-0.282	0.038 [f]	0.137	0.087	0.749	0.161
N		156			78			108			66			30	
Adjusted R^2		0.701			0.665			0.507			0.363			0.222	

Notes:
- [a] Dependent variable is the Log10 transformed sales price
- [b] Base variable for is the first auction
- [c] Base variable is "independent inventors"
- [d] Measured as minimum remaining protection time
- [e] Indicates 10% significance
- [f] Indicates 5% significance
- [g] Indicates 1% significance

Thus, in Table 10.31, when systematically excluding the high value technologies, I can conclude that, in most of the models, the coefficients show the same direction as in the full model. While, in general, the models confirm the results found for the hypothesized variables in the full model, the effects appear less significant among low value technologies. In the mid-range of technologies, the effect of value perception is found to be positive and highly significant.

To summarize, from the results displayed in Table 10.30 and Table 10.31, I can conclude that the log-linear OLS regression model better predicts the sales prices of high value technologies. $Model_2$ has the overall highest adjusted R^2 value of 0.771, however, it only has four significant effects. The model also seems to predict the sales prices for mid-range technologies fairly well with the second highest adjusted R^2 of 0.665. While various models confirm the effects of the hypothesized variables found in the full model, all effects appear jointly only in two models ($Model_1$ and $Model_3$).

b) Robustness across auctions

Similar to the logistic model, in further sensitivity analyses, I investigated how the log-linear OLS regression model behaves if computed according to four auction sub-sets. Owing to too small case numbers and thus overfitting problems with the model, it could hardly be computed for individual auctions. Thus, Table 10.32 reveals the results from the analyses of two pairs of sub-samples. $Model_9$ and $Model_{10}$ represent the results from analyses of only the US auctions compared to the European auctions in Germany and the UK. $Model_{11}$ and $Model_{12}$ present the results of first auction in each country (US, Germany, and the UK) compared to the subsequent auctions. As mentioned above, those are available only for the US in my dataset.

To a large extent, the results presented in Table 10.32 confirm the results found in the full model. The effect of value perception is significant and positive in all models at the one per cent level. The effect of technological impact is positive in all models and highly significant at the one per cent level in three models ($Model_9$, $Model_{10}$, and $Model_{11}$). Similar to the effect found in the full model, the effect of technological complexity is negative in all models and significant in three models (at the one per cent level in $Model_9$, $Model_{11}$, and at the ten per cent level in $Model_{12}$). In addition, $Model_9$ shows negative significant effects of technological uncertainty at the one per cent level and technology quality at the ten per cent level. $Model_{11}$ also reveals a significant effect of technology quality, however, with a positive prefix.

Table 10.32 - Log-linear OLS regression models for auction sub-sets

Models [a]	**Log Transformed Model** All Auctions			**Model$_9$** First Auctions (A1; A4; A5)			**Model$_{10}$** Only Follow Up Auctions (A2; A3; A6)			**Model$_{11}$** US Auctions (A1 A2; A3, A6)			**Model$_{12}$** EU Auctions (A4; A5)		
	Stand. Coef.	**Sig.**	**SE**	**Stand. Coef.**	**Sig.**	**SE**	**Stand. Coef.**	**Sig.**	**SE**	**Stand. Coef.**	**Sig.**	**SE**	**Stand. Coef.**	**Sig.**	**SE**
(Constant)		0,001 [g]	0,300		0,000 [g]	0,291		0,559	0,319		0,293	0,347		0,009 [g]	0,368
Control Variables															
Auction 2 [b]	0,553	0,000 [g]	0,102							0,547	0,000 [g]	0,125			
Auction 3	0,473	0,000 [g]	0,107				-0,121	0,196	0,111	0,483	0,000 [g]	0,132			
Auction 4	0,232	0,027 [f]	0,201	0,450	0,000 [g]	0,156									
Auction 5	0,499	0,000 [g]	0,123	0,806	0,000 [g]	0,092							0,917	0,000 [g]	0,140
Auction 6	0,474	0,000 [g]	0,106				-0,104	0,293	0,113	0,464	0,000 [g]	0,132			
Univ. / research institute [c]	-0,057	0,308	0,148	0,045	0,496	0,133	-0,035	0,657	0,287	-0,121	0,069 [e]	0,292	0,288	0,001 [g]	0,102
SME	0,115	0,045 [f]	0,086	0,100	0,151	0,109	0,156	0,089 [e]	0,108	0,024	0,759	0,119	0,217	0,018 [f]	0,142
Large firm	0,091	0,164	0,092	0,251	0,003 [g]	0,108	0,090	0,347	0,122	-0,023	0,796	0,126	0,163	0,072 [e]	0,125
Patent_nationality	0,158	0,096 [e]	0,172	0,266	0,007 [g]	0,131							-0,080	0,448	0,132
Hypothesized Variables															
Technological_complexity	-0,279	0,000 [g]	0,101	-0,667	0,000 [g]	0,104	-0,062	0,508	0,149	-0,274	0,001 [g]	0,141	-0,111	0,085 [e]	0,121
Technological_uncertainty [d]	-0,049	0,359	0,126	-0,205	0,001 [g]	0,117	0,048	0,625	0,205	-0,064	0,407	0,192	0,037	0,549	0,184
Technological_impact	0,289	0,000 [g]	0,076	0,304	0,000 [g]	0,090	0,314	0,003 [g]	0,106	0,302	0,000 [g]	0,103	0,051	0,522	0,096
Technology_quality	0,016	0,758	0,153	-0,097	0,091 [e]	0,176	0,137	0,118	0,199	0,152	0,040 [f]	0,215	-0,036	0,597	0,180
Value_perception	0,535	0,000 [g]	0,073	0,636	0,000 [g]	0,084	0,557	0,000 [g]	0,097	0,488	0,000 [g]	0,096	0,390	0,000 [g]	0,160
N	156			62			94			120			36		
Adjusted R^2	0,701			0,868			0,469			0,560			0,909		

Notes:

[a] Dependent variable is Log10 of sales price
[b] Base variable is the first auction
[c] Base variable is "independent inventors"
[d] Measured as minimum remaining protection time
[e] Indicates 10% significance
[f] Indicates 5% significance
[g] Indicates 1% significance

With regard to the covariates, Table 10.32 shows various significant effects for all models except $Model_{10}$. Similar to the full model, $Model_{10}$ and $Model_{12}$ reveal a significant positive effect of technologies developed by SME. In $Model_{11}$ and $Model_{12}$, technologies developed by universities / research institutes also have a significant impact, although this impact is negative in $Model_{11}$ and positive in $Model_{12}$. In $Model_9$ and $Model_{12}$, technologies developed by large firms also have a positive impact on the sales price. $Model_9$ confirms the positive effect of patent nationality. Across all models, the average adjusted R^2 is 0.702, which is thus very close to the adjusted R^2 of the full model.

With regard to Table 10.32, with a few exceptions, I can conclude that the effects found for the hypothesized variables in the full model appear robust across different auction sub-samples independent of whether the auctions took place in the US or Europe or whether it was the first or a subsequent auction. In the subsequent auctions ($Model_{10}$), the effect of technological complexity has a similar prefix (negative) to that of the full model, but is not significant. The effect of technological impact has a positive prefix in all models, but is not significant in the EU auctions ($Model_{12}$). Furthermore, in the first auctions ($Model_9$), technological uncertainty and technology quality have negative significant effects. In the US auctions however, technology quality has a positive effect as predicted by the corresponding hypothesis. For the first auctions ($Model_9$), all the hypothesized variables have significant effects. The effect of technology quality however has the opposite direction than predicted by the corresponding hypothesis. With regard to the covariates, no clear picture emerges from the results displayed in Table 10.32.

10.3.3 Summary of Major Regression Results

To address my second research question, I extracted insights from the available literature (see Part II), which I subsequently combined with firsthand knowledge generated from qualitative pre-studies (see particularly Chapter 9) to understand the transaction governance structure of auctions. Based on the literature and the pre-studies, hypotheses were generated (Section 9.3). In order to test these, measures were developed in Chapter 8 to compute variables from a patent dataset primarily applying or referring to measures used in econometric patent (value) studies. The operationalized measures were used to translate the conceptual model (see Figure 10.3) into analytical, multivariate regression models. Key results from running various multivariate logistic and log-linear OLS regression models are presented in the multiple tables above. In the following, I synthesize the results and discuss whether, or which of the hypotheses can be confirmed or need to be rejected for the five hypothesized variables (the covariates are excluded). Table 10.33 summarizes the results with regard to the hypotheses displaying the effect directions predicted by the hypotheses and combines these with the observed coefficients and significance levels of the full logistic and log-linear OLS regressions as well as the multiple sensitivity analyses.

Table 10.33 - Summary of logistic and log-linear OLS regression results

Hypothesized Variables	Predicted Effect		Full Model		Sub-sample *															
					Young Technologies (3 Models)								Old Technologies (3 Models)							
			Coef.	Sig.	Coef.		Sig.						Coef.		Sig.					
	ID	Direction			pos.	neg.	pos.			neg.			pos.	neg.	pos.			neg.		
							1%	5%	10%	1%	5%	10%			1%	5%	10%	1%	5%	10%
Technological Complexity	H_{1a}	-	-	1%																
	H_{1b}	+	-	1%		3						1		3				3		
Technological Uncertainty	H_{2a}	-	+																	
	H_{2b}	-	-			3				1	1		3		2					
Technological Impact	H_{3a}	+	+	10%																
	H_{3b}	+	+	10%	3								3		3					
Technology Quality	H_{4a}	+	+	5%																
	H_{4b}	+	+	5%		3					1	1	3		3					
Value Perception	H_{5a}	-	-																	
	H_{5b}	+	+		3		3						3		3					

Hypothesized Variables		Sensitivity Analyses															
		Value Ranges (8 Models)								Auction-sets (4 Models)							
		Coef.		Sig.						Coef.		Sig.					
	ID	pos.	neg.	pos.			neg.			pos.	neg.	pos.			neg.		
				1%	5%	10%	1%	5%	10%			1%	5%	10%	1%	5%	10%
Technological Complexity	H_{1a}		8				3	1	2		4				1	2	
	H_{1b}	2	6			1	1	3			4				2		1
Technological Uncertainty	H_{2a}	6	2							2	2						
	H_{2b}	3	5			1				2	2				1		
Technological Impact	H_{3a}	6	2	3	1	1				4		2					
	H_{3b}	8		1	1	1				4		3					
Technology Quality	H_{4a}	8		1	3	3				4			3	1			
	H_{4b}	4	4						1	2	2		1				1
Value Perception	H_{5a}	3	5	1			3			3	1						
	H_{5b}	6	2	5				1		4		4					

Note: * Models for the sub-samples were only presented for Log-linear OLS models with the dependent variable being the Log10 transformed sales price. Logit models (grey shaded) were computed, but did not show any significant effects for technological uncertainty.

Table 10.33 indicates that, in the full models, six hypothesized variables (H_{1a}, H_{1b}, H_{3a}, H_{3b}, H_{4a}, and H_{4b}) show significant effects. Two effects are each significant at the one per cent, five per cent, and ten per cent levels. From these variables however, the effect of technological complexity on the sales price (H_{1b}) has the opposite direction than predicted by the corresponding hypothesis. The effects of three of the remaining four variables (H_{2b}, H_{5a}, and H_{5b}) show similar directions than hypothesized, but appear insignificant in the full models.

However, to investigate whether the observed effects are robust, the analyses of various models with sub-samples revealed additional results. The results from the sensitivity analyses also revealed further insights into the effects that did not appear significant in the full models.

With regard to the effects on the *sales probability*, the sensitivity analyses revealed the following picture. The negative effect of technological complexity must be regarded as absolutely robust. In all 12 models, the coefficient is always negative – similar to the hypothesized direction. In nine models, the effect of technological complexity is significant. Four models show significant effects at the one per cent level, three at the five per cent level, and two at the ten per cent level. Similarly, the positive effect of technology quality should be considered robust. In all 12 models, its coefficient is positive. This effect is significant in 11 models. In one model it is significant at the one per cent level, in six models at the five per cent level, and in four models at the ten per cent level. The positive effect of *technological impact* is similarly robust. Among these 12 models, technological impact has positive coefficients in ten models, similar to the hypothesized direction. In five models, the effect was significant at the one per cent level, while, in one model, it was significant at the five per cent level and, in another, at the ten per cent level. Regarding the two models, in which technological impact has a negative coefficient, those effects are not significant. Although the negative impact of *value perception* was found in the full model, the effect was not significant. Among the twelve models, the effect was also found to be negative, but significant. In three models the effect appears significant at the one per cent level. However, in one model, the effect also appears to be positive at the one per cent significance level. A closer look at the results from the sensitivity analyses reveals that the effect can be considered robust for technologies with moderate and low values. Finally, the effect of *technological uncertainty* reveals a mixed picture. While eight models show a positive effect, four models also reveal a negative effect. As none of the models reveal any significance levels, I must conclude that I could not prove any effect of technological uncertainty.[31]

The ranking of the effect sizes of the three variables with significant effects are similar in both the untransformed and transformed models. In both models, technology quality has the strongest effect on sales probability. Technological complexity has the second strongest effect and technological impact has the third strongest effect on sales probability. In absolute terms, all three effect sizes are sufficiently large to be regarded as having an impact.

With regard to the effects on the *sales price*, the sensitivity analyses revealed the following picture. The effect found for *technological complexity* must be considered robust, however, in the opposite direction than hypothesized. The results of ten of the twelve models reveal a negative effect that is significant in seven models. In three models, the effect is significant at the one per cent level, in three models at the five per cent level, and in one model at the ten per cent level. In one of the two models, where the effect is positive, it is significant at the ten per cent level. A similar picture separately reveals the additional analyses for two groups of young or old technologies. In each of these six models, the effect was found to be negative and, in four models, it was considered significant. The effect is significant at the one per cent level in three of these models and at ten per cent in one model. The positive effect of *technological impact* should be considered similarly robust. In the full model, the effect is significant at the ten per cent level. Among all twelve models in the sensitivity analysis, the effect has a positive direction; it is considered significant in six models. In four models, the effect is significant at the one per cent level, while, in one model, it is significant at the five per cent, and for another at the ten per cent level. The effect direction is also positive in the two groups of young or old technologies. The positive effect of *value perception* must also be considered robust. As hypothesized, the effect is positive in the full model, however, it is not significant. A closer look at the results from the sensitivity analyses reveals that the effect remains positive in ten models, being highly significant at the one per cent level in nine of them. The effect appears negatively significant at the five per cent level only in one model. Moreover, among young and old technologies, the effect of value perception shows a positive direction (as hypothesized) and is highly significant at the one per cent level in all six models. The positive effect of *technology quality* must also be considered robust, although to a lesser extent. Similar to the hypothesized direction, the effect is positive and significant at the five per cent level in the full model. It is further positive in six of the models from the sensitivity analyses and is significant at the five per cent level in one model. However, in contrast, the effect is negative significant at the ten per cent level in two models. Among the old technologies, the effect is strongly positive at the one per cent significance level in all models. However, among the young technologies, the effect is negatively significant in two models: one at the five per cent level and one at the ten per cent level. Finally, the direction of the effect of *technological uncertainty* was hypothesized to be negative. The full model confirms the direction of the effect, but does not show any significance. A closer look at the sensitivity analyses reveals that the effect is negative in seven models, but only significant in one model, although at the one per cent level. In contrast, the effect is positive in five models, but only significant at the ten per cent level in one model. The separate analyses of the two groups reveal a clear picture. In all three models for young technologies the effect of technological uncertainty is negative and significant in two models at the one per cent and five per cent levels respectively. In the group of old technologies, the effect is always positive for all three models, being highly significant at the one per cent level in two models. Thus, I conclude that the effect of technological uncertainty is robust, although only for young technologies.

From the three variables with significant effects in the transformed and untransformed models, the value perception has the strongest effect on the sales price. In the transformed model, technological impact has the second strongest impact on the sales price, while it has the third strongest impact on the untransformed model. Consequently, in the untransformed model, the technological complexity has the third strongest impact on the sales price, while it ranks second in the untransformed model. In absolute terms, all three effect sizes are sufficiently large to be considered to have an impact.

To conclude, the following picture emerged from various multiple regression analyses. The multiple logistic regression analyses proved four effects of the *sales probability*. First, an increasing technological complexity has a negative effect on the sales probability (H_{1a}). Second, an increasing technological impact has a positive effect on the sales probability (H_{3a}). Third, an increasing technology quality has a positive effect on the sales probability (H_{4a}). Fourth, an increasing value perception has a negative effect on the sales probability, however this only applies to technologies with moderate and low values (H_{5a}). No effect could be proven for technological uncertainty (H_{2a}).

The multiple log-linear OLS regression analyses revealed four effects on the *sales price*. First, an increasing technological impact has a positive effect on the sales price (H_{3b}). Second, an increasing technology quality has a positive effect on the sales price (H_{4b}). Third, an increasing value perception of the seller has a positive effect on the sales price (H_{5b}). Fourth, an increasing technological complexity has a negative effect on the sales price (H_{1b}). The direction of this effect is in contrast to the hypothesized direction. Finally, an increasing technological uncertainty also has an effect, but with different directions, on young and old technologies (H_{2b}). With regard to young technologies, an increasing technological uncertainty has a negative effect on the sales price. In contrast, concerning old technologies, an increasing technological uncertainty has a positive effect on the sales price. These solid, but changing directions of this effect among the two groups might actually indicate a u-shaped relationship to the sales price in contrast to the assumed linear relationship.

NOTES

[1] Recognized patents comprise a rich source of data for the economic analysis of innovation. Hall et al. (2001: 4) point out that 'indeed, there are numerous advantages to the use of patent data... Each patent contains highly detailed information on the innovation itself, the technological area to which it belongs, the inventors (for example, their geographical location), the assignee, etc.' According to OECD (2009: 27), 'each patent document contains detailed information on the invention process: a reasonably complete description of the invention, the technology field concerned, the inventors (name, address), the applicant (owner), citations to previous patents, and scientific articles to which the invention relates, etc.'.

[2] According to Hall et al. (2001) patent data have been investigated primarily since they were computerized. As soon as the large amount of data became available in digital/electronic databases, they could be analyzed more economically using standard PC software. The idea of using patent data on a large scale for economic research, however, dates back at least to Schmookler (1966), followed by Scherer (1982), Griliches (1984), and Narin and Olivastro

(1988). Since then, a research stream has emerged in industrial organization that has by now identified, analyzed, and discussed various value indicators.

[3] Based on van Zeebroeck (2007) in OECD (2009), Table 8.1 provides a comprehensive overview of the different types of indicator classes. Hall et al. (2001) also provide an overview that was helpful in this section. The seminal work of Griliches (1990) is often cited as having first provided a comprehensive review of indicators.

[4] The duration of this period is specified by the auction firms who claim a right to be involved in and compensated for any transaction that was not closed during the live auction event but during a specified period after the auction event. The following argument justifies this rule: Sellers can specify an own MRP that might remain unknown to the bidders and thus might not be reached during the bidding, even if the final bid comes close to the MRP. In this case, the technology is not sold during the auction. However, the seller and the interested bidders might come to an agreement on a reasonable price acceptable for both parties during post-auction negotiations, thus closing a transaction that was prepared by the auction firm to a large extent.

[5] With regard to the transactions, one might question whether they actually constitute auctions (that is competitive bidding situations) as they are a result of direct renegotiations between the seller and buyer. However, it appears reasonable to assume that these transactions can also be regarded as auctions, since the auction firm has provided the basis for the transaction. It is quite likely the final bids on the live auction event did not reach the MRP, that is the highest bid was eventually a little below the MRP. Regarding transactions that were closed after $S_{2a,}$ one might very well question whether they are closed as auctions or rather bargaining negotiations. However, those transactions are not included in the post-auction press release and were thus not considered in the dataset.

[6] It would have been practically difficult to collect reliable and comprehensive data on transactions closed after the publication of the press release, since no collective information was published providing a comprehensive and systematic overview of post-auction sales.

[7] For instance, the comprehensive compendium of indicators derived from patent data does not provide a measure of complexity OECD (2009).

[8] According to Haupt et al. (2007), several problems appear when attempting to assess the position of a technology within its technology lifecycle using patent data. The measurement of patent applications (in general, of a patent activity) requires the complete statistical survey of all patent applications and applicants of the considered technological field. With regard to most technologies, this survey is difficult or even impossible – despite modern patent databases. The reason is that the international system of patent classification (IPC) does not offer classes exactly corresponding to a certain, product-related technology. Moreover, most technologies cannot be identified by simply using a clearly defined set of technical search terms. To calculate such a relative indicator Haupt et al. (2007) propose indices. Although much simpler than full scale approaches, these indices appeared unsuitable for this study.

[9] In December 1994, the US congress passed the Uruguay Round Agreements Act (URAA) with several provisions affecting the US patent law. Relevant to this analysis, the URAA changed the term of patent protection from 17 years from the grant date to twenty years from the filing date for applications first filed on or after 8 June 1995 (US legislation number 35 U.S.C. §154(a)(2)). In this way, the US law became comparable with patent laws in the EU and other countries. In the calculation of this study's measure, I accounted for this change on the US patent legislation.

[10] Not to be confused with the citations, a patent document contains what are called 'backward citations' (Lanjouw and Schankerman (2004), and Harhoff and Reitzig (2004)). Additionally, the literature differentiates between the types of documents that are cited, most prominently into other patent documents or non-patent documents, that is scientific literature (Harhoff and Reitzig (2004), and Harhoff et al. (2003)). Moreover, the literature identifies the person who is cited. This can either be another person or the same inventor (Lanjouw and Schan-

kerman, 2001). For this study, data are only available on forward citations, thus I do not discuss the other citation types any further.

[11] Forward citation data, although available, for example, through the PATSTAT database, were not available when the dataset was compiled.

[12] According to Tijssen (2002: 512) 'the inventors listed on patents are the key information source on the history of the R&D process related to the technical invention'.

[13] For an overview of different strategies that have been used to approximate patent values see, for example, Sapsalis et al. (2006).

[14] Patents filed at the German patent office were selected because the German DPMA has the highest maintenance fees in Europe which become increasingly expensive the later a patent is held. Thus, it was assumed that firms must have a reason to pay the full maintenance fees, which were thought to indicate that the patents have at least a certain value to its owner.

[15] One has to be careful when interpreting the results from these studies. The values cannot be generalized to each and every patent, but rather represent the upper value limit because the included patents were held full term. An average patent is not renewed for more than 10 years and only about 10 pe cent of all patents survive the whole statutory term (Schankerman and Pakes, 1986).

[16] Gambardella et al. (2006) provide a further explanation for and discussion on this value measure. Accordingly, the 'value of a patent is different from the value of the patented invention. It is the latter net of the value of the invention if the inventor had no patent on it. This nails down the specific value of having a patent rather than the value of the invention per se. Some authors have called it the "patent premium".' However, Gambardella et al. (2006: 3) understand the value of a patent as the value of an asset and defined it as 'the sum of the annual discounted profits obtained from the invention when patented net of the same discounted stream if there was no patent on it'.

[17] According to OT, the source of 'estimated or anticipated selling price or value of any lot found in this catalogue is the seller and any written or oral estimate, appraisal or other statement made by OT with respect to such estimated or anticipated selling price or value is only a statement of opinion and shall not be relied upon by bidder as a prediction or guarantee of any lot's actual selling price or value. OT makes no representations or warranties of any kind to bidder with respect to any lot or its value or selling price'.

[18] As the share of missing data for expected values amounted to less than 25 per cent, I imputed the missing data using the Expectation-Maximization algorithm (EM), one of the likelihood approaches available in SPSS following Horton and Kleinman (2007). See also the discussion about missing values in Section 8.2.

[19] According to Giuri et al. (2007), the business sector and, in particular, large firms are the most common source of inventions. The business sector accounts for about 93 per cent of all PatVal patents. Universities account for 3.2 per cent and other public research institutions for 2 per cent.

[20] This accounts for a total number of 1.461 patents, from which 77.1 per cent were US patents, 7.3 per cent were EPO patents, 5.7 per cent were German patents, and 1.7 per cent were Japanese patents with the remaining patents scattered across 20+ countries.

[21] This technology (Lot 19c) entitled 'methods for internet shopping with a one-stop shopping cart' was sold by Avias, Inc. in the area of 'web based services / e-commerce'.

[22] For the regression analyses the missing data were imputed using values available for the opening bids of A_4, which were also available for A_5 that could be used to test their correlation.

[23] It is important to note that the values I found in my study are close to those reported in a large scale study by Wuchty et al. (2007). They found average team sizes for the development of inventions in the range from 1.7 to 2.3, with a positive trend towards larger teams.

[24] I refrained from using stepwise procedures to find the best suitable model. Preferably, systematic regression should only be used in the exploratory phase of research or for purposes of pure prediction, not for theory testing. According to Menard (1995:117) 'there appears to be

general agreement that the use of computer-controlled stepwise procedures to select variables is inappropriate for theory testing because it capitalizes on random variations in the data and produces results that tend to be idosyncratic and difficult to replicate in any sample other than the sample in which they were originally obtained'.

[25] For details on different auction types see Chapter 5.

[26] Using square root transformations for the variables, the Nagelkerke R^2 value increases to 0.206.

[27] The covariate patent nationality had to be omitted from the models as in certain auctions, no technologies with non-US patents were offered.

[28] Thus, although sellers tend to overvalue the technologies they own in absolute monetary terms (see Table 10.9), they tend to correctly predict the relative order of technology values, that is to identify which technologies have higher values than others.

[29] A similar analysis for the logistic regression model did not reveal any significant effect of technological uncertainty on the sales probability, thus is not presented in this chapter.

[30] It seems that, for extreme values or at the tails of the distribution (that is for old technologies with their patents close to the expiry date or technologies that were developed very recently), the measure that I applied for technological uncertainty appears problematic.

[31] This finding is actually in line with the transaction theory literature. As pointed out in the literature review by Macher and Richman (2008), the results of prior empirical studies found inconsistent results with regard to uncertainty as a determinant for the choice of governance structure.

11. Discussion of Results

This chapter discusses selected aspects of this study's results, particularly in relation to transaction cost (TC) theory. The research conducted during the course of the present study has centered on determining under which circumstances and particularly depending on which technology properties, auctions represent a preferred governance structure for transactions on the markets for technologies and ideas (MfTI). This question was broken down into two specific research questions. The first research question paved the way for the second one. Hence, the following discussion primarily addresses aspects related to the second research question, but also broadens the discussion beyond that question to account for aspects within the overarching theme of open innovation.

Which technologies should be auctioned?

To a large extent, the results of the regression analyses confirmed the hypotheses derived in Section 9.3, which were based on the qualitative pre-studies and are in line with the predictions from TC theory. Hence, technologies have a higher probability to be auctioned, if they have a lower technological complexity, but a higher technological impact and higher technology quality. Technologies with a higher technological impact and technology quality also reach higher sales prices at the auctions. Furthermore, the sellers' value perception is also positively associated with the sales price.[1] However, the results of three technological properties are not so obvious and need further discussion.

First, in contrast to the arguments provided in Section 9.3, technological complexity has a negative relationship with the sales price. A possible explanation for this effect is prospective buyers' restricted possibilities to conduct the due diligence. Within the due diligence, two aspects need further mentioning that constrain prospective buyers and prohibit them from sufficiently conducting the due diligence for complex technologies. First, the use of virtual data rooms might be less suitable to analyze complex technologies. Investigating complex technologies also appears to be a more complex task. Prospective buyers need to validate the legal, economic, and technical properties of all patents and their patent families. Moreover, they need to validate whether complex technologies are complete or whether certain necessary IP assets are missing, that might be held by other companies. To validate that a technology is not infringing upon others' IP assets is already a resource consuming task for just one patent, but becomes increasingly difficult if a technology consists of multiple patents. Furthermore, to understand the functioning of a complex technology, with all the interactions between its different components, requires considerably more (that is most likely overproportional) time than for discrete technologies only. This also relates to the second aspect within the due diligence period that constrains the analysis of complex technologies. Second, the limited due diligence period constrains prospective buyers, particularly if they are not large firms who might have sufficient resources to dedicate to this task. Hence, prospective buyers might not be able to complete a sufficiently holistic due diligence within the time span provided by the auction firms. The remaining uncertainty then results in a negative impact on the prospective buyers' Willingness-to-pay for highly complex technologies. Third, the pre-studies showed that the auction design represents a transaction model that enables prospective buyers to remain anonymous. Hence, although I have not been able to study the buyer side (for which this study's results should not be overinterpreted), this mechanism attracts buyers that do not want to reveal their identity. If these buyers have a reason to remain anonymous not only in the due diligence phase, but also during the bidding on the live auction event, they might not be interested in revealing their identity after they have purchased any technology. Hence, they are unlikely to purchase complex technologies, which might well require the transfer of complementary tacit know-how to enable utilizing a technology after the auction. Therefore, this effect limits the number of bidders for complex technologies, which has a reverse impact on the price (that is empirical auction studies showed that a higher number of bidders has a positive impact on the price – see Chapter 5).

Second, with regard to technological uncertainty's impact on the sales price, the results of the quantitative study reflect the hypothesis only to a certain extent. I was only able to observe a relationship when grouping the technologies into old and young technologies.[2] Accordingly, the results show that, in the group of young technologies, an increasing technological uncertainty has a negative impact on the sales price. However, in the group of old technologies, increasing technological uncertainty has a positive impact on the sales price. In Figure 11.1, these results are depicted by the bold black straight lines.

There are two possible explanations for this observation. First, the technological uncertainty of very young technologies (T_{VY}) (that is those invented very recently with their patents showing a remaining protection of almost 20 years) appears likely to be very high. Hence, from the date of the invention to the auction, too little time had passed for (tacit) knowledge about this technology to diffuse from the inventors' minds to the expert community. Thus, the expert community will have difficulties to evaluate a very young technology, particularly during a short due diligence period as is characteristic of a technology auction. Second, the distinct characteristics of the due diligence as discussed above (for example, virtual data rooms) further complicate the evaluation of such very young technologies. While virtual data rooms might not be suitable for the evaluation of highly uncertain technologies, the due diligence period might be too limited as well. Thus, risk averse bidders will have a lower Willingness-to-pay for highly uncertain technologies and particularly for very young technologies, which they can hardly evaluate sufficiently by means of the due diligence process provided by the auction firms.

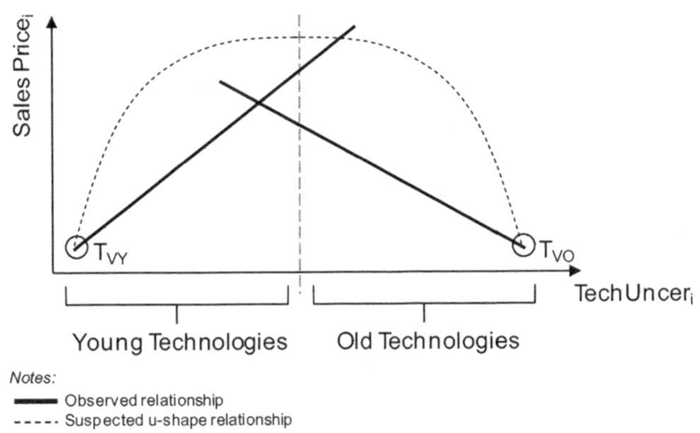

Figure 11.1 - Suspected u-shaped relationship of technological uncertainty and sales price

The proven relationship in the group of old technologies can be explained as follows. Chapter 6 shows that, if a patent is very close to its expiry (that is with a remaining protection close to zero), it often has very little value (T_{VO}). If a technology has become widely used, whether only within the owner's products or by others through licensing, after almost twenty years, the knowledge to utilize such a technology can be assumed to have widely diffused through the expert community. Therefore, firms other than the owner will have little difficulty in imitating the technology. The only element that might prevent them from doing so will be the patent protection. However, as soon as the patents expire, their last remaining value (that is the patent premium) will diminish. Many examples of this phenomenon can be found in the pharmaceutical industry where, commonly, generic versions appear on the market on the day of the patent expiry (for example, Raasch (2006)). As the results of this study show that buyers at technology auctions are less interested in the knowledge to utilize a technology (that is none of the buyers in the four case studies requested any additional knowledge transfer after the auction) but rather in the IP assets, they will have a strongly decreasing Willingness-to-pay for technologies close to their expiry dates (T_{VO}).

To summarize, instead of the suspected linear relationship, the results reveal a u-shaped (unimodal) relationship between technological uncertainty and the sales price with strongly decreasing effects at both flanks, as depicted by the dotted line in Figure 11.1. This relationship can also be explained as a square function that results from multiplying the two linear functions of the expected value, with a given level of uncertainty (that is a fixed probability that a technology will work) and the linear function of patents' remaining protection time and their value. Instead of the measure applied in this study, which only uses one indicator (that is patent age), future studies should rather use a multi-dimensional measure.

Third, the results of this study prove an effect of the sellers' value perceptions on the sales probability of auctioned technologies. However, the (negative) direction of the effect was in contrast to the direction formulated in the hypothesis. This result can be explained in terms of the following argument. The descriptive statistics provided in Table 10.9 indicate that the endowment effect is present in auctions that have been well substantiated in the literature. Comparing actual realized sales prices with the sellers' expected sales prices revealed that the actual realized sales prices are considerably lower than the values expected from the sellers. Thus, at the auctions, sellers seem to be positively biased toward the value of their technologies. This may result from the uncertainty regarding realistic prices in the MfTI. According to Chesbrough (2006: 146), there is 'no systematic reporting of previous prices paid for external technologies and their associated IP. This makes it hard for sellers to know what price to expect to receive or what price would be reasonable, given similar transactions in the past.' Thus, sellers might be overambitious about the prices they can expect. I have already mentioned that sellers tend to apply the cost-based approach when valuing technologies, at least to determine their 'floor price level' (Chapter 3) and recapture the development costs of a technology. However, buyers, who tend to rather apply a market based valuation approach, will be hardly willing to pay for the development costs of a technology, if the technology's market potential is lower than the cost. If sellers tend to overvalue a technology, they will likely set an MRP above the prospective buyers' Willingness-to-pay, wherefore the technology will not be sold.

How can the auction design be improved?

In this chapter, I discuss the current design of technology auctions and how it can be improved. Following Klemperer (2004: 122) 'anyone setting up an auction would be foolish to blindly follow past successful designs; auction design is not 'one size fits all'... In the practical design of auctions, local circumstances matter and the devil is in the details'. Hence, I discuss six elements of the technology auction design, including the selection of the prospective buyers, the minimum reserve price (MRP), the bidding procedure, the commission fees, the payment scheme, and the due diligence.

Prospective buyers

Within the observed open auction designs, sellers are not able to exclude certain prospective buyers from acquiring technologies. However, the case studies revealed that sellers perceive this as a disadvantage. Thus, firms might refrain from choosing auctions as external technology exploitation channels in certain cases, for instance if they want to prohibit a direct competitor from acquiring one of their technologies. If the auction design had a mechanism that would allow the sellers to specify prospective buyers they would want to exclude, the auctions might become more attractive to a wider range of sellers. Three possibilities appear feasible. First, the auction firms can provide the sellers with a list of registered bidders from which the sellers mark those that should not be allowed to bid for certain technologies. However, this approach might conflict with the auction firm's policy to protect bidders' anonymity. Second, auction firms could ask the sellers to specify industry classes so that all firms within the scope of the named industry would automatically be excluded from the bidding for certain technologies. However, this approach bears the risk that a too wide range of firms will be excluded from the bidding, thus the auction would unnecessarily lose bidders that would otherwise have contributed to the competitive bidding. This approach is also difficult to implement since large firms often have a number of business units with operations across a range of industries. Third, the auction firms could allow sellers to specify firms that should be excluded from the bidding of certain technologies, without knowing whether any of these firms had registered as a bidder. This approach would rely on the assumption – which seems reasonable – that sellers know which firms they do not want to acquire a particular technology. In order to avoid excluding too many sellers, the auction firms can charge an extra fee for each firm named by the seller to be excluded. Thereby, the auction firms create an incentive to limit the list to only a few firms. However, the auction firms should also ensure that prospective buyers do not circumvent this approach and hire other intermediaries to bid on their behalf.

None of these approaches solves the problem of prospective buyers hiring an intermediary to act on their behalf in order to circumvent this barrier (for example, a patent law firm). Alternatively, following the design of public technology auctions, sellers can also organize own auctions that take place in closed settings, where bidders are only allowed to participate if they fulfill certain criteria.

The issue of bidder exclusion or the number of bidders is directly linked to auctions' success (Chapter 5). Klemperer (2004: 122) argues that the 'most important properties of an auction are its robustness against collusion and its attractiveness to potential bidders.' Similarly, Bulow and Klemperer (1996) argue that, for the success of any auction, a major area of concern is to attract bidders, since an auction with too few bidders risks being unprofitable for the auction firm. With an increasing number of bidders, the competitive bidding element of auctions becomes more and more advantageous over bargaining negotiations. Kagel and Levin (1986) consider the number of bidders large if higher than six and small if between three and four. While attending the two auctions (A_4, A_5), I observed that the bidding for most technologies occurred between two or three bidders. Thus, the auction design is particularly important for such auctions with a low number of bidders.[3]

Although – for the viability of the auction model – it is crucial to attract additional bidders by improving the auction design, it seems difficult, particularly in the thin MfTI.[4] However, one can expect that technology auctions gain more attention by potential bidders over time when they become more popular and continue to prove their concept. Furthermore, one possibility to attract more bidders is to adjust the commission fee structure (see separate discussion below), which sellers considered too high, but which buyers also have to pay. Future research is needed to gain a better understanding of the nature of buyers. Most likely, independent inventors and universities rather act on the supply side and therefore sell technologies on the auctions, but hardly acquire technologies. Moreover, it is doubtful whether SME purchase technologies on the MfTI. Quite likely, the majority of prospective buyers will be large, multinational firms (that actually might act as sellers and buyers) or TMI (for example, patent trolls). For instance, Ewing (2010a) finds that Intellectual Ventures (IV) are among the large buyers at the technology auctions. Correcting for the asymmetric empirical base (that is constrained access to buyer data) is left to future research.

Minimum reserve price

To account for the risk of too few bidders, auction firms employed the MRP property. In general, the MRP specifies a minimum price below which assets remain unsold, thus minimizing the risk to the sellers. In addition, the MRP is a common property used in auctions in which the number of bidders is limited, since it serves as an extra bidder (Berz, 2007). The auction firm thus participates in the bidding until the MRP is reached. Until this level, the asset will be 'bought' by the auction firm. If the price climbs above the MRP, the auction firm drops out as bidder and the asset is sold to the bidder who placed the highest bid.

In the observed technology auctions, the MRP is either fixed to a minimum level pre-determined by the auction firm or pre-determined by the sellers if they were willing to pay an extra fee. However, the results from my study indicate that sellers have a positively biased price expectation (Chapter 6). The high expected sales prices of the sellers indicate that they might set MRPs too high. Sellers should rather understand that bids will not primarily reflect the development costs of a technology but rather that the 'true value of any auction object is the price assigned to the object by a bidder, in case full information regarding the auction object is available to the bidder,' that is the current market prices of the technology are known (Berz, 2007: 104).

By letting sellers pre-determine the MRP for a relatively low fee (although IPA linked it to the sales price), the auction firms bear the risk that technologies remain unsold. Thus, auction firms should find a way to incentivize sellers to set MRPs conservatively, that is to determine realistic MRP levels. The data revealed from this study already contribute to a more realistic and increased understanding of technology values for future auctions. Hence, increasing the transparency of technology values, as suggested by, for instance, Chesbrough (2006), can help to establish 'value corridors' for certain technology types as value guidelines ('rules of thumb') for future transactions.

Bidding procedure

The auction literature (Chapter 5) shows that the bidding procedure is a particularly important element in the auction design. All of the observed technology auctions applied English (that is ascending) bidding procedures. However, ascending auctions have particular weaknesses. According to Klemperer (2002: 111), they are 'vulnerable to a lack of entry and to rule breaking (collusion) by the bidders since they necessarily pass through a stage where there is just one (or a few) excess bidders and the ascending structure allows a cheat time to assess the success of its strategy'. IPA also attempted to use Dutch (that is descending) auction style. However, none of the technologies were sold. Although Dutch auctions are known to accelerate negotiations (that is as the first bid wins), accelerated negotiations do not seem to be crucial for technology auctions. English auctions might also not be optimal for technology auctions, since they favor the winner's curse effect, particularly if technologies are auctioned with a large common value element (for example, standard technologies, such as the XML technology in the CommerceOne case). In such cases, sealed bid auctions might be a better solution and also avoid collusion. Nevertheless, the winner's curse effect is optimal for the sellers, since it increases sales prices, as well as for the auction firms, who calculate the commission fees based on a fixed rate of the sales price. In contrast, buyers who may be largely affected by the winner's curse, may disfavor English auctions and even refrain from participating in such auctions, which ultimately results in sub-optimal resource allocations. However, as argued above, sellers and the auction firms likewise, have little incentive to adjust the auction design to minimize the winner's curse effect.

Thus, while auction firms need to think about additional means to facilitate entry and avoid rule breaking, the English auction nature bears one further problem. The auction literature has shown that English auctions do not necessarily reveal the true market prices or the buyers' maximum Willingness-to-pay (indifference price). This problem has been shown to result from the fact that the winning bidder does not necessarily pay his maximum Willingness-to-pay, but rather needs to pay only a small fraction more than the second highest bid (Milgrom (2005), Berz (2007)). The auction literature has shown that second price auctions (for example, Vickrey auctions) avoid this effect and reveal true market prices. Although, in such auction designs, the winner does not pay more than in English auctions, or pays a little less (that is the second highest bid), the winner will reveal his or her maximum Willingness-to-pay. However, the auction firms might not have such an incentive to favor Vickrey auctions, as they will not raise the auction prices. Instead, as the auction firms benefit from high sales prices through the commission fees, they will not mind a winner's curse, from which sellers also benefit, but not the buyers. To determine an optimal auction design is a complex question, which cannot be discussed exhaustively in this study. It remains for future research to determine which price levels lead to efficient resource allocations on the MfTI.

Commission fees

Commission fees represent the major determinant for sellers' and buyers' TC, since they constitute a fixed share of the sales price. As sellers and buyers are also subject to other costs (for example, having to prepare the due diligence), the commission fee's share of the total TC is higher for high value technologies. Although the fee structure of technology auctions is currently simple, standardized, and similar to those of art auctions, the commission fees of technology auctions should rather be adjusted if auction firms want to compete with bargaining negotiation schemes for high value technologies (see also Figure 11.3). The fees should either be kept as a fixed share of the sales price, but be limited to a maximum level, or a declining share should be defined if sales prices pass certain threshold levels.

Payment scheme

All observed technology auctions only allowed lump sum payment schemes for the payment of the auctioned technology. This decision seems reasonable if one has the intention to standardize and thus reduce the TC, particularly for complex technologies with high uncertainty regarding future applications and market potential. However, such a simple payment scheme could be problematic. If lump sum payments are the preferred payment scheme for technology transactions, why would the literature (and associations, practitioners, etc.) have developed an intense body of writings and concepts for advanced licensing schemes?

If the intention of auction firms is to establish their model as exploitation channels for low value technologies only, that is those technologies that firms exploit externally because they have no use for it anymore or do not believe in their future potential (that is only for monetization motives), then one might keep the current auction payment scheme. In order to further develop the auction design to better fit the needs of transactions of high value technologies, one would have to adapt this payment scheme to allow for more flexibility. This is a difficult issue, since the introduction of lump sum payments is likely to have the highest effect on the standardization of the auction TGS. At this point, the development of solutions for the conflicting relationship between lump sum payments and the need for transactions of highly complex, promising (uncertain), and highly valuable technologies and the degree of standardization that reduces TC, needs to be left for future research.

Due diligence

In general, the sellers reported few disadvantages in the due diligence process. The results of this study indicate that auctions accelerate the transaction speed, at least taking into account that they parallel different transactions. However, the results show that the TGS of auctions seems less feasible for technologies with high technological complexity and uncertainty.

One reason for this is the limited due diligence duration. The due diligence seems to be too short to sufficiently evaluate the legal, economic, and technical properties of more complex technologies with higher uncertainty. Hence, the auction firms should reconsider the length of the due diligence period. Furthermore, the virtual data rooms seem to perform fairly well for the evaluation of the technologies through prospective buyers, although additional properties should be embedded to enable the sellers and buyers to communicate more personally.

Which decisions do firms face in technology transactions?

Before continuing, I reflect on a set of decisions that firms face if they want to shift from a closed to an open innovation approach when engaging in technology market transactions. During the course of this study, it appeared that firms face three critical decisions when they want to actively practice open innovation and hence they have to consider how to design the governance of transactions.

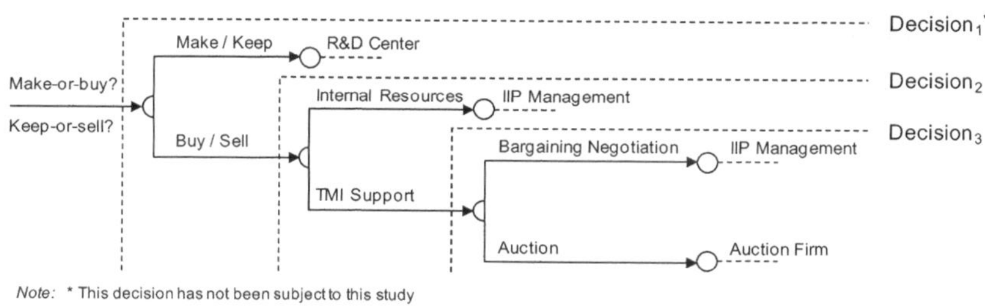

Figure 11.2 - Firms' decision sequence in technology market transactions

The *first decision* is the most prominent in the literature and concerns a firm's technology base. If innovation and IP (IIP) management faces the need for a technology, it has to decide whether the firm should develop the technology internally in the firm's own R&D centers or acquire it on the MfTI (make-or-buy decision). If a firm owns a technology, the IIP management has to decide whether it wants to keep the technology to itself or whether it will allow the technology to be exploited externally on the MfTI (keep-or-sell decision). As this decision has been discussed by others (for example, by Tschirky and Koruna (1998)), this study has waived to also deal with it in a similar way as with the motives behind this decision. This study assumes that firms choose between the 'buy' or 'sell' options instead of the 'make' or 'keep' options, which are related to the closed innovation paradigm where firms develop and exploit technologies solely internally.

If a firm's IIP management decides on either the buy or the sell option because it wants to innovate openly and source / exploit technologies on the MfTI, it faces a *second decision*. Having decided to innovate beyond its own boundaries, the firm needs to engage in technology market transactions. In such a case, the IIP management has to decide whether it wants to govern a transaction with internal resources or seek TMI support. This study is primarily concerned with transactions in which IIP management have decided to engage TMI. Although, for simplicity reasons, I have assumed that this decision is of a dichotomous nature, this decision might appear more complex for IIP management. For instance, regarding the different managerial tasks of an external exploitation project illustrated in Lichtenthaler (2006a) ETE process model, IIP management can choose to fully disintegrate (that is outsource the complete transaction governance to a TMI) or it can choose to outsource only specific tasks to a TMI (for example, the search for a prospective buyer in the planning phase). Reflecting on the auction model, it appears that this model must be characterized as fully integrated and is suitable for firms that want to outsource by far most of their transaction governance to a TMI. The auction firm governs the transaction as it keeps control of the process. The firm can be reactive to a large extent and supply only the necessary material for the due diligence requested by the auction firm. This study shows that firms often decide to use a TMI because they lack internal resources and managerial capabilities that prohibit them from governing transactions internally (for example, in SME).

Whether the IIP management decides to seek the support of a TMI or to govern a transaction internally, it faces a subsequent *third decision*. IIP management has to decide on the most economical acquisition / exploitation strategy (see Figure 3.4).[5] Applying the TC perspective in this study, I have argued that different acquisition / exploitation strategies are characterized by different TGS. In this study, again for simplicity reasons, I have assumed a dichotomous set of options for this decision. Firms either decide on the auction model operated by TMI (that is auction firms) or to govern transactions with internal resources.

If they decide on the latter option, I have further assumed that firms tend to use bargaining negotiations for which they design idiosyncratic TGS on a case by case basis. This assumption may only be realistic when open innovation is new to a firm. As soon as firms have engaged in a number of transactions over time, IIP management might rather refrain from designing each transaction fully idiosyncratically, and recycle certain elements of previous transactions instead. For instance, by means of a cumulative learning process, IIP management will acquire knowledge over time with regard to contract design. Hence, IIP management will recycle contract modules in future transactions, ultimately even establishing internal standardized contracts that can be used in every future transaction. However, if firms have chosen not to use internal resources to govern the transaction, aside of the auction model that was subject to this study, various other transaction models have recently been developed by TMI.[6] If not assuming the option set for the decision to be dichotomous, one would also have to consider that firms have to choose between a variety of other transaction models. In this study, however, I have concentrated only on the in-depth investigation of the auction model, its TGS, and the technology properties that favor auctioning. Studying the other models is beyond the scope of this study.

In the following, I discuss when auctions are the superior choice over bargaining negotiations (that is the third decision in Figure 11.2) by applying arguments of the TC theory. Thereafter, again applying arguments from TC theory, I broaden the discussion of the decision between auctions and bargaining negotiations towards the second decision in Figure 11.2. Consequently, I discuss when firms prefer to use TMI and when they decide to govern transactions using internal resources.

When are auctions a superior transaction model?

Having investigated the TGS of technology auctions in two qualitative studies, I concluded in Section 5.3 that the TGS of technology auctions are situated close to the right extreme on the TGS continuum (see Figure 7.1) and hence almost represent spot market transactions. The results of both studies further show that the observed technology auctions' major cost drivers are the commission fees that are proportional (but not limited) to the sales price of successfully auctioned technologies. Therefore, as commission fees are set at a considerable fixed share (25 per cent) of the sales price, they can be assumed to grow linearly with the bid price.[7] Thus, when auctioning a technology with an infinite final bid, the seller and buyer will have to pay the fixed share of the infinite bid. Consequently, auctioned technologies' TC to firms (buyers and sellers) are assumed to follow a straight line in relation to the transaction volume (see Figure 11.3).

In idiosyncratically designed bargaining negotiations, the nature of the TC is different. First, one can assume that, even for low priced technologies, the firm will face certain fixed costs. Second, with an increasing transaction volume, TC are likely to drop off. For instance, the marketing efforts to advertise a technology on the MfTI will not increase proportionally if a technology is marketed in 10 or 11 countries, as firms will be able to reuse some of their marketing material. Moreover, the number of negotiation rounds (and its associated resources) is unlikely to increase proportionally to the transaction volume when the number of prospective buyers doubles. Thus, the TC for internally governed transactions (that is even with support of a PLF) are likely to show a marginally decreasing shape in relation to the transaction volume (that is a technology's sales price). In Figure 11.3, the straight line depicts the comparison between the TC of technology auctions ($TC_{Technology\ auction}$) and the marginally decreasing TC shape of bargaining negotiations ($TC_{Bargaining\ negotiation}$). This figure can be further compared to

Figure 9.4, which depicts the development of TC for different governance models.[8]

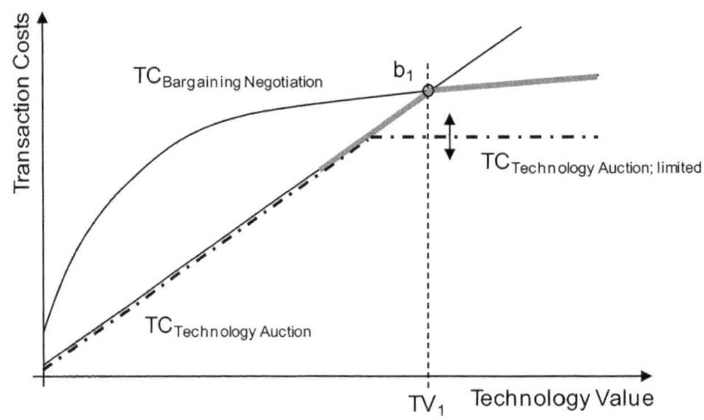

Figure 11.3 - Transaction costs of auctions and bargaining negotiations

Figure 11.3 illustrates that, until there is a cut-off technology point (TV_1), firms will favor technology auctions over bargaining negotiations. Regarding particularly high value technologies that are above the cut-off technology point (TV_1), firms will rather favor bargaining negotiations, due to their marginally decreasing $TC_{hierarchy}$.

In addition to the TC argument, one has to take an additional qualitative effect into account. From both pre-studies (see Section 9.3) I concluded that the auction design imposes a high degree of standardization onto the TGS, thus restricting the buyers' and sellers' possibilities to influence the transactions. Hence, auctions not only have higher TC for high value technologies than bargaining negotiations but they restrict sellers' and buyers' flexibility to influence the transaction (for example, to negotiate a royalty scheme). The effects of standardization are manifold. The seller and buyer might expect that a higher degree of flexibility will enable them to better leverage their advantages. However, prospective buyers with weak bargaining power can benefit from the standardized auction design although auctions are also vulnerable to the strong positions of advantaged bidders. Consequently, for sellers and buyers, the current technology auction design bears the risk that they will pay higher TC_{market} for high value technologies (above TV_1) than they would in bargaining negotiations ($TC_{hierarchy}$). Furthermore, the standardized TGS of technology auctions constrain the possibilities for the seller and buyer to influence the transaction.

Auction firms should thus account for these two effects in their auction design. To ensure that the TC_{market} remains below the $TC_{hierarchy}$, for internally governed bargaining negotiations, appears relatively simple. The most obvious possibility would be to introduce a maximum commission fee ($TC_{Technology\ auction;\ limited}$). Figure 11.3 illustrates such an effect. From a TC perspective, a technology auction would then remain cost efficient over bargaining negotiations even for high value technologies. However, determining the maximum level is not trivial as it depends on the buyer's and seller's Willingness-to-pay, which fluctuates across different firms and industries.

Accounting for the lack of flexibility is more difficult. Auction firms can reduce the degree of standardization and adjust the auction design so that it allows for more flexibility or for the TGS to be more influenced by the sellers and buyers. However, flexibility should not be given away for free. Similar to the MRP feature, which sellers are allowed to pre-determine at an extra fee, sellers might be given the opportunity to 'buy' extra flexibility for the payment of additional fees (for example, an option to include grant back clauses into the contract). Thus, sellers should have clear incentives to use the highest degree of standardization to the extent possible and only consider 'buying' more flexibility if really necessary. A possibility would also be to design packages that range from a low fee package with a high degree of standardization (similar to the currently observed TGS) to a package that allows a high degree of flexibility (for example, more complex contracts, longer due diligence period, the possibility to exclude certain bidders, etc.) for higher fees. Instead of adjusting the auction design, another possibility would be to set the maximum fee level below the level at TV_1. Thereby, firms would have a monetary (cost saving) incentive to use auctions despite the high level of standardization.

Having discussed how the TC of technology auctions (TC_{market}) relate to the TC of internally governed bargaining negotiations ($TC_{hierarchy}$), which determine firms' choice of technology auctions or internally governed transactions, in the following section, I broaden the discussion to the second decision of the framework illustrated in Figure 11.2.

Should firms use TMI or govern technology transactions internally?

Although open innovation, in general, and external technology exploitation/acquisition are not specifically new concepts in the literature, in practice,

these concepts have just recently started diffusing among firms and across industries. Hence, in many firms, internal capabilities to manage technology transactions on the MfTI must still be regarded as weak. Within this market setting, TMI have recently started to develop novel transaction models to facilitate transactions. Firms now face the question of whether (and under which conditions) they should hire TMI to govern a transaction (that is disintegrate) or rather govern a transaction with internal resources (for example, establish a licensing department). Applying the arguments of TC theory, the theory of the firm, as well as Stigler's (1951) arguments to this empirical phenomenon, one can argue that firms tend to disintegrate (that is outsource) the governance of technology transactions to TMI if they provide lower TC than if the firms govern transactions internally. Hence, firms disintegrate if they expect lower TC from an externally governed transaction (see also
Figure 9.4).[9]

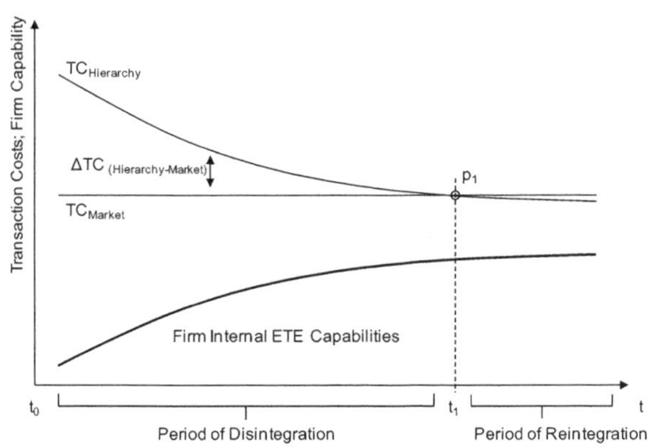

Figure 11.4 - Dynamic TC as reason for disintegration / reintegration phases

When firms want to start open innovation activities, one can assume that they do not possess the internal capabilities to govern transactions internally or without substantially increasing the TC. In such situations, it might seem reasonable that TMI are often founded and managed by experienced IIP managers (for example, the former licensing council of a large firm) who can, due to their experience, establish transaction models for lower TC than if firms internally govern transactions. In Figure 11.4, this situation refers to t_0 where the $TC_{hierarchy}$ is above the TC_{market}.

However, as discussed in Chapter 7, one needs to consider the dynamic nature of TC.[10] Following Jacobides's (2008) arguments, this relates to firm capabilities. In situations in which innovations occur (that is if the open innovation paradigm is considered a managerial innovation, firms commonly lack internal capabilities (that is many firms do not have own licensing departments). Hence, in such a situation, firms face high $TC_{hierarchy}$ when attempting to govern technology transactions on the MfTI. However, if they continuously participate in transactions, they build up their internal capabilities to govern technology transactions. For instance, although not educated as a lawyer, a firm employee will gain insights into designing contracts if he or she continuously participates in technology transactions. As illustrated in Figure 11.4, the increase in internal capabilities to govern technology transactions results in a decreasing $TC_{hierarchy}$ wherefore firms govern technology transactions using internal resources.

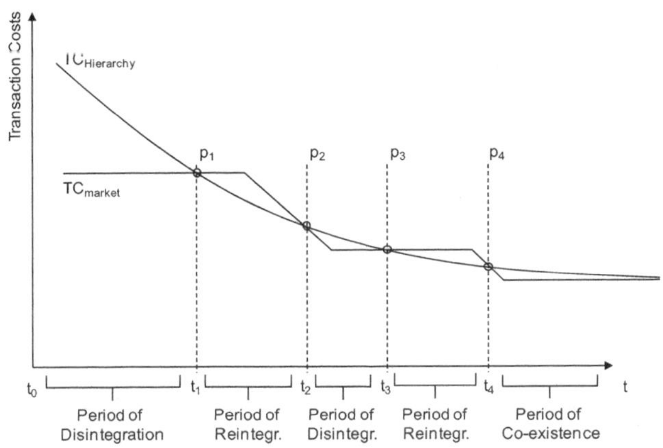

Figure 11.5 – Alternation of disintegration and reintegration phases

Similarly, while firms may gain insights into transactions over time, TMI also ride down the learning curve and improve their transaction models. However, it is reasonable to assume that, even if they decrease their internal costs, TMI will most likely not pass on the decreasing internal costs to their clients, that is the firms. Instead, TMI are likely to capture the internal cost gains as additional profits, at least until they face price competition. Hence, the TC_{market} remains constant over time for the firms that use TMI. Consequently, over time, the TC advantage of market transactions (that is using TMI) over hierarchically governed transactions (that is with an own licensing department) will decrease, which is represented by the decreasing $\Delta TC_{(hierarchy\text{-}market)}$ in Figure 11.5.

At a certain point in time (t_1), the $TC_{hierarchy}$ of internally governed transactions will ultimately fall below the TC_{market} of transactions governed by TMI. Hence, beyond p_1, it becomes cost-effective for firms to govern the transactions themselves. Consequently, following a period of disintegration in which firms use TMI to govern technology transactions on the MfTI, we might experience a period of reintegration in the future where firms start to prefer internally governed transactions. While firms naturally build up capabilities to govern technology transactions over time, they might even accelerate the process towards t_1 by acquiring external capabilities or ultimately certain TMI, which have proven to be a sustainable business model.[11] For instance, firms can organize their own technology auctions in closed environments in which they combine the advantages of the observed technology auctions with the flexibility to have control over the TGS design (for example, the way in which the contracts are designed) if they can create enough market thickness (for example, by forming a consortium with other firms or universities). Hence, we might observe an opposite pattern to the one observed by Granstrand (2004a), who reported cases in which large firms had spun-off their technology trade units to operate independently on the MfTI, but largely failed to survive.

However, TMI might react to this development as soon as they realize that firms increasingly use internal resources to govern transactions, by entering into a cost competition (Figure 11.5). Having realized internal cost reduction through learning, but having not yet passed on these benefits to their clients, TMI might start to lower their fees to compete with the internal firm departments. By engaging in a cost competition with the firms, TMI might be able to move p_1 further down in time (at least to some extent). How quick and to what extent TMI will be able to optimize their transaction models and how fast firms will build up internal capabilities will then determine the outcome of that competition.

However, other developments also impact the firms' future preferences for internally or TMI-governed transactions. The pre-studies showed that information and communication technologies (ICTs) play an important role in the TGS of technology auctions. For instance, the auction firms use virtual data rooms for prospective buyers to conduct their due diligence prior to the auction. Virtual data rooms are less costly than real world data rooms, and hence have a positive impact on the TC_{market}, favoring TMI-governed transactions. Further developments which support technologies in general and ICT technologies in particular can be expected to further lower the TC.[12] However, these developments can be twofold and promote market transactions through lower TC_{market} and vertically integrated (hierarchical) transactions by means of a lower $TC_{hierarchy}$. For instance, management support systems (for example, software tools to identify prospective buyers through a patent analysis) might be developed to assist firms in governing transactions internally that lower the $TC_{hierarchy}$ and favor internally governed transactions. In contrast, for instance, if Yet2.com, an online marketplace for IP transactions, constantly improves its community software, the success rate of seller–buyer matching increases, and the TC_{market} decreases, hence favoring TMI-governed transactions. Consequently, particularly ICT developments and the resulting cost reductions for internally and TMI-governed transactions can become of vital importance for the shift of firms' future preferences for internally and TMI-governed transactions. As ICT developments will remain uncertain and ongoing over time, the 'preference pendulum' can subsequently swing back and forth between both options.

Over time, we may thus expect firms' preferences to fluctuate, which could result in alternating periods of disintegration and reintegration as depicted in Figure 11.5. During a certain period, technology developments may positively impact the ICT systems for IIP management, so that the $TC_{hierarchy}$ decreases below the TC_{market} level. During another period, technology developments may result in ICT software that positively impact the TC_{market}, wherefore firms might favor disintegration (that is outsourcing to TMI). Although ICT developments will remain ongoing and result in continuous improvements of both sides' systems, learning behavior is cumulative and one-directional (for example, firms continuously build up own internal capabilities to govern technology transactions). As this effect speaks in favor of internally governed transactions, TMI will be forced to lower the TC_{market} faster than firms in order to maintain competitiveness in the long run. If TMI and internally governed transactions reach an equilibrium stage (that is $TC_{market} = TC_{hierarchy}$) the two models can coexist. For some technologies, firms might then use TMI while they will govern transactions of other technologies internally.

Having discussed in which conditions firms should use TMI to govern technology transactions or rather govern these with internal resources, I subsequently discuss whether the auction model is viable for technology transactions.

Is the auction model viable?

In general, good arguments hold for the use of auctions to determine the market values of technologies and patents in particular, which has often been quoted as a major obstacle to efficient clearing on the MfTI. Auctions are among the classical price negotiation mechanisms that have proven their validity for the trade of a variety of highly specialized goods. Regarding transacting technologies, since I investigated the six auctions, various subsequent auctions have already taken place.

However, before discussing this question in detail, I reflect on the success measure of judging auctions as a novel transaction model, which was applied in this study. This study investigates auctions in the context of open and distributed innovation processes, but hardly with regard to innovation activities inside the firms (intra-firm). Instead, this study looks into the transaction of technologies between firms (inter-firm) on the MfTI. It was not the aim of this study to empirically measure the overall effect of the complete innovation process. However, the effect should be discussed. This study has shown that auctions can have three effects on inter-firm technology transactions on the MfTI.

First, to acquire technologies, firms can either use auctions instead of other transaction models (for example, bargaining negotiations). If auctions provide a means to acquire technologies for lower TC than through bargaining negotiations they contribute positively to lowering the total costs of the distributed innovation process. Hence, the innovation process becomes more cost-efficient. Second, firms can purchase a technology on the MfTI instead of developing the technology internally. If the sum of the acquisition costs, which consist of the payments to the seller, and the TC_{market} are lower than the internal R&D costs and managerial costs (that is $TC_{hierarchy}$), then the firm saves resources and the complete distributed innovation process becomes more cost-effective. When technologies are patented, this situation appears particularly likely. Internal R&D costs are likely higher if the first-best solution to a technical problem has already been invented by others who have taken out patents (maybe even with a broad patent scope) and the firms have to take additional efforts to invent around it and find a second best solution. Third, through technology acquisitions from the MfTI, firms not only save resources (that is increase the innovation efficiency), they also accelerate the innovation process. This argument is similar to Braunerhjelm and Svensson (2010) licensing arguments.[13] If a firm's cost to acquire a technology (that is payment to the seller and TC_{market}) and the related implementation costs are below the internal R&D costs, firms can develop an innovation more quickly, hence accelerating the innovation speed if that acquisition is faster than the own development (that is despite various obstacles, for example, the well-known not-invented-here syndrome). However, the determinants for the transaction time and the relations to TC and the trade-offs between them should be studied in future research.

Aside from the success measure discussed from the demand perspective, auctions contribute to distributed innovation processes on the supply side as well. Auctions offer the possibilities (similar to some other TMI-developed transaction models for firms that lack internal ETE capabilities) for SME and independent inventors to supply the MfTI. If technology owners can find partners that possess better resources and capabilities to exploit a technology, auctions contribute to improved resource allocations for the efficient creation of innovations. At the same time, if a technology is exploited externally, the R&D RoI increases. For instance, concerning a patent family, separate patents can be auctioned for certain country designations, in which the technology owner has no operations. In this way, the sale of a patent can be interpreted similar to an exclusive, non-redeemable license for a certain region. However, to sell patents for specific applications, similar to licenses, appears hardly possible.

From the supply perspective, one might further notice that auctions can improve distributed innovation processes if a buyer finds a technology through an auction that he or she would not have found through other acquisition channels. For instance, the auction firms send their auction catalogues to the firms in their network (push marketing approach). A firm which they would perceive as interested in a certain technology would receive the catalogue, scan through it, and become aware of the availability of a technology as a solution to another problem the firm has. This firm might then acquire a technology it would have never considered before because it would have faced infinite TC if it would have searched for it through other (conventional) acquisition channels.

Through their marketing approach, the auction firms also bridge industrial barriers. However, due to the limited possibilities to exclude certain buyers, sellers are likely not to sell sensitive technologies, which they do not want competitors to acquire, although they would sell them to other firms from other industries or in countries where they do not have own operations.

In this study, I have primarily discussed the first and second effects focusing on the TC arguments. Hence, the comparative discussion of auction and bargaining negotiations refers primarily to the success criteria, efficiency and efficiency gains, in distributed innovation processes. From this study, I can hardly say anything about the quality of the outcome of a distributed innovation process where auctions have been used for technology transactions. For instance, although a distributed innovation process might be more cost efficient using auctions, the innovative product resulting from it might not necessarily be automatically more radical and widely accepted by the users or consumers. Moreover, from this study, I can hardly make any judgments on the quality and improvements of knowledge flows to accelerate innovation processes, although it became obvious that the auction firms act as network hubs that bridge industrial boundaries.

Furthermore, the study's results show that, in contrast to bargaining negotiations, the sellers experience relatively few problems with the observed technology auctions (Section 9.2). The case study firms agreed that the standardized contracts are a major advantage of auctions. Furthermore, the standardized transaction procedure, the increased transparency, the acceleration of the transaction process, and the increased quality of the technology valuation are considered advantageous in comparison to bargaining negotiations. The firms also agreed that the difficulties appearing when searching for prospective buyers (what has commonly been cited as obstacle to technology trade), appear resolved though the auction model. Sellers do not consider it problematic that they have limited information on the buyer of their technology. Although the sellers pointed out few disadvantages of the auction model, they perceive the disclosure of sensitive firm data and the inability to select / exclude certain buyers in general as disadvantageous. Particular sellers perceived the overall costs of the transactions as too high.

In general, the sellers also appreciated the role of the auction firm in supporting the transactions and indicated the importance of the auction firm's role with regard to various tasks. For instance, the sellers highly valued the support of the auction firm's governance of the transactions. The auction firms further play a major role in closing the transactions, providing a standardized legal framework (including the contracts) for the transaction, and in supporting the information exchange between prospective buyers and the sellers in the negotiation phase of the transaction. Furthermore, the sellers acknowledged the role of the auction firm in the general process administration, that is the auction firm as process owner. This appreciation is further reflected by the fact that the firms acknowledged the convenience of the transaction as a particular success measure.

However, one observation leads me to question whether TMI will be able to maintain the auction model on the MfTI. Across the six auctions in my dataset, the overall observed sales ratio was only 39.7 per cent, although this ratio slightly increased from auction to auction. This sales ratio remains far below those of art auctions (that is commonly found to be higher than 70 per cent[14]) that share certain properties with regard to their auction process structure (for example, primarily English bidding procedures) and the transacted assets (for example, art pieces are commonly unique and also difficult to value). However, the sales ratio of the investigated technology auctions appears to be in the same range as other types of technical auctions. For instance, Genesove (1995) finds that average sales ratios in wholesale, used automobile auctions vary between 32 per cent and 42 per cent. One also has to consider that the six auctions included in the dataset of this study include three first auctions in the US, Germany, and the UK. Having conducted a number of auctions, the auction firms will encounter learning effects that will help them to improve the auction design to make it more attractive to sellers and buyers, but also to improve their internal cost structure.

This study also shows that the current auction model is feasible for simple technologies with moderate values and for buyers and sellers that are mainly interested in a transfer of ownership rights of the related patents. Auction firms can survive if they ensure an internal and profitable cost structure and transaction possibilities for moderate value technologies at reasonable TC_{market} (that is currently mainly driven by the commission fees) that are below the $TC_{hierarchy}$ of internally governed transactions. If auction firms rather want to profit from selling small amounts of high value technologies than large amounts of moderate value technologies, they will need to adjust their current auction design and their TGS (for example, by limiting their commission fees; see Figure 11.3).

However, as argued above (see Figure 11.4 and 11.5), auction firms will likely face a cost competition with firms that continuously build up their transaction management capabilities through learning over time. To what extent the auction firms will be able to compete with them and reduce the TC_{market} in the future remains to be seen. This also depends on the auction firms' internal cost structure, which was not investigated in this study. Hence, in terms of this study, it appears hard to tell whether the auction model can compete in the long run with transactions that are internally governed by firms' IIP departments. However, even if auction firms do not survive by offering the auction model to firms, the auction model as such seems likely to survive to transact technologies. For instance, as soon as firms have built up enough internal transaction capabilities, hence have reduced their $TC_{hierarchy}$, they might try to design own auctions in rather closed settings that allow for more flexibility and more control of the TGS similar to, for instance, the CommerceOne auction.

To conclude, at least due to a particularly low TC_{market}, technology auctions seem to have opened up the MfTI by unlocking a previously unexplored technology segment. At least moderate value technologies can be auctioned, which are not profitably tradable via bargaining negotiations that are governed internally by firms due to a high $TC_{hierarchy}$. According to Tao et al. (2005), the segment of moderately valuable technologies is considerably large (see Figure 6.2) and accounts for approximately 50 per cent of all patents. Whether auction firms will survive offering their auction models depends on future developments. However, the auction model as such has proven its viability. Hence, through an increasing number of transactions of moderately valuable technologies on the MfTI, firms can avoid redundant R&D efforts and overcome the IP assembly problem, for which increased market clearing will contribute to the acceleration of firms' innovation processes.

However, the previous discussion on future MfTI developments has largely assumed that everything else remains equal. Relaxing this assumption, one should further consider that other developments take place, which impact the MfTI development. Whether or not governments will be able limit the increasing number (and size) of infringement suits against primarily large multinational firms from non-practicing entities (that is patent trolls) will certainly have an impact on the use of patents in the future. Governments also contribute to further marketing developments through policy measures on the one hand, and technology procurement programs on the other. In recent years, governments in various countries have invested heavily in the creation of TTO at universities. As governments have been dedicated to such investments for many years, TTO are likely to stay. However, they are still searching for their roles in the institutional framework of the MfTI. How will the roles of government funded TTO differ from those of privately funded TMI? Moreover, exogenous events, such as the financial crisis might trigger future MfTI developments. For instance, firms' IIP management might face considerable resource constraints during a financial crisis, and subsequently be forced to innovate even more economically. However, the question remains: Will the crisis lower business opportunities for TMI if firms innovate less in general or push the firms to outsource even more technology transactions using TMI, hence creating further business opportunities for novel transaction models?

NOTES

1 The effect sizes of these variables that were found in the regression models are all in a sufficiently strong range, and are hence not further discussed here.
2 Two equal sub-samples were formed with each group, representing 50 per cent of the number of cases. The cut-off point was an average remaining protection time of 12.17 years.
3 Klemperer (2004: 113) similarly argues that 'auction design may not matter very much when there is a large number of potential bidders for whom entry to the auction is easy'.
4 Related is the small numbers problem. Another example of a thin matching market is a marriage market with highly idiosyncratic assets.
5 This decision might actually not be strictly sequential as the decision between a TMI and transaction model might not be independent. For instance, some TMIs might offer unique transaction models, so that a firm would face extreme difficulties (and costs) running that transaction model itself.
6 To my knowledge, Mortara (2010) provided the most recent presentation of TMI developed transaction models.
7 Sellers pay a premium of 15 per cent on the sales prices and buyers are charged ten per cent on the sales price. However, sellers and buyers have additional costs (for example, the sala-

ries of employees to conduct the due diligence), wherefore a strict linearity assumption might not be realistic, but sufficient for the purpose of this discussion.

[8] See also the discussion of the notions, 'governance cost' and 'transaction costs', in Chapter 7.

[9] It should be noted that, in my discussion of integration, I restrict my discussion only to the TC, although there are other lines of arguments, which might very well lead to different conclusions.

[10] The variables used in my analytical models for the regression analyses have been entered as static variables. However, at least some of them are also dynamic. For instance, in Chapter 6, I explain why technological uncertainty changes over time.

[11] See also the organizational integration literature, rooted in the organizational theory literature, for instance Barki and Pinsonneault (2005), Schweizer (2005), Ettlie and Reza (1992), and Hamel (1991).

[12] Granstrand (2000b: 1064) argues that 'as infocom technologies advance and transactional information can be collected and processed more efficiently, it might be expected that various transaction costs (including search costs) could be lowered'.

[13] Edquist et al. (2000) also studied the effect of technology acquisition through governments on innovation (that is technology procurement).

[14] For instance, Ashenfelter and Graddy (2003) ran a study of sales rates and prices of impressionist and contemporary art based on an extensive dataset covering 150 auctions with more than 20 thousand items for sale from different departments at Christie's in London in 1995 and 1996. Although they find that sales rates vary tremendously over time and also systematically across different types of auctions and for different assets, the sales rates are considerably higher. In total, 96 per cent of items put up for sale in auctions of arms and armor were sold, 89 per cent of wine was sold, and 71 per cent of impressionist and modern art items were sold. On average, they computed a sales rate of 71 per cent for impressionist and modern art and an average sales rate of 77 per cent for contemporary art.

12. Conclusions, Implications, and Research Recommendations

12.1 CONCLUSIONS

For various reasons (for example, the increasing complexity of technical products and intensified competition), firms have recently started to open up their innovation processes beyond the own firm boundaries. When innovating openly, firms engage in transactions on the MfTI where they trade ideas (that is in an early stage of the innovation process) as well as patents, licenses, and technologies. Some firms act on the demand side and acquire technologies externally, other firms exploit their technologies externally, acting on the supply side, and some firms might do both. However, when entering the MfTI, firms often lack the internal capabilities to govern transactions efficiently. Furthermore, the institutional framework is still rich with many obstacles that prohibit efficient transactions, which leads to high transaction costs. This study is concerned with novel transaction models developed by technology market intermediaries (TMI) and whether or not they can improve the efficiency of technology transactions primarily through the reduction of transaction costs. This study particularly emphasizes the auction model.

The results of two qualitative studies show that the structure of technology auction processes fits the three core phases of the Lichtenthaler (2006b) model for external technology exploitation projects. Accordingly, three phases can be distinguished in the auction process structure. Each is distinguished by a key event:

1. Pre-auction (planning) phase: Initiated by the auction announcement. It includes the registration of sellers and the submission of their technology proposals, as well as the internal evaluation of technologies through the auction firm.
2. Bidding (negotiation) phase: Initiated by the auction catalogue publication. It includes the registration of prospective buyers, the due diligence of technologies through prospective buyers and the bidding at the live auction event.

3. Post-auction (realization) phase: Initiated after the live auction event. It includes the formal transfer of the payments from the buyer to the seller in exchange for the ownership rights (that is intellectual property rights), eventually a transfer of complementary (tacit) knowledge from seller to buyer, the payment of commission fees to the auction firm from the buyers and sellers, and post-auction negotiations of unsold assets that have been withdrawn from the auction.

When auctions are compared to bargaining negotiations (for example, with idiosyncratically designed contracts) the results of the present study show that sellers save resources, particularly during the negotiation and realization phases of the transaction process. Hence, in contrast to bargaining negotiations, the auction process design accelerates the transaction speed.

Furthermore, the results of the present study reveal that six elements induce standardization in the TGS of technology auctions. These refer to standardized participation fees from the sellers, joint marketing activities for all technologies, standardized buyer participation fees, virtual data rooms, a standardized legal framework, and a fixed commission fee rate. Thus, the finding of this study leads to the conclusion that the TGS of the investigated technology auctions are situated close to the right extreme on the continuum of transaction governance structures (TGS) (Figure 7.1), for which technology auctions almost represent spot market transactions.

This study finds that the standardization of TGS together with the economies of scale reduce TC for technology transactions. Auction firms realize economies of scale by means of bundling marketing expertise and capitalizing on the auction firms' customer networks across a wide range of industries.

However, even if auctions reduce the TC of technology transactions, sellers and buyers will still be liable for certain TC. This study's results reveal that, in technology auctions, the major share of TC is determined by the commission fees that sellers and buyers have to pay to the auction firm. As the commission fees are charged as a fixed share of the sales price for high value technologies, the current auction design appears inferior to bargaining negotiations, at least from a TC perspective. In the observed auction design, sellers and buyers are largely restricted in their possibilities to adjust the rules governing the transaction and have little flexibility to adjust the TGS to their individual preferences. These restrictions also disfavor transactions of technologies with high degrees of technological complexity and uncertainty.

This study tested which technologies appear particularly suitable for auctioning. Based on the pre-studies' results and the framework for the choice of TGS proposed by Williamson (1979), hypotheses were developed for five technological properties. These include the technological complexity, technological uncertainty, technological impact, technology quality, and the owner's perceived technology value. The hypotheses were tested with 393 technologies (156 (39.7 per cent) were auctioned and 237 (60.3 per cent) remained unsold) using various multivariate logistic and log-linear OLS regression models and validated using a range of sensitivity analyses. The analyses reached a mean sales price of €194.15 thousand (median of €77.23 thousand) and confirmed the hypotheses to a large extent.

The results show that technologies are more likely to be auctioned if they are characterized by low complexity, high technology quality, and technological impact. Technologies for which the sellers have medium / moderate value expectations are more likely to be auctioned. The results show also that technologies reach higher sales prices if they have a higher technological complexity, higher technological impact, and higher technology quality. The sales price also appears to be positively related to the value perception of the sellers. The results of this study indicate that, due to a particularly low TC_{market}, technology auctions seem to open up the MfTI by unlocking a previously unexplored technology segment. Moderate value technologies can be auctioned, which cannot be transacted profitably due to a high $TC_{hierarchy}$ caused by bargaining negotiations that are governed internally by firms.

12.2 IMPLICATIONS FOR INNOVATION AND IP MANAGEMENT

When firms open up innovation processes beyond the own firm's boundaries, they can improve innovation outcomes in terms of lower development costs, higher innovation speed, and increased innovation quality through the acquisition and exploitation of external technologies from and on the MfTI. To what extent firms can realize the benefits of open and distributed innovation activities largely depends on firms' innovation and IP (IIP) management capabilities to govern technology market transactions. The results of this study support firms' IIP management in various facets.

This study's results imply that firms or their IIP management, face three decisions in the context of technology transactions (see the decision framework in Figure 11.1). First, firms need to decide whether they want to develop a technology internally or acquire it on the MfTI. This decision is related to the widely discussed make-or-buy decision in the industrial organization literature, which deals with vertical integration questions. Similarly, firms that own technologies have to decide whether they want to exploit them solely internally or beyond the firms' boundaries on the MfTI. These firms face a keep-or-sell decision. As this study is related to the open innovation approach, it is concerned with the sell and buy options. Following this decision, IIP management has to decide on how to govern the technology market transactions.

As shown in the decision framework (Figure 11.1), IIP management subsequently needs to decide on whether to govern transactions internally or seek the support of a TMI. The results of this study imply that, when making this decision, IIP management should evaluate the firm's internal costs for governing a transaction ($TC_{hierarchy}$) and compare these to the TC_{market} they would face if engaging a TMI in order to govern a transaction. Hence, firms should prefer internally governed transactions if the associated $TC_{hierarchy}$ is below those of TMI governed transactions (TC_{market}). When firms start to engage in open innovation activities, they are likely to lack the internal capabilities to efficiently manage technology transactions. Thus, internally governed transactions will likely not be cost-efficient.

However, the results of this study also show that TC are not static but change over time. By participating in technology market transactions, IIP management will learn and establish internal transaction management capabilities cumulatively (Figure 11.4). Hence, IIP management will be able to constantly reduce the $TC_{hierarchy}$ and force TMI into a cost competition. TMI will then have to react to firms' increased capabilities by lowering their TC_{market}.

The decision framework presented in Figure 11.4 further implies that IIP management faces a third decision. IIP management has to select the most promising transaction model for any acquisition or exploitation project. Although there are various transaction models, from this study's results, implications can be derived to support those of firms, particularly when considering technology auctions as possible transaction model. The results of this study imply that the observed TGS of technology auctions favors the auctioning of particular technologies. IIP management should primarily consider technologies that are characterized by low complexity, high technology quality, and technological impact as well as moderate values if auction firms do not improve their design. Particularly for high value technologies, the advantages of the auction model dimish due to a constant commission fee rate that accounts for a major share of the TC_{market}. For high value technologies, IIP management should dare to develop idiosyncratic bargaining negotiations, maybe seeking the support of a PLF. In bargaining negotiations, the sellers can maintain control of the TGS and use the transaction model's flexibility to leverage their bargaining power to optimize prices.

In exceptional cases, where IIP management lacks internal capabilities, IIP management can also consider auctions as an option for high value technologies, particularly if many prospective buyers are interested in acquiring such a technology. The winner's curse effect, which is associated with the English bidding procedure can be optimized for sellers, especially for technologies with a strong common value element. However, if IIP managers struggle to identify prospective buyers by themselves, they should be careful to use auctions. Although auction firms can draw on their networks that might reach beyond the network of a technology owner, few prospective buyers are likely to compete and thus the firm is unlikely to achieve high sales prices. In addition, if sellers would bear high costs to supply necessary information to prospective buyers (that is due to high technological uncertainty and a large share of tacit know-how), they should prefer bargaining negotiations.

In addition to the TC perspective, IIP management should evaluate whether the firm can reach higher sales prices if a technology is auctioned instead of using bargaining negotiations (for example, through the competitive bidding element and the winner's curse effect). Furthermore, in general, IIP management should always evaluate whether TMI-developed transaction models possess distinct advantages which firms will find difficult to replicate. IIP management should prefer TMI-governed transactions if the specific transaction model possesses a particular advantage through which higher sales prices can be achieved (that is higher than what can be achieved through own negotiation capabilities). IIP management should then accept an even higher TC_{market}, at least up to a level where additional profits can be realized. Auctions are also advantageous if firms attempt to shorten transaction times, but do not allow sellers to exclude certain prospective buyers from the bidding (for example, direct competitors).

As soon as sellers have established internal capabilities (that is by learning from constant participation in technology market transactions), firms should consider conducting their own auctions. The results of this study imply that the auction model in general has various advantages over bargaining negotiations. However, the auction firms have compromised some of the advantages through a high degree of standardization, thereby lowering the TC_{market}. If sellers conduct their own auctions, they can still be designed to benefit from the competitive bidding element (and the winner's curse effect) in order to optimize their outcome. Moreover, the auction design can be adapted to fit the auction firm's preferences (for example, to develop case-by-case contracts and include royalty schemes). Sellers can also exclude certain prospective buyers, to whom they do not wish to sell a particular technology (for example, direct competitors) by conducting private auctions in a closed setting with only pre-selected prospective buyers. Thus, firms that have established internal governance capabilities for technology market transactions can combine the advantages of competitive bidding situations of auctions with idiosyncratically designed bargaining negotiations into semi-standardized auctions. The CommerceOne case showed that a semi-standardized model can work quite successfully.

If individual firms cannot create sufficient market thickness to conduct an own auction, they can form consortia with other firms as well as university sponsored TTO. Through further ICT developments (for example, the internet in general and patent management software in particular), firms will increasingly be able to widely diffuse the auction announcement and subsequently reach more prospective buyers. In addition, further developed software tools will support them in identifying prospective buyers, who they can then invite to the auction (for example, the availability of semantic patent analysis methods will improve the match-making).

Aside from the implications regarding firms' decisions in the context of technology market transactions, the results of this study imply that technology owners are positively biased when evaluating expected prices for technologies that they own. This study's results indicate that owners apply cost-based valuation approaches, which lead to higher expected sales prices than realized sales prices. The results show an average sales price of €194.15 thousand for a technology. However, the sellers expected an average sales price of €525.56 thousand. When firms determine the minimum reserve price at auctions, depending on their internal valuation, technologies for which internal cost-based valuation approaches lead to distinctively higher results than those of market valuation approaches remain unsold at auctions. Hence, the results of this study imply that technology owners should refrain from valuing technologies using cost-based approaches, but accept that they can only expect prices that represent market prices; even if these prices do not allow them to recapture their (sunk) R&D costs to develop the technology. The results of this study offer a guideline for sales prices of auctioned technologies, at least if their transaction governance structures are designed similarly.

These implications should be relevant for large firms that might possess the internal resources (although maybe not yet the capabilities) to govern complex technology transactions internally, as well as for SME, TTO from universities / research institutes, and independent inventors who need to rely more on the external support and expertise of TMI to govern transactions.

12.3 IMPLICATIONS FOR THE DESIGN OF TECHNOLOGY AUCTIONS

The results of this study imply that the design of technology auctions has proven its viability in general. On average, each of the observed auctions generated about €5 million in sales. Each auction sold an average of 26 technologies in a reasonably short timeframe. The sales ratios are comparable to other technical machinery auctions in which unique (used) assets are auctioned. Although the costs of organizing technology auctions remain unknown, the fact that various auctions have taken place after the six auctions that were investigated in this study indicates that the auction model can be sustainably maintained.

However, the results of this study show that the TMI-developed auction models compete with bargaining negotiations that are internally governed by technology-based firms. Over time, particularly firms that engage in open innovation activities will establish internal transaction management capabilities through learning. Therefore, auction firms will face an increasing (cost) competition with firm-governed transactions. Hence, auction firms should adjust the auction model's TGS in order to remain competitive. Resulting from this study, I propose measures to improve the auction design. These measures should also be considered by technology-based firms that plan their own technology auctions.

This study's results imply that, by standardizing the TGS, the auction firms have developed a transaction model which allows them to transact technologies at a lower TC_{market} compared to idiosyncratic, bargaining negotiations. However, the auction firms have standardized the TGS to such a degree that it constrains transaction possibilities for certain technologies. The current auction design – in line with the theoretical predictions derived from TC theory – appears preferable mainly for technologies with low complexity, with low technological uncertainty, and moderate values.

To favor transactions of highly complex and uncertain technologies (that is particularly as these are increasingly developed and needed by technology-based firms to economically innovate increasingly complex products, processes, and services) auction firms should allow for more flexibility in the TGS of technology auctions. Sellers (as well as buyers) should be able to influence the TGS, at least to a limited extent, to account for the idiosyncratic nature of technologies. In particular, auction firms should allow for alternative payment schemes (that is royalties) than simple lump sum payments. Moreover, they should adjust the standardized contracts (for example, to include grant back clauses). Furthermore, this study's results imply that the auction design should be adjusted to give sellers the possibility to exclude certain prospective buyers (for example, direct competitors). The results of this study imply that the preferred approach is to ask sellers to specify firms that should be excluded from bidding on certain technologies, without knowing if any of these firms has registered as a prospective buyer. This approach relies on the reasonable assumption that the sellers know which firms they do not want to acquire a particular technology. In order to avoid excluding too many sellers, the auction firms should charge a fee for each firm that is named by the seller to be excluded. Thereby, the auction firms create an incentive to limit the list to only a few firms. This approach allows the prospective buyers to remain anonymous, since the auction firms do not reveal their names to the sellers, thereby also avoiding the disadvantages of a broad approach, in which sellers specify industry classes from which they do not want firms to purchase one of their technologies. However, the auction firms should ensure that prospective buyers do not circumvent this approach and hire intermediaries to bid on their behalf.

The results of this study imply that auction firms should not give away additional flexibility for free. Similar to the minimum reserve price, which sellers are allowed to pre-determine at an extra fee, sellers should be given the possibility to purchase extra flexibility at an additional fee. This will create an incentive to keep the TC_{market} as low as possible. Moreover it will provide an incentive for sellers to use the highest degree of standardization and only consider buying more flexibility if necessary. A further advanced possibility to allow more flexibility is to design packages that range from a low fee package with a high degree of standardization (similar to the currently observed TGS) to a package that allows a high degree of flexibility (for example, more complex contracts and a longer due diligence period) at a higher fee.

This study's results imply another measure to make the auction model more attractive for transactions of high value technologies. The auction firms should limit the commission fee payments to a maximum level or at least specify a lower commission fee rate above a certain sales price. Furthermore, the results of this study imply that auction firms should also consider other types of bidding procedures than the English auction, which is vulnerable to the winner's curse effect. In particular, as the observed auctions have suffered from a low number of bidders, the auction firms should raise the attractiveness of technology auctions for prospective buyers. Despite the fact that the winner's curse effect is favorable for sellers, who benefit from high sales prices, and the auction firms, which benefit from the sales prices through a fixed commission fee rate, auction firms should make technology auctions more attractive for prospective buyers by means of minimizing the risk of the winner's curse effect.

12.4 RESEARCH RECOMMENDATIONS

While various aspects were researched in this study, many had to be excluded to keep the scope of the study manageable. In addition, new aspects also arose during this study. In the following section, I present selected aspects that should receive further research to generate relevant insights for firms' innovation and IP management (IIP), technology market intermediaries (TMI), the markets for technologies and ideas (MfTI), auction design and theory, as well as for transaction cost (TC) theory.

Traditional patent law firms, university (and hence government) sponsored technology transfer offices (TTO), brokers, as well as dealers of IP assets and technologies are hardly new phenomena and have already received considerable research attention. However, recently TMI have emerged on the MfTI that develop novel transaction models (for example, auctions) that largely rely on new ICT technologies (for example, virtual data rooms). Although previous research has developed a conceptual basis for intermediaries, in the specific context of technology market transactions, they have only been studied to a limited extent. In addition to the results generated by this study, further research is needed to support IIP management to substantiate why, when, and how to use which transaction model to improve distributed innovation processes (that is accelerate them through transaction time reduction, increase cost efficiency through TC reduction, and improve the quality of the innovation outcomes).

Obviously, when firms have to select a transaction model for each acquisition / exploitation project, the IIP management needs to understand the different novel transactions models developed by TMI. Although this study's results support them, the decision had to be simplified in this study in order to be dichotomous. This study has assumed that firms decide between auctions and bargaining negotiations. Future research should distinguish between more than these two extreme choices (that is spot market auctions vs. idiosyncratic bargaining negotiations) and further conceptualize and empirically test a multi-dimensional choice among more alternatives. So far, research has been hindered due to the early stage nature of this phenomenon. As TMI still experiment with manifold transaction models the empirical field remains diverse, heterogeneous, opaque, and dynamic. New TMI constantly enter the MfTI while others with unsustainable transaction models exit them. In such a situation (that is where the number of observations is still limited), explorative single case studies appear as suitable approaches for further research. For instance, investigating the business model of Intellectual Ventures (IV), which has recently emerged as a large player in the patent business, purchasing large amounts of patents backed up by large firms, appears to be an interesting single case study.[1] Hence, it seems likely that IV – due to its purchasing power – will have a sustainable impact on the MfTI developments. Within a single case study, further research should investigate how IV will shape the MfTI. Furthermore, IV's business model is built on the ownership concept (that is IV acts as dealer), which is in contrast to the auction model, which is built on the broker concept. Another interesting concept that should receive further research is patent pools. Patent pools have shown to be a possible solution to the IP assembly problem, for example, in the DVD or MPEG2 cases (Andersson and Phillips, 2005), despite the conflict potential with antitrust laws. Furthermore, results from additional case studies of individual transactions via auctions or other transaction models should also be investigated, such as the CommerceOne case. Preferably, the case studies should combine data from the sellers, TMI, and buyers to avoid biased results due to an asymmetric empirical basis.

To support firms' decisions on the most suitable transaction model, the case studies should then be compiled into a typology of the transaction models. Such a typology should at least attempt to quantify the TC (and transaction time) of the transaction models, on a relative scale. This process is similar to the ranking of acquisition / exploitation strategies according to the degree of organizational integration between the Granstrand framework (Figure 3.4) and the continuum of governance structures, thereby ordering the transaction models of TMI between both extremes of spot market and hierarchically governed transactions (Figure 7.1). The typology proposed by Tietze (2010) as well as the recently released study by Mortara (2010) can serve as a starting point. As illustrated by the IV example, such a typology should also distinguish between the broker and dealer concept (that is dealers take ownership and brokers do not). When developing the typology, results from the marketing field should be used, particularly with regard to the choice of distribution channels, although the different nature of tangible products and intangible IPR needs to be considered.

Later, when this phenomenon (that is TMI governed transaction models) has become more stable and firms (sellers and buyers) have gained sufficient experience, quantitative survey studies can follow the qualitative case studies. Ultimately, research is needed to address three questions:

Which of the existing novel transaction models will survive? Which new TMI types will emerge and what novel transaction models will they develop? Can there be any permanently superior model for technology transactions?

Furthermore, this study demonstrated that auction results can be used to study determinants of technology and patent values. As it turns out, the use of auction data is particularly advantageous because the technology (and patent) values can be measured in terms of the real prices paid on the MfTI instead of proxies, which were commonly used in previous studies. However, further research is needed to better understand what value types the prices paid at auctions reflect in comparison to other patent value measures, for instance, the measure used in the PatVal studies. Moreover, it should be researched to what extent auction prices can be generalized to the population of technologies. This study's results only show that the auctioned technologies (even if they constitute 3.7 patents) have lower values than the PatVal patents. To explain this difference, for instance, regression analyses can be computed in which properties of individual patents from this dataset (that is single patent technologies) are compared with sub-samples of PatVal patents.

In addition, five research recommendations are derived from this study with regard to methodological aspects. First, the analytical models should be substantiated through larger datasets (for example, derived from the PatStat database) allowing researchers to compute models with a larger number of indicators that also account for interaction effects (for example, moderator and mediator effects). Second, regressions should be computed so that models are not only linear. For instance, the discussion of the regression results shows that the relationship between technological uncertainty and the sales price appears to be u-shaped rather than linear. Third, it should be researched whether Heckman selection models can confirm the results of separate logistic and OLS regression models as used in this study, and additional results should be revealed. Fourth, further research should put more emphasis on the coefficients and the elasticity of technological properties. Fifth, despite the amount of empirical studies that use patent data, and hence a variety of measures for different concepts (for example, technological impact) on different levels (for example, industry, firm), the models can be improved through further research to develop more advanced measures.

Concerning these measures, particularly three research recommendations can be formulated. First, using patent data, future research should develop measures to determine whether transacted technologies are product, process or product–process technologies, especially since firms often perceive the latter ones to have a higher value. Second, measures need to be developed in order to operationalize technology properties that combine multiple dimensions (or indicators) to strengthen the reliability of the empirical results applicable for patent data. Third, measures need to be developed further in order to allow for the measurement of individual patents' as well as patent bundles' properties. For instance, the distance of IPC classes can be counted for individual patents as an indicator of a patent's breadth; however, this measure needs to be extended in order to be applicable for a set of patents.

Furthermore, the results of this study show that sellers tend to overvalue the technologies they own (that is endowment effect). Therefore, they are likely to specify a minimum reserve price (MRP), which is above the technology's market value and, hence, these technologies remain unsold. Subsequently, it should be further researched how sellers can determine the optimal minimum reserve price (MRP) – a subject that has received considerable attention in previous auction theory studies (for example, Ashenfelter et al. (2003) but not yet in technology auctions. Hence, additional models should be computed to better quantify the positive seller bias (that is the relationship between the expected sales price and the realized sales price) and discriminate it between different seller types. For instance, one can argue that large firms might be more advanced in technology valuations and hence be more capable of improving their R&D return on investments; therefore, this effect might be more prominent among SME.

During this study, I noticed that the literature provides only a few techniques to identify components of technologies from patent data. This study's dataset (or the auction catalogues) can serve as a valuable data source to develop automatized techniques that use patent databases to identify the individual components of technologies or the related patents. The results from an automatized algorithm can be compared to the data used in this study (that is the patent bundles as listed in the auction catalogue) in order to validate such techniques' performance.

Although this study reveals the TGS of technology auctions, the design of technology auctions needs to be further researched in order to determine how the design of technology auctions can be altered to improve resource allocations between buyers and sellers (that is market matching). Ultimately, it also needs further research to determine which price levels represent efficient resource allocations on the MfTI. For instance, which price levels and auction designs contribute to the alignment of buyers' and sellers' objective functions and hence eliminate principal–agent problems. When studying technology auctions, future research should also correct for this study's asymmetric empirical base and include the buyers' perspective (for example, motives and Willingness-to-pay). For instance, access to the prospective buyers' bidding data would allow simulation studies to be conducted to determine the outcome of auctions that use different bidding procedures (for example, Dutch or Vickrey auctions). Moreover, auctions for art (and antiques), which share certain properties with technologies and also prefer to use English bidding procedures, were found to be susceptible to bidder collusion (Beckmann, 2004). Consequently, it should be studied if and to what extent collusion and rings take place in technology auctions and have an impact on the auction outcomes. Future research should also investigate whether technology auctions are subject to the winner's curse effect. Prospective buyers' bidding data should be used to determine whether this effect is present and, if so, to determine the size of it.

Aside from the study of rather specific technology auctions, I also recommend developing a generic typology that clusters auction designs according to its properties. Although the auction literature has seen a recent growth with the rise of the internet (for example, since large datasets from online auctions have become accessible), it seems that a comprehensive typology of auction design is missing, particularly one that can also be applied by non-auction theorists. Commonly, the auction literature distinguishes between auction types according to the properties of the bidding procedure (for example, ascending / descending). However, I have found that auctions should not only be discriminated according to bidding procedure properties, but also properties related to their governance structure. At least two properties have emerged from this study that should be considered in such a typology. First, it should be considered whether any bidder should be allowed to participate or whether participation is restricted (open / closed). Second, the typology should consider whether the bidding as well as the due diligence should take place online or offline.

This study was primarily based on TC theory and elements from the capability-based approach to explain recent and predict future developments of the MfTI, particularly with regard to firms' use of TMI. These arguments were used in the previous chapter to discuss how firms disintegrate and reintegrate transaction activities to and from TMI over time and how TMI alter the division of labor between the actors on the MfTI. In addition, it should be researched whether other theoretical arguments also support this study's findings. For instance, researchers should determine whether the arguments from the literature dealing with institutional change (developed after the seminal contribution of North (1971)) and those dealing with vertical integration (following the notion of the division of labor pioneered by Stigler (1951)) can be used to explain the emergence of TMI and predict their impact on the MfTI. Moreover, future studies should investigate whether the arguments provided by the endogenous growth theory (for example, associated with Paul Romer, Philippe Aghion, Chad Jones, and

Peter Howitt) and evolutionary growth models (for example, associated with Richard Nelson, Sidney Winter, Giovanni Dosi, and Sten Metcalfe) can be applied to explain the emergence of TMI, future developments (that is alternating reintegration and disintegration periods), and the recently observed growth of the MfTI.[2]

Further research is necessary to better understand the impact of TC on the choice of transaction models, but also to identify other determinants that affect this decision. I have argued that TMI reduce search costs, in particular, to find prospective buyers by providing a consolidated marketing approach for various technologies. However, future research should examine to what extent these costs are reduced. More importantly, future research should determine how efficient the auction firms' advertising efforts are or how important sellers regard those costs as contributions to the decision of transaction models – although there are some implications that the search costs constitute a large share. In addition to TC, there are other determinants for this decision in firms. For instance, firms might consider the transaction time an important determinant, particularly if they face competition with other firms to bring a product to a market first (that is accelerate innovation speed to reduce market lead times). Moreover, future research should investigate to what extent the function of a technology in the firms' portfolios (that is strategic vs. blocking patent) determines the transaction model. Furthermore, for a holistic examination, future research should not only consider the contributions of the independent determinants for the selection of transaction models, but also their interrelations (for example, the transaction time and its relations to TC as well as the trade-offs between them). Firms' decisions are determined by micro-level determinants (that is firm preferences) as well as macro-level factors, which comprise, for instance, ICT developments as discussed in Chapter 11, as well as other contextual factors (for example, governmental policy) and exogenous events (for example, the recent financial crisis).

I also recommend compiling another typology that clusters obstacles which prohibit efficient resource allocations on the MfTI. To date, different scholars have reported various obstacles, which are scattered across different studies. Based on the literature presented in Part II, through a more advanced literature review, these obstacles can be consolidated. Subsequently, the obstacles should be clustered in a typology, which measures each obstacle's impact on market efficiency (although maybe only on a relative scale, that is ranking from high to low impact). Such a typology would not only support firms in their decisions, but also governments to develop appropriate policy measures to tackle the obstacles in a decreasing impact order.

Although this study provides arguments that speak in favor of TMI's contribution to the efficiency of distributed innovation processes, this discussion needs to be extended and substantiated through further research. Other valid arguments need to be considered. For instance, in contrast, one could argue that, following the emergence of a new type of institution on the MfTI (that is additional actor), the institutional complexity increases, which might negatively impact innovation outcomes. Investigating whether and under which environmental conditions TMI improve the efficiency of transactions and the general functioning of the MfTI was beyond the scope of this study, but should be considered a relevant topic for future research. Future research should empirically assess the contributions of TMI in general and of different transaction models in particular, not only to assess their efficiency, but also to determine the qualitative outcomes of distributed innovation processes. Although these two effects are difficult to measure and quantify, a relative assessment approach complemented by a theory-guided discussion can shed some light on this topic. A possible approach to quantitatively determine the efficiency of auctions could be to compare technology auctions' results with empirical results from auctions that sell assets with similar properties. For instance, this study's results of technology auctions can be compared to the results of empirical studies on art auctions, as art pieces share certain properties with technologies (for example, uniqueness and difficulty to value). Results from econometric studies of art auctions (for example, Ashenfelter and Graddy (2003), Frey and Pommerehne (1990), Anderson (1974)), for instance, reveal a number of determinants for pricing art. Investigating whether these determinants are valid for and can be transferred to technology auctions is open to further research.

Hence, to gain a better understanding of additional determinants for firms' decisions of transaction models that are directly linked to the efficiency of MfTI and distributed as innovation processes remains a multifaceted challenge for future research. This research will assist firms, IIP management, TMI and governments to understand which determinants they can influence and to what extent this can be done, in order to increase transaction efficiency through the reduction of transaction obstacles. In turn, this will lead to improved resource allocation, increased innovation efficiency, as well as better innovation outcomes.

NOTES

[1] See further the report by Ewing (2010b).

[2] For instance, Aghion (2005), developed a notion showing how different types of institutions or policies maximize growth at different stages of technological development.

Annexes

ANNEX I - EVOLUTION OF TMI

This data was collected by means of a pyramiding exercise by experts from the TMI community. I cannot confirm this sample's completeness, since the firms are so widely spread across the globe, the majority is small, and the market is currently subject to a strong dynamic and fluctuation. However, it could be claimed that the sample includes the most important firms and covers at least 80 to 90 per cent of the firms up to December 2006. Further details on the sampling procedure can be found in Tietze and Barreto (2007).

Foundation year	Newly founded TMIs annually	Accumulated	Growth ratio to y-1	Non- US based	US based	Share of US based
1948	1	1	na		1	100.00%
1968	1	2	100.00%		1	100.00%
1972	1	3	50.00%		1	100.00%
1975	1	4	33.33%	1	0	0.00%
1982	1	5	25.00%		1	100.00%
1984	3	8	60.00%		3	100.00%
1985	1	9	12.50%		1	100.00%
1989	5	14	55.56%	2	3	60.00%
1990	4	18	28.57%	1	3	75.00%
1992	4	22	22.22%	1	3	75.00%
1994	1	23	4.55%		1	100.00%
1996	3	26	13.04%	1	2	66.67%
1997	2	28	7.69%		2	100.00%
1998	6	34	21.43%	1	5	83.33%
1999	1	35	2.94%		1	100.00%
2000	3	38	8.57%		3	100.00%
2001	8	46	21.05%	2	6	75.00%
2002	4	50	8.70%		4	100.00%
2003	4	54	8.00%		4	100.00%
2004	3	57	5.56%		3	100.00%
2005	1	58	1.75%	1	0	0.00%
2006	2	60	3.45%	2	0	0.00%
Total	60		23.52%	12	48	78.86%

ANNEX II - CONCORDANCE TABLE FOR APPLICATION AREAS

ID #	Consolidated application	Original application area
1	Automotive Systems & Devices / Industrial Machinery	Automotive
		Automotive / Industrial Machinery
		Automotive Systems and Devices
		Automotive Systems and Devices
2	Aviation & Aerospace	Aerospace
		Aviation & Aerospace
3	Business Methods / Financial Services	Business Methods / Data Systems
		Business Methods / Financial Services
		Financial Services
4	Chemical Engineering / Applied Sciences / Advanced Materials / Nanotech.	Advanced Material and Chemicals
		Applied Sciences
		Chemical Engineering
		Chemicals/ Materials
		Material Science /Nanotechnology
5	Consumer Products / Electronics	Consumer Products / Electronics
6	Digital & Home Media / Entertainment	Digital and Home Media
		Digital Media / Entertainment
		Digital Media Systems and Management
7	Display / Multimedia / User Interface Technologies	Display / Multimedia Technologies
		Display / Multimedia Technologies
		Display Technology
		Display Technology and Digital Imaging
		User Interface Technology
8	Electrical / Mechanical Engineering / Process Automation	Electrical Engineering
		Mechanical Engineering
		Process Automation
9	Energy / Utilities / Power Technologies / Oil & Gas	Energy / Utilities
		Oil and Gas
		Power Technologies
10	Green Energy / Technology / Industrial Systems	Green Energy and Technology
		Green Industrial Systems
		Green Technology

Source: OT and IPA catalogues

ID #	Consolidated application	Original application area
11	Internet / Web Services / E-Commerce / Internet TV	Internet
		Internet / Web Services
		Internet TV Network
		Online Advertising and Content
		Social Networking
		Web Based Services / E-Commerce
12	Location-based Systems / Applications / Logistics	Location-Based Applications
		Location-Based Systems / Applications
		Location-Based Systems and Logistics
		Logistics
		Manufacturing / Logistics
13	Medical & Life Sci / Devices / Telemedicine / Bio-chip / Bio-tech	Biotech / Life Sciences
		Medical
		Medical / Bio-chip/ Bio-tech
		Medical / Life Sciences / Telemedicine
		Medical Science
		Telemedicine / Medical Devices
14	(Micro-) Electronics / Computer Systems & Technologies	Computer-Based Systems and Technologies
		Computer Systems and Data Storage
		Computer Systems and Electronics
		Data Storage
		Electronics and Handheld Devices
		Integrated Circuits and Memory
		Memory Technology
		Microelectronics / IT
		Semiconductor / MEMS / Memory
		Semiconductor / Semiconductor Manufacturing
		Software / Messaging / Computer Systems
15	Telecommunications (Wireless, Cellular, Optical & Other)	Bar Code / RFID Technologies / Smartcard
		Communications
		Telecommunications
		Wireless / Cellular / Optical and Other Telecom
16	Miscellaneous	Entertainment and Gaming
		Food Services
		Miscellaneous

ANNEX III - CORRELATION TABLE OF VARIABLES

		Sales status	Sales price [T€]	Technological complexity	Technological complexity (Lg10)	Technological uncertainty	Technological uncertainty (Lg10)	Technological impact	Technological impact (Lg10+1)	Technology quality	Technology quality (Lg10)	Expected value [T€]	Expected value [T€] (EM Imputation)
Sales status	Pearson Correlation	1											
	Sig. (1-tailed)												
	N	393											
Sales price [T€]	Pearson Correlation	.[a]	1										
	Sig. (1-tailed)	,000											
	N	156	156										
Technological complexity	Pearson Correlation	-,185**	-,038	1									
	Sig. (1-tailed)	,000	,321										
	N	393	156	393									
Technological complexity (Lg10)	Pearson Correlation	-,218**	-,021	,837**	1								
	Sig. (1-tailed)	,000	,400	,000									
	N	393	156	393	393								
Technological uncertainty	Pearson Correlation	-,018	-,133*	-,137**	-,165**	1							
	Sig. (1-tailed)	,362	,049	,003	,001								
	N	393	156	393	393	393							
Technological uncertainty (Lg10)	Pearson Correlation	-,004	-,136*	-,119**	-,147**	,944**	1						
	Sig. (1-tailed)	,468	,045	,009	,002	,000							
	N	392	155	392	392	392	392						
Technological impact	Pearson Correlation	,035	,388**	,154**	,248**	-,414**	-,363**	1					
	Sig. (1-tailed)	,243	,000	,001	,000	,000	,000						
	N	390	156	390	390	390	389	390					
Technological impact (Lg10+1)	Pearson Correlation	,018	,350**	,206**	,347**	-,534**	-,463**	,829**	1				
	Sig. (1-tailed)	,362	,000	,000	,000	,000	,000	,000					
	N	390	156	390	390	390	389	390	390				
Technology quality	Pearson Correlation	,035	-,009	,168**	,134**	,028	,036	-,048	,000	1			
	Sig. (1-tailed)	,247	,457	,000	,004	,287	,240	,171	,497				
	N	393	156	393	393	393	392	390	390	393			
Technology quality (Lg10)	Pearson Correlation	,045	-,019	,216**	,202**	,001	,008	-,016	,045	,933**	1		
	Sig. (1-tailed)	,186	,407	,000	,000	,493	,436	,376	,187	,000			
	N	393	156	393	393	393	392	390	390	393	393		
Expected value [T€]	Pearson Correlation	-,021	,568**	,529**	,411**	-,088	-,094	,127*	,151**	,019	,038	1	
	Sig. (1-tailed)	,357	,000	,000	,000	,065	,053	,014	,005	,371	,259		
	N	299	112	299	299	299	298	296	296	299	299	299	
Expected value [T€] (EM Imputation)	Pearson Correlation	-,047	,664**	,523**	,406**	-,092*	-,096*	,183**	,191**	,030	,041	1,000**	1
	Sig. (1-tailed)	,176	,000	,000	,000	,034	,029	,000	,000	,276	,210	,000	
	N	393	156	393	393	393	392	390	390	393	393	299	393

References

Abernathy, W. J. (1978). *The Productivity Dilemma: Roadblock to Innovation in the Automobile Industry*. Baltimore, Johns Hopkins University Press.

Abernathy W. J. and K. B. Clark (1985). 'Innovation: Mapping the winds of creative destruction.' *Research Policy* **14**(1), 3-22.

Abernathy, W. J. and J. M. Utterback (1978). 'Patterns of industrial innovation.' *Technology Review* **80**(7), 40-47.

Aggarwal, N. and E. A. Walden (2009). 'Intellectual Property Bundle (IPB) theory: Managing transaction costs in technology development through network governance.' *Decision Support Systems* **48**(1), 23-32.

Aghion, P. (2005). 'Growth and institutions.' *Empirica* **32**(1), 3-18.

Aghion, P., C. Harris, P. Howitt and J. Vickers (2001). 'Competition, imitation and growth with step-by-step innovation.' *The Review of Economic Studies* **68**(3), 467-492.

Agrawal, A. and R. Henderson (2002). 'Putting patents in context: Exploring knowledge transfer from MIT.' *Management Science* **48**(1), 44-60.

Agresti, A. (2010). *Analysis of Ordinal Categorical Data, 2nd edition*. Hoboken, N.J., John Wiley & Sons.

Ahuja, G. (2003). 'When Atlas shrugged: Preemption, complexity and division of labor in a theory of appropriability.' *Ann Arbor* **1001**, 48109-1234.

Akerlof, G. A. (1970). 'The market for "lemons": Quality uncertainty and the market mechanism.' *The Quarterly Journal of Economics* **84**(3), 488-500.

Albers, S. (2007). *Methodik der empirischen Forschung*. Wiesbaden, Betriebswirtschaftlicher Verlag Gabler.

Alcácer, J. and M. Gittelman (2004). How do I know what you know? Patent examiners and the generation of patent citations. *Working Paper*. New York University.

Alcácer, J., M. Gittelman and B. Sampat (2009). 'Applicant and examiner citations in US patents: An overview and analysis.' *Research Policy* **38**(2), 415-427.

Alchian, A. A. (1950). 'Uncertainty, evolution, and economic theory.' *The Journal of Political Economy* **58**(3), 211-221.

Allain, M. L., E. Henry and M. Kyle (2008). The timing of licensing: theory and empirics. *Working paper*. Sciences Po - Département d'économie,

Allen, D. W. (1999). 'Transaction Costs', in. B. Bouckaert and G. D. Geest (eds), *Encyclopedia of Law and Economics*, Cheltenham, UK and Northampton, MA, USA, Edward Elgar Publishing, 893-926.

Allen, R. C. (1983). 'Collective invention.' *Journal of Economic Behavior & Organization* **4**(1), 1-24.

Almeida, P. and B. Kogut (1999). 'Localization of knowledge and the mobility of engineers in regional networks.' **45**(7), 905-917.

Anand, B. N. and T. Khanna (2000). 'The structure of licensing contracts.' *Journal of Industrial Economics* **48**(1), 103-135.

Anderson, B. A. (1979). 'Acquiring and selling technology - marketing techniques.' *Research Management* **22**(2), 26-28.

Anderson, E. (1985). 'The salesperson as outside agent or employee: a transaction cost analysis.' *Marketing Science* **4**(3), 234-254.

Anderson, E. and D. C. Schmittlein (1984). 'Integration of the sales force: an empirical examination.' *The Rand Journal of Economics* **15**(3), 385-395.

Anderson, R. C. (1974). 'Paintings as an investment.' *Economic Inquiry* **12**(1), 13-26.

Andersson, C. and M. Phillips (2005). Patent Pooling: A multiparty licensing mechanism as a response to a world of ever-increasing number of nestled patents. Göteborg, Chalmers University of Technology.

Arévalo, J. J. (2004). 'Gradual Nash bargaining with endogenous agenda: a path-dependent model.' *Colombian Economic Journal* **2**(1), 191-212.

Ariely, D. and I. Simonson (2003). 'Buying, bidding, playing, or competing? Value assessment and decision dynamics in online auctions.' *Journal of Consumer Psychology* **13**(1), 113-123.

Arora, A. and R. Merges (2004). 'Specialized supply firms, property rights and firm boundaries.' *Industrial and Corporate Change* **13**(3), 451-475.

Arora, A., M. Ceccagnoli and W. M. Cohen (2004). 'R&D and the patent premium.' *International Journal of Industrial Organization* **26**(5), 1153-1179.

Arora, A., A. Fosfuri and A. Gambardella (2001). *Markets for Technology: The Economics of Innovation and Corporate Strategy*. Cambridge, Mass., MIT Press.

Arranz, N. and J. C. F. Arroyabe (2008). 'A model to analyse governance structures in technological networks.' *Strategy and Governance of Networks*, 249-268.

Arrow, K. J. (1962). 'Economic welfare and the allocation of resources for innovation', in. R. Nelson (eds), *The Rate and Direction of Inventive Activity*, Princeton, NJ, Princeton University Press, 609-625.

Arrow, K. J. (1963). 'Uncertainty and the welfare economics of medical care.' *The American Economic Review* **53**(5), 941-973.

Arrow, K. J. (1996). 'The economics of information: an exposition.' *Empirica* **23**(2), 119-128.

Arrow, K. J. (1968). 'The economics of moral hazard: Further comment.' *American Economic Review* **58**(3), 537-539.

Arundel, A. and I. Kabla (1998). 'What percentage of innovations are patented? Empirical estimates for European firms.' *Research Policy* **27**(2), 127-141.

Ashby, W. R. (1960). *Design for a Brain: The Orgin of Adaptive Behaviour*. London, Chapman & Hall.

Ashenfelter, O. (1989). 'How auctions work for wine and art.' *Journal of Economic Perspectives* **3**(3), 23-36.

Ashenfelter, O. and K. Graddy (2002). Art Auctions: A Survey of Empirical Studies. *Center for Economic Policy Research, Discussion Paper*. London.

Ashenfelter, O. and K. Graddy (2003). 'Auctions and the price of art.' *Journal of Economic Literature* **41**(3), 763-787.

Ashenfelter, O. and K. Graddy (2005). 'Anatomy of the rise and fall of a price-fixing conspiracy: auctions at Sotheby's and Christies's.' *Journal of Competition Law and Economics* **1**(1), 3-20.

Ashenfelter, O., K. Graddy and M. Stevens (2003). A Study of Sale Rates and Prices in Impressionist and Contemporary Art Auctions. *Centre for Economic Policy Research (CEPR), Discussion Paper*. London.

Aslesen, H. W. and A. Isaksen (2007). 'New perspectives on knowledge-intensive services and innovation.' *Geografiska Annaler: Series B, Human Geography* **89**(s1), 45-58.

Athreye, S. and J. Cantwell (2007). 'Creating competition? Globalisation and the emergence of new technology producers.' *Research Policy* **36**(2), 209-226.

Atteslander, P. and J. Cromm (2006). *Methoden der empirischen Sozialforschung*. Berlin, Schmidt.

Ausubel, L. M. and P. Cramton (1995 (revised 22 Jul 2002)). Demand Reduction and Inefficiency in Multi-Unit Auctions. *Papers of Peter Cramton*. University of Maryland, Department of Economics.

Azasu, S. (2006). Auctions in the Real Estate Market - A Review. *Department of Real Estate and Construction Management, Royal Institute of Technology, Working Paper*. Stockholm.

Backhaus, K., B. Erichson, W. Plinke and R. Weiber (2006). *Multivariate Analysemethoden : eine anwendungsorientierte Einführung*. Berlin, Springer.

Bagehot, W. (1873). *Lombard Street: A Description of the Money Market (1962 edition)*. Homewood, IL, Richard D. Irwin.

Bajari, P. and A. Hortacsu (2002). The Winner's Curse, Reserve Prices and Endogenous Entry: Empirical Insights from E-Bay Auctions. *SIEPR Discussion papers*. Stanford.

Bajari, P., S. Houghton and S. Tadelis (2007). Bidding for Incomplete Contracts: An Empirical Analysis. *NBER Working Paper*, Cambridge, Mass.

Bajari, P., R. McMillan and S. Tadelis (2008). 'Auctions versus negotiations in procurement: An empirical analysis.' *Journal of Law, Economics, and Organization* **25**(2), 372-399.

Balakrishnan, S. and B. Wernerfelt (1986). 'Technical change, competition and vertical integration.' *Strategic Management Journal* **7**(4), 347-359.

Barki, H. and A. Pinsonneault (2005). 'A model of organizational integration, implementation effort, and performance.' *Organization Science* **16**(2), 165-179.

Bates, T. (1990). 'Entrepreneur human capital inputs and small business longevity.' *The Review of Economics and Statistics* **72**(4), 551-559.

Bauman, Y. and G. Klein (2010). *The Cartoon Introduction to Economics - Volume 1: Microeconomics*. New York, NY, Macmillan.

BBN Corporation. (1997, March 18). "PROPHET StatGuide." from http://www.basic.northwestern.edu/statguidefiles/n-ist_alts.html#Transformations.

Beckert, J. (1996). 'What is sociological about economic sociology? Uncertainty and the embeddedness of economic action.' *Theory and Society* **25**(6), 803-840.

Beckmann, M. (2004). 'Art auctions and bidding rings: Empirical evidence from German auction data.' *Journal of Cultural Economics* **28**(2), 125-141.

Behnken, E. (2005). The Innovation Process as a Collective Learning Process. *11th International Conference on Concurrent Enterprising*. Munich, Germany: 219-226.

Bekkers, R., B. Verspagen and J. Smits (2002). 'Intellectual property rights and standardization: the case of GSM.' *Telecommunications Policy* **26**(3-4), 171-188.

Belz, C., M. Reinhold and H. Tschirky (2000). 'Technologiemarketing II – Formen, Strategien und Fallstricke der Vermarktung von Technologien.' *Management Zeitschrift* (6), 40-48.

Benassi, M. and A. Di Minin (2009). 'Playing in between: patent brokers in markets for technology.' *R&D Management* **39**(1), 68-86.

Benassi, M., D. Corsaro and G. Geenen (2010). Are patent brokers a possible first best? *Department of Economics University of Milan, Italy, Departemental Working Papers*.

Bensaou, M. and E. Anderson (1999). 'Buyer-supplier relations in industrial markets: when do buyers risk making idiosyncratic investments?' *Organization Science* **10**(4), 460-481.

Berz, G. (2007). *Spieltheoretische Verhandlungs- und Auktionsstrategien : mit Praxisbeispielen von Internetauktionen bis Investment Banking*. Stuttgart, Schäffer-Poeschel.

Bessant, J. and H. Rush (1995). 'Building bridges for innovation: the role of consultants in technology transfer.' *Research Policy* **24** (1), 97-114.

Bessen, J. (2004). 'Holdup and licensing of cumulative innovations with private information.' *Economics Letters* **82**(3), 321-326.

Bessen, J. and E. Maskin (2009). 'Sequential innovation, patents, and imitation.' *The RAND Journal of Economics* **40**(4), 611-635.

Bessy, C. and E. Brousseau (1998). 'Technology licensing contracts features and diversity.' *International Review of Law and Economics* **18**(4), 451-490.

Bessy, C., E. Brousseau and S. Saussier (2008). Payment schemes in technology licensing agreements: A transaction cost approach. *SSRN*,

Bikhchandani, S. and J. G. Riley (1991). 'Equilibria in open common value auctions.' *Journal of Economic Theory* **53**(1), 101-130.

Birkenmeier, B. (2003). Externe Technologie-Verwertung: Eine komplexe Aufgabe des Integrierten Technologie-Managements. *No. 15240*. Zurich, ETH.

Black, J. (2002). *Oxford Dictionary of Economics*. Oxford, Oxford: Oxford University Press.

Bonaccorsi, A., T. P. Lyon, F. Pammolli and G. Turchetti (2003). Auctions vs. Bargaining: An Empirical Analysis of Medical Device Procurement. *Department of Economics, Working Paper*. University of Washington,

Bonoma, T. V. (1985). 'Case research in marketing: opportunities, problems, and a process.' *Journal of Marketing Research* **22**(2), 199-208.

Borch, K. H. and K. Henrik (1968). *The Economics of Uncertainty*. Princeton, NJ, Princeton University Press.

Bortz, J. and N. Döring (2006). *Forschungsmethoden und Evaluation für Human- und Sozialwissenschaftler: Für Human-und Sozialwissenschaftler*, Springer.

Bourgeois III, L. J. and K. M. Eisenhardt (1988). 'Strategic decision processes in high velocity environments: Four cases in the microcomputer industry.' *Management Science* **34**(7), 816-835.

Boyens, K. (1998). Externe Verwertung von Technologischem Wissen. Wiesbaden, Deutscher Universitätsverlag.

Bozeman, B. (2000). 'Technology transfer and public policy: a review of research and theory.' *Research Policy* **29**(4-5), 627-655.

Braun, V. and C. Herstatt (2007). 'Barriers to user innovation: moving towards a paradigm of licence to innovate?' *International Journal of Technology, Policy and Management* **7**(3), 292-303.

Braunerhjelm, P. and R. Svensson (2010). 'The inventor's role: was Schumpeter right?' *Journal of Evolutionary Economics* **20**(3), 413-444.

Brett, J. M. and T. Okumura (1998). 'Inter-and intracultural negotiation: US and Japanese negotiators.' *Academy of Management Journal* **41**(5), 495-510.

Brewer, J. and A. Hunter (1989). *Multimethod Research: A Synthesis of Styles*. Newbury Park, CA, Sage.

Brockhoff, K. (1992). 'R&D cooperation between firms - A perceived transaction cost perspective.' *Management Science* **38**(4), 514-524.

Brockhoff, K., A. K. Gupta and C. Rotering (1991). 'Inter-firm R&D cooperations in Germany.' *Technovation* **11**(4), 219-230.

Brodbeck, H. (1999). *Strategische Entscheidungen im Technologie-Management : Relevanz und Ausgestaltung in der unternehmerischen Praxis*. Zürich, Verl. Industrielle Organisation.

Brosius, F. (1998). *SPSS 8.0 : Professionelle Statistik unter Windows*. Bonn, MITP-Verlag.

Brousseau, E., R. Coeurderoy and C. Chaserant (2007). 'The governance of contracts: Empirical evidence on technology licensing agreements.' *Journal of Institutional and Theoretical Economics JITE* **163**(2), 205-235.

Budd, C., C. Harris and J. Vickers (1993). 'A model of the evolution of duopoly: Does the asymmetry between firms tend to increase or decrease?' *The Review of Economic Studies* **60**(3), 543-573.

Bullock, R. J. and M. E. Tubbs (1987). 'The case meta-analysis method for OD.' *Research in Organizational Change and Development* **1**, 171-228.

Bulow, J. and P. Klemperer (1996). 'Auctions vs. negotiations ' *American Economic Review* **86**(1), 180-194.

Burgelman, R. A. and R. S. Rosenbloom (1989). 'Technology strategy: An evolutionary process perspective. Research on technological innovation.' *Management and Policy* **4**(1), 1-23.

Burns, R. B. (2000). *Introduction to Research Methods*. London, Sage.

Campagnolo, D. and A. Camuffo (2010). 'The concept of modularity in management studies: A literature review.' *International Journal of Management Reviews* **12**(3), 259-283.

Campbell, R. S. (1983). 'Patent trends as a technological forecasting tool.' *World Patent Information* **5**(3), 137-143.

Canning, D. (1989). 'Bargaining theory', in. F. Hahn (eds), *The Economics of Missing Markets, Information and Games*, Oxford, Oxford University Press, 163-187.

Capen, E., R. Clapp and W. Campbell (1971). 'Competitive bidding in high-risk situations.' *Journal of Petroleum Technology* **23**(6), 641-653.

Cardullo, M. W. (1996). *Introduction to Managing Technology*. Taunton, Somerset, England and New York, Research Studies Press.

Carlton, D. W. and J. M. Perloff (2008). *Modern Industrial Organization*. Boston, Mass., Pearson/Addison Wesley.

Carter, A. P. (1989). 'Know-how trading as economic exchange.' *Research Policy* **18**(3), 155-163.

Cassady, R. (1980). *Auctions and Auctioneering*. Berkeley, University of California Press.

Cassing, J. and R. W. Douglas (1980). 'Implications of the auction mechanism in baseball's free agent draft.' *Southern Economic Journal* **47**(1), 110-121.

Caves, R. E., H. Crookell and J. P. Killing (1983). 'The imperfect market for technology licenses.' *Oxford Bulletin of Economics and Statistics* **45**(3), 249-267.

Ceccagnoli, M., S. J. H. Graham, M. J. Higgins and J. Lee (2009). Firm Reliance on External Technology in the Pharmaceutical Industry. SSRN

Cesaroni, F., A. Gambardella and W. Garcia-Fontes (2004). *R&D, Innovation and Competitiveness in the European Chemical Industry*. Boston, Mass., Kluwer Academic.

Chang, H. F. (1995). 'Patent scope, antitrust policy, and cumulative innovation.' *The RAND Journal of Economics* **26**(1), 34-57.

Chatterjee, S. and A. S. Hadi (1988). *Sensitivity Analysis in Linear Regression*. New York, John Wiley & Sons.

Chen, E. K. Y., P. Enderwick, S. J. Rubin, J. H. Dunning and D. Wallace (1994). *Technology Transfer to Developing Countries*. London, Routledge.

Chesbrough, H. (2003a). 'The era of open innovation.' *MIT Sloan Management Review* **44**(3), 35-41.

Chesbrough, H. (2003b). 'The logic of open innovation: managing intellectual property.' *California Management Review* **45**(3), 33-58.

Chesbrough, H. (2003c). *Open Innovation. The New Imperative for Creating and Profiting from Technology*. Boston, Harvard Business School Press.

Chesbrough, H. (2006). Emerging secondary markets for Intellectual Property, Research Report to National Center for Industrial Property Information and Traning (NCIPI).

Chiesa, V., R. Manzini and E. Pizzurno (2003). 'The market for technolgical intangibles: a conceptual framework for the commcerical transactions.' *Proceedings of the R&D Management Conference*, 1-10.

Christensen, C. M. (1997). *The Innovator's Dilemma: When New Technologies Cause Great Firms to Fail*. Boston, Mass., Harvard Business Press.

Clark, D. J. and K. A. Konrad (2008). 'Fragmented property rights and incentives for R&D.' *Management Science* **54**(5), 969-981.

Clark, K. B. and S. C. Wheelwright (1993). *Managing New Product and Process Development: Text and Cases*. New York, Free Press.

Cline, D. (2010). "Limitations, Delimitations." Retrieved 16.07.2011 from http://education.astate.edu/dcline/Guide/Limitations.html.

Coase, R. H. (1937). 'The nature of the firm.' *Economica* **4** (16), 386-405.

Cohen, W. M., R. R. Nelson and J. P. Walsh (2000). Protecting their intellectual assets: Appropriability conditions and why US manufacturing firms patent (or not). *NBER working paper*

Coles, M. and R. Wright (1994). Dynamic bargaining theory. *Research Department Staff Report No.172*. Minneapolis, Federal Reserve Bank of Minneapolis.

Collins, H. M. (1974). 'The TEA set: tacit knowledge and scientific networks.' *Social Studies of Science* **4**(2), 165-185.

Colombo, M. G. (2003). 'Alliance form: A test of the contractual and competence perspectives.' *Strategic Management Journal* **24**(12), 1209-1229.

Commons, J. R. (1931). 'Institutional economics.' *American Economic Review* **21**(4), 648-657.

Contractor, F. J. (1981). *International Technology Licensing: Compensation, Costs, and Negotiation.* Lexington, Lexington Books.

Cooper, R. G. (2008). *Winning at New Products: Accelerating the Process from Idea to Launch.* New York, Basic Books.

Cowan, R. and D. Foray (1997). 'The economics of codification and the diffusion of knowledge.' *Industrial and Corporate Change* **6**(3), 595-622.

Czarnitzki, D., G. Licht, C. Rammer and A. Spielkamp (2001). 'Rolle und Bedeutung von Intermediären im Wissens- und Technologietransfer.' *ifo Schnelldienst* **4**, 40-49.

Dahlander, L. and M. G. Magnusson (2005). 'Relationships between open source software companies and communities: Observations from Nordic firms.' *Research Policy* **34**(4), 481-493.

Dahlin, K., M. Taylor and M. Fichman (2004). 'Today's Edisons or weekend hobbyists: Technical merit and success of inventions by independent inventors.' *Research Policy* **33**(8), 1167-1183.

Dahlman, C. J. (1979). 'The problem of externality.' *Journal of Law and Economics* **22**(1), 141-162.

Davila, T. (2000). Performance and the Design of Economic Incentives in New Product Development. *Stanford University, Graduate School of Business Research Paper No. 1647*

Davis, J. L. and S. S. Harrison (2001). *Edison in the Boardroom: How Leading Companies Realize Value from their Intellectual Assets.* New York, John Wiley & Sons.

De Smith, M. J., M. F. Goodchild and P. Longley (2007). *Geospatial Analysis: A Comprehensive Guide to Principles, Techniques and Software Tools.* Leicester, Troubador Publishing.

de Vries, D. (2011). *Leveraging Patents Financially – A Company Perspective.* Wiesbaden, Betriebswirtschaftlicher Verlag Gabler.

DeGennaro, R. P. (2005). 'Market imperfections ' *Journal of Financial Transformation* **14**, 107-117.

Delmas, M. A. (1999). 'Exposing strategic assets to create new competencies: the case of technological acquisition in the waste management industry in Europe and North America.' *Industrial and Corporate Change* **8**(4), 635-671.

Delphion (2004). Glossary of Patent Terms

Demchak, C. C. (1992). 'Complexity, rogue outcomes and weapon systems.' *Public Administration Review* **52**(4), 347-355.

Demsetz, H. (1964). 'The exchange and enforcement of property rights.' *Journal of Law and Economics* **7**(Oct.), 11-26.

Demsetz, H. (1968). 'The cost of transacting.' *The Quarterly Journal of Economics* **82**(1), 33-53.

Demsetz, H. (1988). 'The theory of the firm revisited.' *Journal of Law, Economics, and Organization* **4**(1), 141.

Den Hertog, P. and R. Bilderbeek (2000). 'The new knowledge infrastructure: the role of technology-based knowledge-intensive business services in national innovation systems', in. M. Boden and I. Miles (eds), *Services and the Knowledge-based Economy*, London, Continuum, 222-246.

Denicolò, V. and P. Zanchettin (2004). Competition and Growth in Neo-Schumpeterian Models. *Papers in Economics, 04/28*. Dept. of Economics, University of Leicester

Dessauer, J. P. (1977). *Book publishing: What it is, What it does*. New York, Bowker.

Dewally, M. and L. H. Ederington (2004). What Attracts Bidders to Online Auctions and What Is Their Incremental Price Impact? SSRN

Diamond Jr, A. M. (2006). 'Schumpeter's creative destruction: A review of the evidence.' *Journal of Private Enterprise* **22**(1), 120-146.

Díez-Vial, I. (2007). 'Explaining vertical integration strategies: market power, transactional attributes and capabilities.' *Journal of Management Studies* **44**(6), 1017-1040.

Dosi, G. (1982). 'Technological paradigms and technological trajectories: A suggested interpretation of the determinants and directions of technical change.' *Research Policy* **11**(3), 147-162.

Dosi, G. and M. Egidi (1991). 'Substantive and procedural uncertainty.' *Journal of Evolutionary Economics* **1**(2), 145-168.

Dosi, G. and R. Nelson (2010). 'Technical change and industrial dynamics as evolutionary processes'. (eds), *Handbook of the Economics of Innovation* Amsterdam, North Holland / Elsevier.

Drucker, P. F. (1954). The Practice of Management, New York: Harper and Row Publishers.

Dyer, J. H. (1997). 'Effective interim collaboration: how firms minimize transaction costs and maximise transaction value.' *Strategic Management Journal* **18**(7), 535-556.

Ebersberger, B., K. Laursen, J. Saarinen and A. Salter (2005). The origins of radicalness and complexity in product innovation - An analysis of the Finnish innovation database. *DRUID Tenth Anniversary Summer Conference*. Copenhagen Business School, Denmark: 19-21.

Edosomwan, J. A. (1989). *Integrating Innovation and Technology Management*. New York, John Wiley & Sons.

Edquist, C., L. Hommen and L. J. Tsipouri, Eds. (2000). *Public Technology Procurement and Innovation*. Dordrecht, Kluwer Academic Publishers.

Eichberger, J. and I. R. Harper (1997). *Financial Economics*. Oxford, Oxford University Press.

Eisenhardt, K. M. (1989). 'Building theories from case study research.' *Academy of Management Review* **14**(4), 532-550.

Eisenhardt, K. M. and M. E. Graebner (2007). 'Theory building from cases: Opportunities and challenges.' *Academy of Management Journal* **50**(1), 25-32.

Elton, J., B. Shah and J. Voyzey (2002). 'Intellectual property: Partnering for profit.' *Mc Kinsey Quarterly*(Special Edition - Technology), 59-67.

Engelbrecht-Wiggans, R. (1980). 'Auctions and bidding models: A Survey.' *Management Science* **26**(2), 119-142.

Engelbrecht-Wiggans, R. and C. M. Kahn (1998). 'Multi-unit auctions with uniform prices.' *Economic Theory* **12**(2), 227-258.

Engelbrecht-Wiggans, R., M. Shubik and R. M. Stark (1983). *Auctions, Bidding, and Contracting: Uses and Theory*. New York, New York University Press.

Englmaier, F. and A. Schmöller (2008). Reserve Price Formation in Online Auctions. *Cesifo Working Paper No. 2374.*

EPO, OECD and UK Patent Office (2006). *Patents: realising and securing value - Executive summary*. EPO-OECD-UKPO Conference, London.

Ernst, D. and Unctad (2005). 'Complexity and internationalisation of innovation: why is chip design moving to Asia?' *International Journal of Innovation Management* **9**(1, special issue in honour of Keith Pavitt), 47-73.

Ernst, H., C. Leptien and J. Vitt (2000). 'Inventors are not alike: the distribution of patenting output among industrial R&D personnel.' *IEEE Transactions in Engineering Management* **47**(2), 184-199.

Escher, J.-P. (2000). The process of external technology exploitation as part of technology marketing: a conceptual framework. *Technolovation*, Technologie- und Innovationsmanagement Inside Report des ETH-Zentrums für Unternehmenswissenschaften.

Escher, J.-P. (2001). 'Process of external technology exploitation as part of technology marketing: a conceptual framework.' *Proceedings of the PICMET conference.*

Escher, J.-P. (2005). Technology Marketing in Technology-based Enterprises – The Process and Organization Structure of External Technology Deployment. Zurich, Swiss Federal Institute of Technology Zurich.

Ettlie, J. E. and E. M. Reza (1992). 'Organizational integration and process innovation.' *Academy of Management Journal* **35**(4), 795-827.

European Commission (2002). 'Étymologie du terme "gouvernance".'

European Commission (2003). *Official Journal of the European Union* **46**(20 May 2003).

European Commission (2008). Intellectual Property Rights and Competitiveness: Challenges for ICT-Producing SMEs Study report Brussels, DG Enterprise and Industry. Report No. 08.

Evans, D. S. and B. Jovanovic (1989). 'An estimated model of entrepreneurial choice under liquidity constraints.' *The Journal of Political Economy* **97**(4), 808-827.

Evenson, R., J. Putnam and S. Kortum (1991). Estimating Patent Counts by Industry Using the Yale-Canada Concordance. *Final Report to the National Science Foundation, Yale University*. Fairbank, JK.

Ewald, A. (1989). *Organisation des strategischen Technologie-Managements: Stufenkonzept zur Implementierung einer integrierten Technologie- und Marktplanung*. Berlin, Schmidt.

Ewing, T. (2010a). 'Inside the world of public auctions.' *Intellectual Asset Management,* (July/August), 63-71.

Ewing, T. L. (2010b). The Intellectual Ventures Portfolio in the United States. Second Edition. Tulsa, Avancept LLC.

Fabozzi, F. J., F. Modigliani and M. G. Ferri (1994). *Foundations of Financial Markets and Institutions*. London, Prentice-Hall International.

Fagerberg, J. and B. Verspagen (2009). 'Innovation studies - The emerging structure of a new scientific field.' *Research Policy* **38**(2), 218-233.

Feller, J., P. Finnegan, J. Hayes and P. O'Reilly (2009). 'Institutionalising information asymmetry: governance structures for open innovation.' *Information Technology & People* **22**(4), 297-316.

Fielding, N. (2001). 'Qualitative Interviewing', in. G. N. Gilbert (eds), *Researching Social Life*, London, Sage Publications.

Fisher, R., W. Ury and B. Patton (1991). *Getting to Yes: Negotiating Agreement Without Giving In*. New York, NY, Penguin Books.

Fleming, L. (2001). 'Recombinant uncertainty in technological search.' *Management Science* **47**(1), 117-132.

Fleming, L. (2007). 'Breakthroughs and the "Long Tail" of innovation.' *MIT Sloan Management Review* **49**(1), 69-74.

Flikkema, M., P. Jansen and L. Van Der Sluis (2007). 'Identifying neo-Schumpeterian innovation in service firms: A conceptual essay with a novel classification.' *Economics of Innovation and New Technology* **16**(7), 541-558.

Flygansvær, B. M., S. A. Haugland and A. I. Rokkan (2002). A discussion of governance under conditions of technological uncertainty-suggesting hypotheses and a research model. *18th IMP-conference*. Dijon, France

Folta, T. B. (1998). 'Governance and uncertainty: The trade-off between administrative control and commitment.' *Strategic Management Journal* **19**(11), 1007-1028.

Fontana, A. and J. H. Frey (1994). 'Interviewing: The art of science', in. N. K. Denzin and Y. S. Lincoln (eds), *Handbook of Qualitative Research*, Thousand Oaks, Sage, 361-376.

Foray, D. (2004). *Economics of Knowledge*. Cambridge, Mass., MIT Press.

Ford, D. (1985). 'The management and marketing of technology.' *Advances in Strategic Management* **21**(5), 85-95.

Ford, D. (1988). 'Develop your technology strategy.' *Long Range Planing* **21**(5), 85-95.

Ford, D. and C. Ryan (1977). 'The marketing of technology.' *European Journal of Marketing* **11**(6), 369-383.

Ford, D. and C. Ryan (1981). 'Taking technology to market.' *Harvard Business Review* **59**(2), 117-126.

Forrester, J. W. (2001). An Introduction to Sensitivity Analysis. MIT System Dynamics in Education Project. Cambridge, Mass., Massachusetts Institute of Technology.

Frey, B. S. and W. W. Pommerehne (1990). *Muses and Markets: Explorations in the Economics of the Arts*. Oxford, Blackwell.

Friedman, D. (1993). 'The double auction market institution: A survey', in. D. Friedman and J. Rust (eds), *The Double Auction Market: Institutions, Theories, and Evidence*, Reading, MA, Addison-Wesley, 3-26.

Friedman, L. (1956). 'A competitive bidding strategy.' *Operations Research* **4**(1), 104-112.

Fröhling, W. (2005). 'Practical Experiences Regarding the Evaluation of Medium-Sized Patent Portfolios.' *presentation at EPO-OECD-BMWA International Conference on Intellectual Property as an Economic Asset: Key Issues in Valuation and Exploitation, 30 June-1 July 2005, Berlin.*

Fu, S. and D. S. Perkins (1995). 'Technology licensors and licensees: who they are, what resources they employ, and how they feel.' *International Journal of Technology Management* **10**(7/8), 907-920.

Furman, J. L. and S. Stern (2006). Climbing atop the shoulders of giants: The impact of institutions on cumulative research. *NBER Working Paper No. 12523.*

Gallini, N. T. (1984). 'Deterrence by market sharing: a strategic incentive for licensing.' *American Economic Review* **74**(5), 931-941.

Gallini, N. T. (2002). 'The economics of patents: Lessons from recent US patent reform.' *Journal of Economic Perspectives* **16**(2), 131-154.

Gambardella, A. (2002). 'Successes and failures in the markets for technology.' *Oxford Review of Economic Policy* **18**(1), 52-62.

Gambardella, A., P. Giuri and A. Luzzi (2007). 'The market for patents in Europe.' *Research Policy* **36**(8), 1163–1183.

Gambardella, A., P. Giuri and M. Mariani (2006). Study on evaluating the knowledge economy : what are patents actually worth? The value of patents for today's economy and society. Brussels, European Commission, Directorate-General for Internal Market.

Gambardella, A., D. Harhoff and B. Verspagen (2008). 'The value of European patents.' *European Management Review* **5**(2), 69-84.

Garson, G. D. (2010, 1/25/2010). 'Statnotes: Topics in Multivariate Analysis.' from http://faculty.chass.ncsu.edu/garson/pa765/statnote.htm.

Garud, R. and P. Karnøe (2003). 'Bricolage versus breakthrough: distributed and embedded agency in technology entrepreneurship.' *Research Policy* **32**(2), 277-300.

Gassmann, O. (2006). 'Opening up the innovation process: towards an agenda.' *R&D Management* **36**(3), 223-228.

Gassmann, O., J.-P. Escher and M. Luggen (2003). 'Technologieverwertung durch Spin-off - Ausgründung aus privatwirtschaftlichen Forschungsinstituten.' *Wissenschaftsmanagement* **9**(5), 24-30.

Genesove, D. (1995). 'Search at wholesale auto auctions.' *The Quarterly Journal of Economics* **110**(1), 23-49.

Geng, X., M. Stinchcombe and A. B. Whinston (2001). 'Radically new product introduction using online auctions.' *International Journal of Electronic Commerce* **5**(3), 169-189.

Gerring, J. (2004). 'What is a case study and what is it good for?' *American Political Science Review* **98**(02), 341-354.

Gerring, J. (2007). 'The Case Study: What it Is and What it Does', in. C. Boix and S. C. Stokes (eds), *The Oxford Handbook of Comparative Politics*, New York, Oxford University Press.

Gerybadze, A. (2004). *Technologie- und Innovationsmanagement: Strategie, Organisation und Implementierung*. München, Vahlen.

Gibbert, M., W. Ruigrok and B. Wicki (2008). 'What passes as a rigorous case study?' *Strategic Management Journal* **29**(13), 1465–1474.

Gilbert, R. J. and M. L. Katz (2007). 'Efficient division of profits from complementary innovations.' *International Journal of Industrial Organization* **29**(4), 443-454.

Giuri, P., M. Mariani, S. Brusoni, G. Crespi, D. Francoz, A. Gambardella, W. Garcia-Fontes, A. Geuna, R. Gonzales, D. Harhoff, K. Hoisl, C. Lebas, A. Luzzi, L. Magazzini, L. Nesta, O. Nomaler, N. Palomeras, P. Patel, M. Romanelli and B. Verspagen (2005). Everything you always wanted to know about inventors (but never asked): Evidence from the PatVal-EU survey. *LEM Working Paper*. Sant'Anna School of Advanced Studies, Pisa.

Giuri, P., M. Mariani, S. Brusoni, G. Crespi, D. Francoz, A. Gambardella, W. Garcia-Fontes, A. Geunac, R. Gonzales, D. Harhoff, K. Hoisl, C. L. Bas, A. Luzzi, L. Magazzini, L. Nesta, O. Nomaler, N. Palomeras, P. Patel, M. Romanelli and B. Verspagen (2007). 'Inventors and invention processes in Europe: Results from the PatVal-EU survey.' *Research Policy* **36**(8), 1107–1127.

Goetzmann, W. N. and M. Spiegel (1995). 'Private value components, and the winner's curse in an art index.' *European Economic Review* **39** (3-4), 549-555.

Goldberg, V. P. (1977). 'Competitive bidding and the production of precontract information.' *Bell Journal of Economics* **8**(1), 250-261.

Goldberg, V. P. (1980). 'Relational exchange: economics and complex contracts.' *American Behavioral Scientist* **23**(3), 337-352.

Goodman, L. A. (1981). 'Three elementary views of log linear models for the analysis of cross-classifications having ordered categories.' *Sociological Methodology* **12**, 193-239.

Grabowski, H. and J. Vernon (1994). 'Innovation and structural change in pharmaceuticals and biotechnology.' *Industrial and Corporate Change* **3**(2), 435-449.

Granstrand, O. (1979). Technology Management and Markets - An invesitgation of R&D and Innovation in Industrial Organizations. Göteborg, Sweden, Chalmers University of Technology.

Granstrand, O. (1982). *Technology, Management and Markets*. London, Frances Pinter.

Granstrand, O. (1998). 'Towards a theory of the technology-based firm.' *Research Policy* **27**, 465-489.

Granstrand, O. (2000a). *The Economics and Management of Intellectual Property: Towards Intellectual Capitalism*. Cheltenham, UK and Northampton, MA, USA, Edward Elgar Publishing.

Granstrand, O. (2000b). 'The shift towards intellectual capitalism - the role of infocom technologies.' *Research Policy* **29**(9), 1061-1080.

Granstrand, O. (2003a). 'Are we on our way in the new economy with the optional inventive steps?' in. O. Granstrand (eds), *Economics, Law and Intellectual Property: Seeking Strategies for Research*, Dordrecht, Kluwer Academic Publishers, 223-259.

Granstrand, O., Ed. (2003b). *Economics, law and intellectual property*. Dordrecht, Kluwer Academic Publishers.

Granstrand, O. (2003c). Innovation and Intellectual Property. *DRUID Summer Conference 2003 on Creating, Sharing and Transferring Knowledge*. Copenhagen.

Granstrand, O. (2004a). 'The economics and management of technology trade: towards a pro-licensing era?' *International Journal of Technology Management* **27**(2/3), 209-240.

Granstrand, O. (2004b). 'From R&D/Technology Management to IP/IC Management', in. European Institute for Technology and Innovation Management (EITIM) (eds), *Bringing Technology and Innovation into the Boardroom*, Hampshire, Palgrave Macmillan, 190-220.

Granstrand, O. (2006). Proper property properties of knowledge. *International IPR conference "Intellectual Property Rights for Business and Society"*. London.

Granstrand, O. (2010). *Industrial Innovation Economics and Intellectual Property*. Göteborg, Svenska Kultukompaniet.

Granstrand, O. (2011 (forthcoming)). *Patents and Innovation for Growth and Welfare*. Cheltenham, UK and Northampton, MA, USA, Edward Elgar Publishing.

Granstrand, O. and M. Holgersson (forthcoming). 'Managing multinational technology and intellectual property (IP) - Is there global convergence?' *International Journal of Technology Management*.

Granstrand, O. and S. Sjölander (1990). 'Managing innovation in multi-technology corporations.' *Research Policy* **19**(1), 35-60.

Granstrand, O., E. Bohlin, C. Oskarsson and N. Sjöberg (1992). 'External technology acquisition in large multi-technology corporations.' *R&D Management* **22**(2), 111–134.

Green, J. R. and S. Scotchmer (1995). 'On the division of profit in sequential innovation.' *The RAND Journal of Economics* **26**(1), 20-33.

Griliches, Z., Ed. (1984). *R&D, Patents, and Productivity*. Chicago, University of Chicago Press.

Griliches, Z. (1990). 'Patent statistics as economic indicators: A survey. ' *Journal of Economic Literature* **28**(4), 1661-1707.

Grindley, P. C. and D. J. Teece (1997). 'Managing intellectual capital - Licensing and cross-licensing in semiconductors and electronics.' *California Management Review* **39**(2), 8-41.

Grown, C. and T. Bates (1992). 'Commercial bank lending practices and the development of black owned construction companies.' *Journal of Urban Affairs* **14**(1), 25-41.

Gruber, M., D. Harhoff and K. Hoisl (2009). Invention Processes and Knowledge Recombination across Technological Boundaries. *Summer Conference 2010*. London.

Gu, F. and B. Lev (2000). Markets in Intangibles: Patent Licensing. *SSRN*.

Guellec, D. and B. P. de la Potterie (2001). *R&D and productivity growth: panel data analysis of 16 OECD countries*, Directorate for Science, Technology and Industry, OECD.

Gujarati, D. N. (2003). *Basic Econometrics*. New York, NY, McGraw-Hill.

Hagedoorn, J. (2003). 'Sharing intellectual property rights - An exploratory study of joint patenting amongst companies.' *Industrial and Corporate Change* **12**(5), 1035-1050.

Hair, J. F. (2006). *Multivariate Data Analysis*. Upper Saddle River, NJ, Pearson/Prentice Hall.

Hall, B. (2005). 'Exploring the patent explosion.' *Essays in Honor of Edwin Mansfield*, 195-208.

Hall, B. H. (2009). *The use and value of patent rights*. UK IP Ministerial Forum on the Economic Value of Intellectual Property.

Hall, B. H. and M. Trajtenberg (2006). 'Uncovering general purpose technologies with patent data', in. C. Antonelli, D. Foray, B. H. Hall and E. Steinmueller (eds), *New Frontiers in the Economics of Innovation and New Technology: Essays in Honor of Paul A. David*, Cheltenham, UK and Northampton, MA, USA, Edward Elgar Publishing, 389-427.

Hall, B. H. and R. H. Ziedonis (2001). 'The patent paradox revisited - An empirical study of patenting in the US semiconductor industry 1979-1995.' *RAND Journal of Economics* **32**(1), 101-128.

Hall, B. H., A. B. Jaffe and M. Trajtenberg (2001). 'The NBER patent citations data file: lessons, insights and methodological tools.'

Hall, B. H., A. Jaffe and M. Trajtenberg (2005). 'Market value and patent citations.' *RAND Journal of Economics* **36**(1), 16-38.

Hall, P. (1994). *Innovation, Economics and Evolution: Theoretical Perspectives on Changing Technology in Economic Systems*. New York, NY, Harvester Wheatsheaf.

Hamel, G. (1991). 'Competition for competence and interpartner learning within international strategic alliances.' *Strategic Management Journal* **12**(S1), 83-103.

Hamilton, A. (1781). *'Banks and Politics in America: From the Revolution to the Civil War'* in. B. Hammond (eds), Princeton, NJ, Princeton University Press.

Hampel, F. R. (1971). 'A general qualitative definition of robustness.' *The Annals of Mathematical Statistics* **42**(6), 1887-1896.

Hansen, M. T. (1999). 'The search-transfer problem: The role of weak ties in sharing knowledge across organization subunits.' *Administrative Science Quarterly* **44**(1), 82-85.

Hargadon, A. and R. I. Sutton (1997). 'Technology brokering and innovation in a product development firm.' *Administrative Science Quarterly* **42**(4), 716-749.

Harhoff, D., K. Hoisl and C. Webb (2006). European patent citations-How to count and how to interpret them. *Unpublished manuscript, University of Munich.*

Harhoff, D., F. Narin, F. M. Scherer and K. Vopel (1999). 'Citation frequency and the value of patented innovation.' *Review of Economics and Statistics* **81**, 511-515.

Harhoff, D. and M. Reitzig (2004). 'Determinants of opposition against EPO patent grants - The case of biotechnology and pharmaceuticals.' *International Journal of Industrial Organization* **22** (4), 443-480.

Harhoff, D. and F. M. Scherer (2000). 'Technology policy for a world of skew-distributed outcomes.' *Research Policy* **29**(4-5), 559-566.

Harhoff, D., F. M. Scherer and K. Vopel (2003). 'Citations, family size, opposition and the value of patent rights.' *Research Policy* **32**(8), 1343-1363.

Harrigan, K. R. (1986). 'Matching vertical integration strategies to competitive conditions.' *Strategic Management Journal* **7**(6), 535-555.

Harris, C. and J. Vickers (1987). 'Racing with uncertainty.' *The Review of Economic Studies* **54**(1), 1-21.

Haupt, R., M. Kloyer and M. Lange (2007). 'Patent indicators for the technology life cycle development.' *Research Policy* **36**(3), 387-398.

Hauschildt, J. and S. Salomo (2007). *Innovationsmanagement*. München, Vahlen.

Hausman, J., B. H. Hall and Z. Griliches (1984). 'Econometric models for count data with an application to the patents-R & D relationship.' *Econometrica: Journal of the Econometric Society* **52**(4), 909-938.

Hawking, S. (2003). *On the Shoulders of Giants: The Great Works of Physics and Astronomy*. Philadelphia, Running Press.

Hayek, F. A. v. (1975). 'Die Anmaßung von Wissen (pretence of knowledge)', (eds), *ORDO. Jahrbuch für die Ordnung von Wirtschaft und Gesellschaft*, Stuttgart, 12–21.

Hayek, F. A. v. (1991). *Die Verfassung der Freiheit*. Tübingen, Mohr.

Heide, J. B. and G. John (1990). 'Alliances in industrial purchasing: The determinants of joint action in buyer-supplier relationships.' *Journal of Marketing Research* **27**(1), 24-36.

Heide, J. B. and A. S. Miner (1992). 'The shadow of the future: Effects of anticipated interaction and frequency of contact on buyer-seller cooperation.' *Academy of Management Journal* **35**(2), 265-291.

Heller, M. A. and R. S. Eisenberg (1998). 'Can patents deter innovation? The anticommons in biomedical research.' *Science* **280**(5364), 698-701.

Henderson, R. M. and K. B. Clark (1990). 'Architectural innovation: the reconfiguration of existing product technologies and the failure of established firms.' *Administrative Science Quarterly* **35**(1), 9-30.

Henke, J. W. (2000). 'Strategic selling in the age of modules and systems.' *Industrial Marketing Management* **29**(3), 271-284.

Hennart, J. F. (1993). 'Explaining the swollen middle: why most transactions are a mix of "market" and "hierarchy".' *Organization Science* **4**(4), 529-547.

Hentschel, M. (2007). *Patentmanagement, Technologieverwertung und Akquise externer Technologien: Eine empirische Analyse*. Wiesbaden, Deutscher Universitätsverlag.

Herman, G. A. and T. W. Malone (2003). 'What Is in the Process Handbook?' in. T. W. Malone, K. Crowston and G. A. Herman (eds), *Organizing Business Knowledge: The MIT Process Handbook*, Cambridge, Mass., MIT Press, 619.

Herriott, R. E. and W. A. Firestone (1983). 'Multisite qualitative policy research: Optimizing description and generalizability.' *Educational Researcher* **12**(2), 14-19.

Herstatt, C. (2007). *Management der Frühen Innovationsphasen: Grundlagen - Methoden - Neue Ansätze*. Wiesbaden, Betriebswirtschaftlicher Verlag Gabler.

Herstatt, C. and B. Verworn (2004). 'Innovation Process Models and Their Evolution', in. European Institute for Technology and Innovation Management (eds), *Bringing Technology and Innovation Management into the Boardroom*, Basingstoke, Palgrave Macmillan, 326-347.

Herstatt, C. and E. von Hippel (1992). 'Developing new product concepts via the lead user method: A case study in a "low tech" field.' *Journal of Product Innovation Management* **9**(3), 213-221.

Hinze, S., T. Reiss and U. Schmoch (1997). Statistical Analysis on the Distance Between Fields of Technology, European Commission TSER Project.

Hipp, C. (1999). 'Information flows and knowledge creation in knowledge intensive business services: Scheme for a conceptualization', in. J. S. Metcalfe and I. Miles (eds), *Innovation Systems in the Service Economy: Measurement and Case Study Analysis*, Boston, Mass, Kluwer Academic Publishers, 149–169.

Hoekman, B. M., K. E. Maskus and K. Saggi (2005). 'Transfer of technology to developing countries: unilateral and multilateral policy options.' *World Development* **33**(10), 1587-1602.

Hoetker, G. (2007). 'The use of logit and probit models in strategic management research: Critical issues.' *Strategic Management Journal* **28**(4), 331.

Holcomb, T. R. and M. A. Hitt (2007). 'Toward a model of strategic outsourcing.' *Journal of Operations Management* **25**(2), 464-481.

Holland, J. H. and J. H. Miller (1991). 'Artificial adaptive agents in economic theory.' *The American Economic Review* **81**(2), 365-370.

Holmström, B. and J. Roberts (1998). 'The boundaries of the firm revisited.' *Journal of Economic Perspectives* **12**(4), 73-94.

Hoppe, T. and A. Sadrieh (2007). An Experimental Assessment of Confederate Reserve Price Bids in Online Auction. *FEMM Working Papers.*

Horstmann, I. J. and C. LaCasse (1997). 'Secret reserve prices in a bidding model with a resale option.' *The American Economic Review* **87**(4), 663-684.

Horton, N. J. and K. P. Kleinman (2007). 'Much ado about nothing.' *The American Statistician* **61**(1), 79-90.

Hosmer, D. W. and S. Lemeshow (1989). *Applied Logistic Regression,* New York, John Wiley & Sons.

Hosmer, D. W. and S. Lemeshow (2000). *Applied Logistic Regression*, New York, John Wiley & Sons.

Hosmer, D. W., T. Hosmer, S. Le Cessie and S. Lemeshow (1997). 'A comparison of goodness-of-fit tests for the logistic regression model.' *Statistics in Medicine* **16**(9), 965-980.

Hosmer, D. W., S. Taber and S. Lemeshow (1991). 'The importance of assessing the fit of logistic regression models: a case study.' *American Journal of Public Health* **81**(12), 1630-1635.

Howard, J. (2005). Knowledge exchange networks in Australia's innovation system: overview and strategic analysis. Canberra, Howard Partners: Department of Education, Science and Training.

Howells, J. (2006). 'Intermediation and the role of intermediaries in innovation.' *Research Policy* **35**(5), 715-728.

Hüner, A., C. Lüthje and C. Stockstrom (2010). Knowledge Base and Technological Impact of User Innovations: Empirical Evidence from the Medical Devices Industry. *17th International Product Development Management Conference.* Murcia, Spain.

Institut der deutschen Wirtschaft Köln (2006). Das Innovationsverhalten der technikaffinen Branchen - Forschung, Patente und Innovationen. Köln, Studie im Auftrag des Verein deutscher Ingenieure (VDI).

IP Bewertungs AG. (2007). "http://www.ipb-ag.de".

IP Law Bulletin. (2005). "Merchant bank puts patents on auction block." Retrieved November, 2007.

IPB. (2005). "Case Studies: Patent Valuation at IP Bewertungs AG (IPB)." Retrieved Aug., 2007.

Jacobides, M. G. (2008). 'How capability differences, transaction costs, and learning curves interact to shape vertical scope.' *Organization Science* **19**(2), 306-326.

Jacobides, M. G. and S. G. Winter (2005). 'Capabilities, transaction costs and evolution: Understanding the institutional structure of production.' *Strategic Management Journal* **26**(5), 395-413.

Jaffe, A. B. and J. Lerner (2004). *Innovation and its Discontents: How our Broken Patent System is Endangering Innovation and Progress, and What to Do About It.* Princeton, NJ, Princeton University Press.

Janzik, L., C. Raasch and C. Herstatt (forthcoming). 'Motivation in innovative online communities: Why join, why innovate, why share?.' *International Journal of Innovation Management.*

Jauch, L. R., R. N. Osborn and T. N. Martin (1980). 'Structured content analysis of cases: A complementary method for organizational research.' *Academy of Management Review* **5**(4), 517-525.

Jensen, R. A., J. G. Thursby and M. C. Thursby (2003). 'Disclosure and licensing of university inventions: The best we can do with the s** t we get to work with.' *International Journal of Industrial Organization* **21**(9), 1271-1300.

Jeppesen, L. B. and L. Frederiksen (2006). 'Why do users contribute to firm-hosted user communities? The case of computer-controlled music instruments.' *Organization Science* **17**(1), 45-63.

John, G. and T. Reve (1982). 'The reliability and validity of key informant data from dyadic relationships in marketing channels.' *Journal of Marketing Research* **19**(4), 517-524.

John, G. and B. Weitz (1989). 'Salesforce compensation: An empirical investigation of factors related to use of salary versus incentive compensation.' *Journal of Marketing Research* **26**(1), 1-14.

Johnson, D. K. N. (2002). The OECD Technology Concordance (OTC): Patents by industry of manufacture and sector of use. *OECD Science, Technology and Industry Working Papers.*

Jones, B. F. (2009). 'The burden of knowledge and the "death of the renaissance man": Is innovation getting harder?' *Review of Economic Studies* **76**(1), 283-317.

Kagel, J. H. (1997). 'Auctions: a survey of experimental research', in. J. H. Kagel and A. E. Roth (eds), *Handbook of Experimental Research*, Princeton, NJ, Princeton University Press, 501-587.

Kagel, J. H. and D. Levin (1986). 'The winner's curse and public information in common value auctions.' *The American Economic Review* **76**(5), 894-920.

Kagel, J. H. and D. Levin (2001). 'Behavior in multi-unit demand auctions: Experiments with uniform price and dynamic Vickrey auctions.' *Econometrica* **69**(2), 413-454.

Kagel, J. H. and D. Levin (forthcoming). 'Auctions: A survey of experimental research, 1995-2008', (eds), *Handbook of Experimental Economics, Vol.2.*

Kahneman, D., J. L. Knetsch and R. H. Thaler (1990). 'Experimental tests of the endowment effect and the Coase theorem.' *Journal of Political Economy* **98**(6), 1325-1348.

Kamien, M. I. and Y. Tauman (1986). 'Fees versus royalties and the private value of a patent.' *The Quarterly Journal of Economics* **101**(3), 471-491.

Kamins, M. A., X. Dreze and V. S. Folkes (2004). 'Effects of seller-supplied prices on buyers' product evaluations: Reference prices in an Internet auction context.' *Journal of Consumer Research* **30**(4), 622-628.

Kamiyama, S., J. Sheehan and C. Martinez (2006). Valuation and Exploitation of Intellectual Property. *OECD (STI WORKING PAPER 2006/5).*

Kash, D. E. and R. Rycroft (2002). 'Emerging patterns of complex technological innovation.' *Technological Forecasting and Social Change* **69**(6), 581-606.

Katila, R. and G. Ahuja (2002). 'Something old, something new: a longitudinal study of search behavior and new product introduction.' *Academy of Management Journal* **45**(6), 1183-1194.

Katok, E. and A. E. Roth (2004). 'Auctions of homogeneous goods with increasing returns: Experimental comparison of alternative "Dutch" auctions.' *Management Science* **50**(8), 1044-1063.

Katz, M. L. and H. S. Rosen (1998). *Microeconomics*. Boston, Mass., Irwin.

Katz, M. L. and C. Shapiro (1986). 'How to license intangible property.' *The Quarterly Journal of Economics* **101**(3), 567-589.

Katz, R. and T. J. Allen (1982). 'Investigating the Not Invented Here (NIH) syndrome: A look at the performance, tenure, and communication patterns of 50 R&D Project Groups.' *R&D Management* **12**(1), 7-20.

Kaufmann, L. (2001). *Internationales Beschaffungsmanagement: Gestaltung strategischer Gesamtsysteme und Management einzelner Transaktionen*. Wiesbaden, Deutscher Universitätsverlag.

Kazumori, E. and J. N. McMillan (2005). 'Selling online versus live.' *The Journal of Industrial Economics* **53**(4), 543–569.

Keep, W. W., G. S. Omura and R. J. Calatone (1994). 'What managers should know about their competitors' patented technologies.' *Industrial Marketing Management* **23**(3), 257-264.

Khalil, T. M. (2000). *Management of Technology: The Key to Competitiveness and Wealth Creation*. Boston, Mass., McGraw-Hill.

Kjerstad, E. (2005). 'Auctions vs negotiations: A study of price differences.' *Health Economics* **14**(12), 1239-1251.

Klein, P. (2005). 'The make-or-buy decision: lessons from empirical studies', in. C. Menard and M. M. Shirley (eds), *Handbook of New Institutional Economics*, Dordrecht, Springer, 435-464.

Klein, S., G. L. Frazier and V. J. Roth (1990). 'A transaction cost analysis model of channel integration in international markets.' *Journal of Marketing Research* **27**(2), 196-208.

Klemperer, P. (1998). 'Auctions with almost common values: The 'Wallet Game' and its applications.' *European Economic Review* **42**(3-5), 757-769.

Klemperer, P. (2002). 'What really matters in auction design.' *The Journal of Economic Perspectives* **16**(1), 169-189.

Klemperer, P. (2004). *Auctions: Theory and Practice*. Princeton, NJ, Princeton Univ. Press.

Klerkx, L. and C. Leeuwis (2008). 'Matching demand and supply in the agricultural knowledge infrastructure: Experiences with innovation intermediaries.' *Food Policy* **33**(3), 260-276.

Kline, D. (2003). 'Sharing the corporate crown jewels.' *MIT Sloan Management Review* **44**(3), 89-93.

Knight, F. H. (1921). *Risk, Uncertainty and Profit*. Boston, NY, Houghton Mifflin.

Knorr-Cetina, K. (1999). *Epistemic Cultures: How the Sciences Make Knowledge*. Cambridge, Mass., Harvard University Press.

Knott, A. M. and H. E. Posen (2005). 'Is failure good?' *Strategic Management Journal* **26**(7), 617-641.

Koch, A. and T. Stahlecker (2006). 'Regional innovation systems and the foundation of knowledge intensive business services. A comparative study in Bremen, Munich, and Stuttgart, Germany.' *European Planning Studies* **14**(2), 123-146.

Kogut, B. and U. Zander (1992). 'Knowledge of the firm, combinative capabilities, and the replication of technology.' *Organization Science* **3**(3), 383-397.

Köhler, W., G. Schachtel and P. Voleske (2007). *Biostatistik: Eine Einführung für Biologen und Agrarwissenschaftler*. Berlin, Springer.

Koruna, S. (2004). 'External technology commercialization: Policy guidelines.' *International Journal of Technology Management* **27**(2-3), 241-254.

Krattiger, A. F. (2004). 'Financing the bioindustry and facilitating biotechnology transfer.' *IP Strategy Today*(8).

Kuhn, T. S. (1962). *The Structure of Scientific Revolutions*. Chicago, University of Chicago Press.

Kumar, M. and S. I. Feldman (1998). Business Negotiations on the Internet. *inet'98*. Geneva Switzerland.

Kumar, N., L. W. Stern and J. C. Anderson (1993). 'Conducting interorganizational research using key informants.' *The Academy of Management Journal* **36**(6), 1633-1651.

Kummer, C. and V. Sliskovic (2007). Do virtual data rooms add value to the mergers and acquisition process? *Institute of Mergers, Acquisitions and Alliances (MANDA)*. Zurich.

Kusunoki, K., I. Nonaka and A. Nagata (1998). 'Organizational capabilities in product development of Japanese firms: a conceptual framework and empirical findings.' *Organization Science* **9**(6), 699-718.

Laffont, J. J. (1989). *The Economics of Uncertainty and Information*. Cambridge, Mass., MIT Press.

Lafontaine, F. and M. E. Slade (forthcoming). 'Inter-firm contracts', in. R. Gibbons and J. Roberts (eds), *Handbook of Organizational Economics*, Princeton, NJ, Princeton University Press.

Lamoreaux, N. and K. Sokoloff (1998). 'Inventors, firms, and the market for technology: US manufacturing in the late nineteenth and early twentieth centuries', in. N. Lamoreaux, D. Raff and P. Temin. (eds), *Learning by Firms, Organizations, and Nations*, Chicago, University of Chicago Press, 19-60.

Lang, J. C. (2001). 'Management of intellectual property rights: Strategic patenting.' *Journal of Intellectual Capital* **2**(1), 8-26.

Lang, S. (1998). 'Who owns the course? Online composition courses in an era of changing intellectual property policies.' *Computers and Composition* **15**(2), 215-228.

Lange, V. (1994). Technologische Konkurrenzanalyse zur Früherkennung von Wettbewerberinnovationen bei deutschen Großunternehmen. *Betriebswirtschaftslehre für Technologie und Innovation*. Wiesbaden, Deutscher Universitätsverlag. **Ph.D.:** XX, 312 S.

Langlois, R. N. (1992). 'Transaction-cost economics in real time.' *Industrial and Corporate Change* **1**(1), 99-127.

Langlois, R. N. (2002). 'Modularity in technology and organization.' *Journal of Economic Behavior & Organization* **49**(1), 19-37.

Langlois, R. N. and N. J. Foss (1999). 'Capabilities and governance: the rebirth of production in the theory of economic organization.' *Kyklos* **52**(2), 201-218.

Lanjouw, J. O. and M. Schankerman (1997). Stylized facts of patent litigation: Value, scope and ownership. *NBER Working Paper No. w6297*.

Lanjouw, J. O. and M. A. Schankerman (1999). The quality of ideas: measuring innovation with multiple indicators. *NBER Working Paper No. w7345*.

Lanjouw, J. O. and M. Schankerman (2001). 'Characteristics of patent litigation: A window on competition.' *RAND Journal of Economics* **32**(1), 129-151.

Lanjouw, J. O. and M. Schankerman (2004). 'Patent quality and research productivity: Measuring innovation with multiple indicators.' *Economic Journal* **114**(495), 441-465.

Lanjouw, J. O., A. Pakes and J. Putnam (1998). 'How to count patents and value intellectual property: The uses of patent renewal and application data.' *Journal of Industrial Economics* **46**(4), 405-432.

Larsson, R. (1993). 'Case survey methodology: Quantitative analysis of patterns across case studies.' *Academy of Management Journal* **36**(6), 1515-1546.

Laursen, K. and A. Salter (2006). 'Open for innovation: the role of openness in explaining innovation performance among UK manufacturing firms.' *Strategic Management Journal* **27**(2), 131-150.

Leffler, K. B., R. R. Rucker and I. Munn (2007). The Choice Among Sales Procedures: Auction vs. Negotiated Sales of Private Timber. *Dept. of Agricultural Economics and Economics, Montana State University.*

Leiblein, M. J. (2003). 'The choice of organizational governance form and performance: Predictions from transaction cost, resource-based, and real options theories.' *Journal of Management* **29**(6), 937-961.

Lemarié, S. (2005). Vertical integration and the licensing of innovation with fixed fees or royalties. *32nd EARIE conference*. Porto.

Lemley, M. A. and C. Shapiro (2005). 'Probabilistic patents.' *Journal of Economic Perspectives* **19**(2), 75-98.

Leonard-Barton, D. (1988). Synergistic design for case studies: Longitudinal single-site and replicated multiple-site. *National Science Foundation Conference on Longitudinal Research Methods in Organizations*. Austin.

Leonard-Barton, D. (1990). 'A dual methodology for case studies: Synergistic use of a longitudinal single site with replicated multiple sites.' *Organization Science* **1**(3), 248-266.

Lettl, C., C. Herstatt and H. G. Gemuenden (2006). 'Users' contributions to radical innovation: evidence from four cases in the field of medical equipment technology.' *R&D Management* **36**(3), 251-272.

Levin, R. C., A. K. Klevorick, R. R. Nelson and S. G. Winter (1987). 'Appropriating the returns from industrial research and development.' *Brookings Papers on Economic Activity* **3**(Special Issue On Microeconomics), 783-820.

Levine, R., N. Loayza and T. Beck (2000). 'Financial intermediation and growth: Causality and causes.' *Journal of Monetary Economics* **46**(1), 31-77.

Lichtenthaler, E. (2004). 'Organising the external technology exploitation process: current practices and future challenges.' *International Journal of Technology Management* **27**(2-3), 255-271.

Lichtenthaler, U. (2005). 'External commercialization of knowledge: Review and research agenda.' *International Journal of Management Reviews* **7**(4), 231-255.

Lichtenthaler, U. (2006a). Leveraging knowledge assets: success factors of external technology commercialization. Wiesbaden, Deutscher Universitätsverlag.

Lichtenthaler, U. (2006b). 'Projekte der externen Technologieverwertung: Ziele, Prozesse und Typen.' *Die Unternehmung* **60**(5), 351-372.

Lichtenthaler, U. (2007). 'Trading intellectual property in the new economy.' *International Journal of Intellectual Property Management* **1**(3), 241-252.

Lichtenthaler, U. and H. Ernst (2006a). 'Attitudes to externally organising knowledge management tasks: A review, reconsideration and extension of the NIH syndrome.' *R&D Management* **36**(4), 367–386.

Lichtenthaler, U. and H. Ernst (2006b). 'Developing reputation to overcome the imperfections in the markets for knowledge.' *Research Policy* **36**(1), 37-55.

Lien, A. P. (1979). 'Acquiring and selling technology: The role of the middleman.' *Research Management* **22**(3), 29-31.

Linder, J. C., S. L. Jarvenpaa and T. H. Davenport (2003). 'Toward an innovation sourcing strategy.' *MIT Sloan Management Review* **44**(4), 43-49.

Lindmark, S. (2002). Evolution of Techno-Economic Systems-An Investigation of the History of Mobile Communications. Göteborg, Chalmers University of Technology.

Lopez, H. and W. Vanhaverbeke (2010). Connecting open and closed innovation markets: A typology intermediaries. *Politecnico di Milano*. Milano.

Lord, L. F. and F. de Mergelina (2004). The Data Room Process, Due Diligence and the Valuation of Financial Entities Operating Manuals 5 and 6. *Inter-American Development Bank*. Washington, D.C.

Lorenz, J. and E. L. Dougherts (1983). Bonus Bidding and Bottom Lines: Federal Offshore Oil and Gas. *58th Annual Fall Technical Conference*. SPE 12024.

Love, J. (2007). *Process Automation Handbook - A Guide to Theory and Practice*. London, Springer.

Loveridge, R. (1990). *The Strategic Management of Technological Innovation*. Chichester, John Wiley & Sons.

Lucas, W. A. (1974). The case survey method: Aggregating case experience. Santa Monica, CA, RAND Corporation.

Lucking-Reiley, D. (2000). 'Auctions on the Internet: What's being auctioned, and how?' *Journal of Industrial Economics* **48**(3), 227–252.

Luderer, B., V. Nollau and K. Vetters (2007). *Mathematical Formulas for Economists*. Berlin, Heidelberg, Springer.

Ludwig-Mayerhofer, W., R. Jacob, W. H. Eirmbter, R. Keller and R. Christian. (2010, 26 Jun). "ILMES - Internet-Lexikon der Methoden der empirischen Sozialforschung.".

Lynn, L. H., N. M. Reddy and J. D. Aram (1996). 'Linking technology and institutions: The innovation community framework.' *Research Policy* **25**(1), 91-106.

Macher, J. T. and B. D. Richman (2008). 'Transaction cost economics: An assessment of empirical research in the social sciences.' *Business and Politics* **10**(1), 1-63.

Macneil, I. R. (1973). 'The many futures of contracts.' *Southern California Law Review* **47**(3), 691-816.

Madhok, A. (2002). 'Reassessing the fundamentals and beyond: Ronald Coase, the transaction cost and resource-based theories of the firm and the institutional structure of production.' *Strategic Management Journal* **23**(6), 535-550.

Magerman, T., B. Van Looy and X. Song (2006). Data production methods for harmonized patent statistics: Patentee name harmonization. *KUL Working Paper No. MSI 0605*.

Mahoney, J. T. and J. R. Pandian (1992). 'The resource-based view within the conversation of strategic management.' *Strategic Management Journal* **13**(5), 363-380.

Malone, M. S. (1995). *The Microprocessor: A Biography*. New York, Springer.

Maltz, A. (1994). 'Private fleet use: A transaction cost approach.' *Transportation Journal* **32**, 46-53.

Manelli, A. and D. Vincent (1995). 'Optimal procurement mechanisms.' *Econometrica* **63**(3), 591-620.

Manning, W. G. and J. Mullahy (2001). 'Estimating log models: to transform or not to transform?' *Journal of Health Economics* **20**(4), 461-494.

Mansfield, E. (1994). Intellectual property protection, foreign direct investment, and technology transfer. *International Finance Corporation, The World Bank*. Washington, D.C.

Mansfield, E. (2000). 'Intellectual property protection, direct investment and technology transfer: Germany, Japan and the USA.' *International Journal of Technology Management* **19**(1), 3-21.

Mansfield, E. and A. Romeo (1980). 'Technology transfer to overseas subsidiaries by US-based firms.' *The Quarterly Journal of Economics* **95**(4), 737-750.

Mantel, S. J. and G. Rosegger (1987). 'The role of third-parties in the diffusion of innovations: a survey', in. R. Rothwell and J. Bessant (eds), *Innovation: Adaptation and Growth*, Amsterdam, Elsevier, 123-134.

March, J. G. (1991). 'Exploration and exploitation in organizational learning.' *Organization Science* **2**(1), 71-87.

March, J. G., H. A. Simon and H. Guetzkow (1958). *Organizations*. New York, John Wiley & Sons.

Marcy, W. (1979). 'Acquiring and selling technology - licensing do's and don'ts.' *Research Management* **22**(2), 18-21.

Mariani, M. and M. Romanelli (2007). '"Stacking" and "picking" inventions: The patenting behavior of European inventors.' *Research Policy* **36**(8), 1128-1142.

Markman, G. D., P. T. Gianiodis, P. H. Phan and D. B. Balkin (2005). 'Innovation speed: Transferring university technology to market.' *Research Policy* **34**(7), 1058-1075.

Markoff, J. (2005). Secretive buyer of some e-commerce patents turns out to be Novell. *The New York Times*. New York.

Maskin, E. and J. Riley (1989). 'Optimal Multi-Unit Auctions', in. F. Hahn (eds), *The Economics of Missing Markets, Information, and Games*, Oxford, Clarendon Press, 312–335.

Masten, S. E. (1984). 'The organization of production: Evidence from the aerospace industry.' *Journal of Law and Economics* **27**(2), 403-417.

Masten, S. E. (1988). 'A legal basis for the firm.' *Journal of Law, Economics, and Organization* **4**(1), 181-198.

Mazzoleni, R. and R. R. Nelson (1998). 'Economic theories about the benefits and costs of patents.' *Journal of Economic Issues* **32**(4), 1031-1033.

McAfee, R. P. and J. McMillan (1987a). 'Auctions and bidding.' *Journal of Economic Literature* **25**(2), 699-738.

McAfee, R. P. and J. McMillan (1987b). 'Auctions with entry.' *Economics Letters* **23**(4), 343-347.

McClure, I. D. (2008). 'Commoditizing Intellectual Property rights: The practicability of a commercialized and transparent international IPR market and the need for international standards.' *Buffalo Intellectual Property Law Journal* **6**, 101-121.

McCutcheon, D. M. and J. R. Meredith (1993). 'Conducting case study research in operations management.' *Journal of Operations Management* **11**(3), 239-256.

McEvily, B. and A. Zaheer (1999). 'Bridging ties: a source of firm heterogeneity in competitive capabilities.' *Strategic Management Journal* **20**(12), 1133-1156.

McGrath, J. E. (1981). 'Dilemmatics: The study of research choices and dilemmas.' *American Behavioral Scientist* **25**(2), 179-210.

McIvor, R. (2009). 'How the transaction cost and resource-based theories of the firm inform outsourcing evaluation.' *Journal of Operations Management* **27**(1), 45-63.

Megantz, R. C. (1996). *How to License Technology*. New York, NY, John Wiley & Sons.

Megantz, R. C. (2002). *Technology Management: Developing and Implementing Effective Licensing Programs*. New York, NY, John Wiley & Sons.

Menard, S. (1995). *Logistic Regression: From Introductory to Advanced Concepts and Applications*. Thousand Oaks, Sage.

Merges, R. P. and R. R. Nelson (1990). 'On the complex economics of patent scope.' *Columbia Law Review* **90**(4), 839-916.

Meyer, M. (2006). 'Are patenting scientists the better scholars? An exploratory comparison of inventor-authors with their noninventing peers in nano-science and technology.' *Research Policy* **35**(10), 1646-1662.

Miles, I., N. Kastrinos, K. Flanagan, R. Bilderbeek, P. Den Hertog, W. Huntink and M. Bouman (1995). Knowledge-intensive business services: Users, carriers and sources of innovation. *European Innovation Monitoring System (EIMS)*, European Commission, CORDIS. **15**.

Miles, M. B. and A. M. Huberman (1994). *Qualitative Data Analysis: An Expanded Sourcebook*. Thousand Oaks, Sage.

Milgrom, P. (1989). 'Auctions and bidding: A primer.' *Journal of Economic Perspectives* **3**(2), 3-22.

Milgrom, P. (2005). *Putting Auction Theory to Work*. Cambridge, Cambridge University Press.

Milgrom, P. R. and R. J. Weber (1982). 'A theory of auctions and competitive bidding.' *Econometrica* **50**(5), 1089-1122.

Miller, D. and H. Peter (1977). 'Strategy-making in context: Ten empirical archetypes.' *Journal of Management Studies* **14**(3), 253-280.

Millien, R. and R. Laurie (2007). Established and Emerging IP Business Models. *The 8th annual Sedona conference on patent litigation*. Sedona, AZ.

Minichiello, V., R. Aroni, E. Timewell and L. Alexander (1990). *In-depth Interviewing: Researching People*. Melbourne, Longman Cheshire.

Mitchell, G. R. (1988). 'Options for the strategic management of technology.' *International Journal of Technology Management* **3**(3), 253-262.

Mittag, H. (1985). *Technologiemarketing - Die Vermarktung von industriellem Wissen unter besonderer Berücksichtigung des Einsatzes von Lizenzen* Bochum, Studienverlag Dr. N. Brockmeyer.

Miyazaki, K. and K. Kijima (2000). 'Complexity in technology management: Theoretical analysis and case study of automobile sector in Japan.' *Technological Forecasting and Social Change* **64**(1), 39-54.

Mokyr, J. (2004). *The Gifts of Athena*. Princeton, NJ, Princeton University Press.

Monteverde, K. and D. J. Teece (1982). 'Supplier switching costs and vertical integration in the automobile industry.' *The Bell Journal of Economics* **13**(1), 206-213.

Morse, J. M., M. Barrett, M. Mayan, K. Olson and J. Spiers (2002). 'Verification strategies for establishing reliability and validity in qualitative research.' *International Journal of Qualitative Methods* **1** (2), 13-22.

Mortara, L. (2010). Getting help with open innovation. Cambridge, UK, Institute for Manufacturing, University of Cambridge.

Mossetto, G. and M. Vecco, Eds. (2002). *Economics of Art Auctions*. Milano, Franco Angeli.

Mowery, D. C. (1983). 'The relationship between intrafirm and contractual forms of industrial research in American manufacturing, 1900-1940.' *Explorations in Economic History* **20**(4), 351-374.

Mowery, D. C., R. R. Nelson, B. N. Sampat and A. A. Ziedonis (2001). 'The growth of patenting and licensing by US universities: an assessment of the effects of the Bayh–Dole act of 1980.' *Research Policy* **30**(1), 99-119.

Mukoyama, T. (2003). 'Innovation, imitation, and growth with cumulative technology.' *Journal of Monetary Economics* **50**(2), 361-380.

Murmann, J. P. and K. Frenken (2006). 'Toward a systematic framework for research on dominant designs, technological innovations, and industrial change.' *Research Policy* **35**(7), 925-952.

Murray, F. and S. O'Mahony (2007). 'Exploring the foundations of cumulative innovation: Implications for organization science.' *Organization Science* **18**(6), 1006-1021.

Muthoo, A. (2000). 'A non-technical introduction to bargaining theory.' *World Economics* **1**(2), 145-166.

Muthoo, A. (2002). *Bargaining Theory With Applications*. Cambridge, Cambridge University Press.

Myerson, R. B. (1981). 'Optimal auction design.' *Mathematics of Operations Research* **6**(1), 58-73.

Napel, S. (2002). *Bilateral bargaining: Theory and applications*. Berlin, Springer.

Narin, F. and A. Breitzman (1995). 'Inventive productivity.' *Research Policy* **24**(4), 507-519.

Narin, F. and E. Noma Ross (1987). 'Patents as indicators of corporate technological strength.' *Research Policy* **16**(2-4), 143-155.

Narin, F. and D. Olivastro (1988). 'Technology indicators based on patents and patent citations', in. A. F. J. v. Raan (eds), *Handbook of Quantitative Studies of Science and Technology*, Amsterdam, North-Holland, 465-507.

Nash, J. F. (1950). 'The bargaining problem.' *Econometrica: Journal of the Econometric Society* **18**(2), 155-162.

Ndonzuau, F. N., F. Pirnay and B. Surlemont (2002). 'A stage model of academic spin-off creation.' *Technovation* **22**(5), 281-289.

Nelson, R. R. (1994). 'The co-evolution of technology, industrial structure, and supporting institutions.' *Industrial and Corporate Change* **3**(1), 47-63.

Nelson, R. R. (2008). 'What enables rapid economic progress: What are the needed institutions?' *Research Policy* **37**(1), 1-11.

Nelson, R. R. and K. Nelson (2002). 'Technology, institutions, and innovation systems.' *Research Policy* **31**(2), 265-272.

Nelson, R. R. and S. G. Winter (1982). *An Evolutionary Theory of Economic Change*. Cambridge, Belknap Press.

Niioka, H. (2006). 'Patent auctions: Business, and investment strategy in IP commercialization.' *Journal of Intellectual Property Law & Practice* **1**(11), 728-731.

Nokia (2009). Form 20-F 2008. *Annual Report*. Espoo, Finland, Nokia Corporation.

Nonaka, I. and H. Takeuchi (1995). *The Knowledge Creating Company*. Oxford, Oxford University Press.

Noordeewier, T. G., G. John and J. R. Nevin (1990). 'Performance outcomes of purchasing arrangement in industrial buyer-vendor relationship.' *Journal of Marketing* **54**(4), 80-93.

Nooteboom, B. (1999). 'Innovation and inter-firm linkages: New implications for policy.' *Research Policy* **28**(8), 793-805.

Nordhaus, W. D. (1967). The optimal life of a patent. *Cowles Foundation Discussion Papers for Research in Economics at Yale University: 241.*

North, D. (1996). 'Institutional change: A framework of analysis', in. S.-E. Sjöstrand (eds), *Institutional Change: Theory and Empirical Findings*, Armonk, NY, Conference on Socio-Economics ; 3 (Stockholm): 1991, 35-47.

North, D. C. (1971). 'Institutional change and economic growth.' *The Journal of Economic History* **31**(1), 118-125.

Nuvolari, A. (2004). 'Collective invention during the British Industrial Revolution: the case of the Cornish pumping engine.' *Cambridge Journal of Economics* **28**(3), 347-363.

NZZ Online (2007). Patente unter dem Hammer; 6. Januar. *Neue Zürcher Zeitung.*

O'Donoghue, T. (1998). 'A patentability requirement for sequential innovation.' *RAND Journal of Economics* **29**(4), 654-679.

Ocean Tomo. (2007a). "http://www.oceantomo.com/events.html." Retrieved 13.11.2007

Ocean Tomo. (2007b). "http://www.oceantomo.com/press_releases.html." Retrieved 15.11.2007

Ocean Tomo. (2007c). "http://www.oceatomo.com/faq." Retrieved 18.11.2007

Ockenfels, A., D. Reiley and A. Sadrieh (2006). Online Auctions. *NBER Working Paper Series.*

OECD (2005). 'Oslo Manual-Guidelines for Collecting and Interpreting Innovation Data.' *Luxembourg: Eurostat.*

OECD (2006). Science, Technology and Industry Outlook (HIGHLIGHTS), OECD.

OECD (2007). *Glossary of Statistical Terms*, Organisation for Economic Cooperation and Development.

OECD (2009). *OECD Patent Statistics Manual*, OECD Publishing.

OECD, BMWI and EPO (2005). *Intellectual property as an economic asset: key issues in valuation and exploitation*, Berlin.

Oriani, R. and Sobrero (2001). *Market Valuation of Firms Technological Knowledge a Real Options Perspective.* 21st Annual International Conference of Strategic Management Society, "Reinventing Strategic Management: Old Truths and New Insights", San Francisco, CA.

Osborne, J. W. and A. Overbay (2004). 'The power of outliers (and why researchers should always check for them).' *Practical Assessment, Research & Evaluation* **9**(6), 1-12.

Osterwalder, A. (2004). The Business Model Ontology - A Proposition in a Design Science Approach. *Ecole des Haute Etudes Commerciales.* Lausanne, Universite de Laussane.

Owen-Smith, J. and W. W. Powell (2003). 'The expanding role of university patenting in the life sciences: assessing the importance of experience and connectivity.' *Research Policy* **32**(9), 1695-1711.

Oxley, J. E. (1999). 'Institutional environment and the mechanisms of governance: the impact of intellectual property protection on the structure of inter-firm alliances.' *Journal of Economic Behavior & Organization* **38**(3), 283-309.

Pakes, A. and M. Schankerman (1984). 'The rate of obsolescence of patents, research gestation lags, and the private rate of return to research resources', in. Z. Griliches (eds), *R&D, Patents, and Productivity*, Chicago, University of Chicago Press, 73-88.

Parchomovsky, G. and R. P. Wagner (2005). 'Patent portfolios.' *University of Pennsylvania Law Review* **154**(1), 1-77.

Parente, D. H., R. Venkataraman, J. F. and and I. Millet (2004). 'A conceptual research framework for analyzing online auctions in a B2B environment.' *Supply Chain Management: An International Journal* **9**(4), 287-294.

Parker, R. S., G. G. Udell and L. Blades (1996). 'The new independent inventor: Implications for corporate policy.' *Review of Business* **17**(3), 7-11.

Parr, C. (2006). 'Due Diligence Worth a Look?' *Business Law Review London* **27**(10), 232-233.

Patent Ratings. (2007). "http://patentratings.com/." Retrieved 13.11.

Pearce, D. W. (1992). *Macmillan Dictionary of Modern Economics*, Basingstoke, UK, Palgrave Macmillan.

Peng, C. Y. J., K. L. Lee and G. M. Ingersoll (2002). 'An introduction to logistic regression analysis and reporting.' *The Journal of Educational Research* **96**(1), 3-14.

Pénin, J. (2008). More open than open innovation? Rethinking the concept of openness in innovation studies. *Working Papers of BETA*.

Penrose, E. (1951). *The Economics of the International Patent System*. Baltimore, MD, Johns Hopkins University Press.

Peppet, S. R. (2003). 'Contract formation in imperfect markets: Should we use mediators in deals.' *Ohio State Journal on Dispute Resolution* **19**(2), 283-367.

Perrow, C. (1984). *Normal Accidents: Living with High-Risk Technologies*. New York, NY, Princeton University Press.

Perrow, C. (1994). 'Accidents in high-risk systems.' *Technology Studies* **1**(1), 1-38.

Perry, M. L., S. Sengupta and R. Krapfel (2004). 'Effectiveness of horizontal strategic alliances in technologically uncertain environments: Are trust and commitment enough?' *Journal of Business Research* **57**(9), 951-956.

Pisano, G. P. (1990). 'The R&D boundaries of the firm: An empirical analysis.' *Administrative Science Quarterly* **35**(1), 153-176.

Pitkethly, R. (1999). The Valuation of Patents: A review of patent valuation methods with consideration of option based methods and the potential for further research. *Judge Institute Working Paper WP 21/97*.

Plott, C. R. and K. Zeiler (2005). 'The willingness to pay-willingness to accept gap, the "endowment effect," subject misconceptions, and experimental procedures for eliciting valuations.' **95**(3), 530-545.

Poddar, S. and U. B. Sinha (2002). The role of fixed fee and royalty in patent licensing. *Department of Economics, National University of Singapore*.

Polanyi, M. (1962). *Personal Knowledge: Towards a Post-Critical Philosophy*. London, Routledge.

Polanyi, M. (1966). *The Tacit Dimension*. Garden City, NY, Doubleday.

Poley, W. L. and Deutscher Wirtschaftsdienst (1981). *Know-How-Export, Lizenzvergabe, Technologie-Transfer: Neue Chancen auf neuen Märkten*. Köln, Deutscher Wirtschaftsdienst.

Pollard, D. (2006). 'Innovation and technology transfer intermediaries: A systemic international study.' *Advances in Interdisciplinary Studies of Work Teams* **12**, 137-174.

Popp, A. (2000). 'Swamped in information but starved of data: Information and intermediaries in clothing supply chains.' *Supply Chain Management Science* **5**(3), 151-161.

Popper, K. R. (1969). *Conjectures and Refutations: The Growth of Scientific Knowledge*. London, Routledge.

Poppo, L. and T. Zenger (1998). 'Testing alternative theories of the firm: Transaction cost, knowledge-based, and measurement explanations for make-or-buy decisions in information services.' *Strategic Management Journal* **19**(9), 853-877.

Powell, W. W. and K. Snellman (2004). 'The knowledge economy.' *Annual Review of Sociology* **30**(2004), 199-220.

Powell, W. W., K. W. Koput and L. Smith-Doerr (1996). 'Interorganizational collaboration and the locus of innovation: Networks of learning in biotechnology.' *Administrative Science Quarterly* **41**(1), 116-145.

Powers, D. A. and Y. Xie (2008). *Statistical Methods for Categorical Data Analysis*. Bingley, Emerald Group Publishing.

Price, S. and R. Erwee (2006). *High technology firms in Canada's description of their competitive advantage–a cross case analysis of a sample of SMEs*. Academy of Management Annual Meeting: Knowledge, Action and the Public Concern, Atlanta, USA, Pace University.

Provan, K. G. and S. E. Human (1999). 'Organizational learning and the role of the network broker in small-firm manufacturing networks', in. A. Grandori (eds), *Interfirm Networks: Organization and Industrial Competitiveness*, London, Routledge, 185-207.

Punch, K. F. (2005). *Introduction to Social Research: Quantitative and Qualitative Approaches*. London, Sage.

PWC (2007). Exploiting intellectual property in a complex world, Price Waterhouse Coopers.

Raasch, A.-C. (2006). Der Patentauslauf von pharmazeutischer Produkte als Herausforderung beim Management des Produktlebenszyklus. Erlangen-Nürnberg, Friedrich-Alexander-Universität.

Raasch, C. and C. Herstatt (forthcoming). 'Product development in open design communities: A process perspective.' *International Journal of Innovation and Technology Management*.

Raasch, C., C. Herstatt and K. Balka (2009). 'On the open design of tangible goods.' *R&D Management* **39**(4), 382-393.

Rasch, B. (2008). *Quantitative Methoden: Einführung in die Statistik*. Heidelberg, Springer.

Reichstein, T. (2009). Economics and Management of Innovation, Technology and Organizational Change. *DRUID-DIME Academy Winter 2009 PhD Conference*. Aalborg, Denmark.

Reid, D., D. Bussiere and K. Greenaway (2001). 'Alliance formation issues for knowledge-based enterprises.' *International Journal of Management Reviews* **3**(1), 79-100.

Reiley, D. H. (2006). 'Field experiments on the effects of reserve prices in auctions: More magic on the internet.' *Rand Journal of Economics* **37**(1), 195-211.

Reinhard Haupt, Karsten Jahn, Marcus Lange and W. Ziegler (2004). 'Der Patentlebenszyklus: Methodische Lösungsansätze der externen Technologieanalyse.' *Arbeits- und Diskussionspapiere der Wirtschaftswissenschaftlichen Fakultät*. Friedrich-Schiller-Universität Jena.

Reitzig, M. (2003). 'What determines patent value? Insights from the semiconductor industry.' *Research Policy* **32** (1), 13-26.

Reitzig, M. (2004). 'Strategic management of intellectual property.' *Sloan Management Review* **45**(3), 35-40.

Reitzig, M., J. Henkel and F. Schneider (2010). 'Collateral damage for R&D manufacturers: How patent sharks operate in markets for technology.' *Industrial and Corporate Change* **19**(3), 947-967.

Rieck, C. (1993). *Spieltheorie Einführung für Wirtschafts- und Sozialwissenschaftler*. Wiesbaden, Betriebswirtschaftlicher Verlag Gabler.

Rihoux, B. (2006). 'Qualitative Comparative Analysis (QCA) and related systematic comparative methods: Recent advances and remaining challenges for social science research.' *International Sociology* **21**(5), 679-706.

Riis, C. and X. Shi (2009). 'Cumulative innovation and exclusive patent policy.' *Department of Economics, Norwegian School of Management, Oslo.*

Riley, J. G. and W. F. Samuelson (1981). 'Optimal auctions.' *American Economic Review* **71**(3), 381-392.

Rindfleisch, A. and J. B. Heide (1997). 'Transaction cost analysis: Past, present, and future applications.' *The Journal of Marketing* **61**(4), 30-54.

Ring, P. S. and A. H. Van de Ven (1989). 'Formal and informal dimensions of transactions', in. A. H. van de Ven, H. L. Angle and M. S. Poole (eds), *Research on the Management of Innovation: The Minnesota Studies*, New York, NY, Oxford University Press, 171-192.

Rivette, K. G. and D. Kline (2000). *Rembrandts in the Attic: Unlocking the Hidden Value of Patents*. Boston, Harvard Business School Press.

Roberts, K. H. (1990). 'Some characteristics of one type of high reliability organization.' *Organization Science* **1**(2), 160-176.

Roberts, K. H. and G. Gargano (1990). 'Managing a high-reliability organization: A case for interdependence', in. M. A. v. Glinow and S. A. Mohrman (eds), *Managing Complexity in High Technology Organizations* New York, NY, Oxford University Press, 146-159.

Robertson, T. S. and H. Gatignon (1998). 'Technology development mode: A transaction cost conceptualization.' *Strategic Management Journal* **19**(6), 515-531.

Rockett, K. E. (1990). 'Choosing the competition and patent licensing.' *Rand Journal of Economics* **21**(1), 161-171.

Roll, R. (1986). 'The hubris hypotheses of corporate takeovers.' *Journal of Business* **59**(2), 197-216.

Romer, P. M. (1990). 'Endogenous technological change.' *Journal of Political Economy* **98**(5), 71-102.

Roos, J., G. Edvinsson, N. C. Dragonetti and L. Edvinsson (1997). *Intellectual Capital - Navigating in the New Business Landscape*. London, Macmillan.

Rosenberg, M. (1968). *The Logic of Survey Analysis*. New York, Basic Books.

Rosenberg, N. (1994). *Exploring the Black Box: Technology, Economics, and History*. Cambridge, Cambridge University Press.

Rosenkopf, L., A. Metiu and V. P. George (2001). 'From the bottom up? Technical committee activity and alliance formation.' *Administrative Science Quarterly* **46**(4), 748-772.

Rosenkranz, S. and P. W. Schmitz (2007). 'Reserve prices in auctions as reference points (Working Paper Version).' *Economic Journal* **117**(520), 637-653.

Rossman, J. and B. S. Sanders (1957). 'The Patent Utilization Study.' *IDEA: The Intellectual Property Law Review* **1**, 74-111.

Roth, A. E. (2008). 'What have we learned from market design?' *The Economic Journal* **118**(527), 285–310.

Rothkopf, M. H. and R. M. Harstad (1994). 'Modeling competitive bidding: A critical essay.' *Management Science* **40**(3), 364-384.

Rothkopf, M. H. and R. M. Stark (1979). 'Competitive bidding: A comprehensive bibliography.' *Operations Research* **27**(2), 364-390.

Rothkopf, M. H., T. J. Teisberg and E. P. Kahn (1990). 'Why are Vickrey auctions rare?' *Journal of Political Economy* **98**(1), 94-109.

Rothwell, R. (1994). 'Towards the fifth-generation innovation process.' *International Marketing Review* **11**(1), 7-31.

Rubenstein, A. H. and A. Ribenstein (1989). *Managing Technology in the Decentralized Firm*. New York, NY, John Wiley & Sons.

Ruzzier, C. A. (2009). Asset Specificity and Vertical Integration: Williamson's Hypothesis Reconsidered. *Harvard Business School Working Paper No. 09-119*.

Rycroft, R. W. and D. E. Kash (1999). *The Complexity Challenge: Technological Innovation for the 21st Century*. London, Pinter Publishers.

Sahal, D. (1981). 'Alternative conceptions of technology.' *Research Policy* **10**, 2-24.

Sahal, D. (1982). 'The form of technology', in. D. Sahal (eds), *The Transfer and Utilization of Technical Knowledge*, Lexington, MA, Lexington Publishing, 125-139.

Sahal, D. (1985). 'Technological guideposts and innovation avenues.' *Research Policy* **14**(2), 61-82.

Salomon Smith Barney (1999). Report on the IBM licensing programme.

Sapsalis, E., B. van Pottelsberghe de la Potterie and R. Navon (2006). 'Academic vs. industry patenting: An in-depth analysis of what determines patent value.' *Research Policy* **35**(10), 1631-1645.

Sauermann, H. (2000). Finanzintermediäre und Kapitalmarkteffizienz. *Lehrstuhl für BWL, Finanzierung und Banken; Prof. Dr. Hummel*, Universität Potsdam.

Scandura, T. A. and E. A. Williams (2000). 'Research methodology in management: Current practices, trends, and implications for future research.' *Academy of Management Journal* **43**(6), 1248-1264.

Schaeffer, N. C. and S. Presser (2003). 'The science of asking questions.' *Annual Review of Sociology* **29**, 65-88.

Schankerman, M. and A. Pakes (1986). 'Estimates of the value of patent rights in European countries during the post-1950 period.' *The Economic Journal* **96**(384), 1052-1076.

Schelling, T. C. (1997). *The Strategy of Conflict.* Cambridge, Mass., Harvard University Press.

Scherer, F. M. (1965). 'Firm size, market structure, opportunity, and the output of patented inventions.' *The American Economic Review* **55**(5), 1097-1125.

Scherer, F. M. (1982). 'Inter-industry technology flows and productivity growth.' *The Review of Economics and Statistics* **64**(4), 627-634.

Schettino, F., A. Sterlacchini and F. Venturini (2008). Inventive Productivity and Patent Quality: Evidence from Italian Inventors. *Munich Personal RePEc Archive.*

Schilling, M. A. (1998). 'Technological lockout: An integrative model of the economic and strategic factors driving technology success and failure.' *The Academy of Management Review* **23**(2), 267-284.

Schilling, M. A. and H. K. Steensma (2002). 'Disentangling the theories of firm boundaries: A path model and empirical test.' *Organization Science* **13**(4), 387-401.

Schlag, P. (1989). 'The problem of transaction costs.' *Southern California Law Review* **62**, 1661-1700.

Schmoch, U., F. Laville, P. Patel and R. Frietsch (2003). Linking Technology Areas to Industrial Sectors - Final Report, European Commission, DG Research.

Schmookler, J. (1966). *Invention and Economic Growth.* Cambridge, MA, Harvard University Press.

Schrader, S. (1991). 'Information technology transfer between firms: Cooperation through information trading.' *Research Policy* **20**(2), 153-170.

Schultz, A.-M. (1980). *Methoden und Probleme der Gebührenbemessung bei internationalen Lizenz- und Know-how-Verträgen*. St. Gallen.

Schumpeter, J. A. (1912). *Theorie der wirtschaftlichen Entwicklung (2006 edition)*. Berlin, Duncker & Humblot.

Schumpeter, J. A. (1949). 'Science and Ideology.' *The American Economic Review* **39**(2), 346-359.

Schweisfurth, T., C. Raasch and C. Herstatt (forthcoming). 'Free revealing in open innovation: A comparison of different models and their benefits for companies.' *International Journal of Product Development*.

Schweizer, L. (2005). 'Organizational integration of acquired biotechnology companies in pharmaceutical companies: The need for a hybrid approach.' *Academy of Management Journal* **48**(6), 1051-1074.

Scotchmer, S. (1991). 'Standing on the shoulders of giants: cumulative research and the patent law.' *The Journal of Economic Perspectives* **5**(1), 29-41.

Scotchmer, S. (2004). *Innovation and Incentives*. Cambridge, Mass., MIT Press.

Seawright, J. and J. Gerring (2008). 'Case selection techniques in case study research: A menu of qualitative and quantitative options.' *Political Research Quarterly* **61**(2), 294.

Sen, D. (2005). 'Fee versus royalty reconsidered.' *Games and Economic Behavior* **53**(1), 141-147.

Severini, T. A. (1996). 'Measures of the sensitivity of regression estimates to the choice of estimator.' *Journal of the American Statistical Association* **91**(436), 1651-1658.

Shapiro, C. (1986). 'Investment, moral hazard, and occupational licensing.' *The Review of Economic Studies* **53**(5), 843-862.

Sheehan, J., C. Martinez and D. Guellec (2004). Understanding Business Patenting and Licensing: Results of a Survey. *Patents, Innovation and Economic Performance*. OECD Conference Proceedings: 89-11.

Shelanski, H. A. and P. G. Klein (1995). 'Empirical research in transaction cost economics: A review and assessment.' *Journal of Law, Economics, and Organization* **11**(2), 335-361.

Shenkar, O. and I. Arikan (2006). 'Auctions in Ohio Real Estate.' *Ohio Real Estate Education & Research Fund*

Shephard, A. (1987). 'Licensing to enhance demand for new technologies.' *The RAND Journal of Economics* **18**(3), 360-368.

Shin, H. D. S. (2003). *The role of uncertainty in transaction cost and resource-based theories of the firm*. The Ohio State University.

Shulman, S. (2003). 'Big ivory takes license.' *Technology Review* **April**, 77.

Silverman, D. (1993). *Interpreting Qualitative Data: Methods for Analysing Talk, Text and Interaction*. London, Sage.

Simon, H. A. (1959). 'Theories of decision-making in economics and behavioral science.' *The American Economic Review* **49**(3), 253-283.

Simon, H. A. (1962). 'The architecture of complexity.' *Proceedings of the American Philosophical Society* **106**(6), 467-482.

Simon, H. A. (1969). *The Sciences of the Artificial*, Cambridge, MA: MIT Press.

Simonin, B. L. (1991). Transfer of knowledge in international strategic alliances: A structural approach. Ann Arbor, University of Michigan.

Simonin, B. L. (1999). 'Ambiguity and the process of knowledge transfer in strategic alliances.' *Strategic Management Journal* **20**(7), 595-623.

Singh, J. (2005). 'Collaborative networks as determinants of knowledge diffusion patterns.' *Management Science* **51**(5), 756-770.

Singh, J. and L. Fleming (forthcoming). 'Lone inventors as sources of breakthroughs: Myth or reality?' *INSEAD Working Paper No. 2009/31/ST.*

Singh, K. (1997). 'The impact of technological complexity and interfirm cooperation on business survival.' *Academy of Management Journal* **40**(2), 339-367.

Singh, N. and S. K. Bhatia (2000). *Industrial Relations and Collective Bargaining: Theory and Practice*. New Delhi, Deep & Deep Publishers.

Smits, R. (2002). 'Innovation studies in the 21st century: Questions from a user's perspective.' *Technological Forecasting and Social Change* **69**(9), 861-883.

Soh, P. H. and E. B. Roberts (2003). 'Networks of innovators: a longitudinal perspective.' *Research Policy* **32**(9), 1569-1588.

Somaya, D. and D. Teece (2000). Combining Patent Inventions in Multi-Invention Products: Transactional Challenges and Organizational Choices. *Haas School of Business CCC Working Paper.*

Somaya, D. and D. J. Teece (2008). 'Patents, Licensing, and Entrepreneurship: Effectuating Innovation in Multi-invention Contexts', in. D. J. Teece (eds), *Technological Know-How, Organizational Capabilities, and Strategic Management: Business Strategy and Enterprise Development in Competitive Environments*, Singapore, World Scientific Publishing, 287-310.

Souder, W. E. and P. Shrivastava (1985). 'Towards a scale for measuring technology in new product innovations.' *Research Policy* **14**(3), 151-160.

Spedding, L. (2008). *Due Diligence Handbook: Corporate Governance, Risk Management and Business Planning*. Oxford, CIMA Publishing / Elsevier.

Spence, M. (1973). 'Job market signaling.' *The Quarterly Journal of Economics* **87**(3), 355-374.

SPSS (2007). *SPSS Missing Values™ 17.0*. Chicago, SPSS Inc.

Stahl, I. and Economic Research Institute (1972). *Bargaining Theory*. Stockholm, EFI.

Stake, R. E. (1994). 'Case studies ', in. N. K. Denzin and Y. S. Lincoln (eds), *Handbook of Qualitative Research*, Thousand Oaks, CA, Sage, 236-247.

Stankiewicz, R. (1995). 'The role of the science and technology infrastructure in the development and diffusion of industrial automation in Sweden', in. B. Carlsson (eds), *Technological Systems and Economic Performance: The Case of Factory Automation*, Dordrecht, Kluwer, 165-210.

Stark, R. M. and M. H. Rothkopf (1979). 'Competitive bidding: A comprehensive bibliography.' *Operations Research* **27**(2), 364-390.

Statistics Solutions. (2010, November 24, 2009). "Assumptions of logistic regression." Retrieved 22.06., 2010.

Steensma, H. K. and K. G. Corley (2000). 'On the performance of technology-sourcing partnerships: The interaction between partner interdependence and technology attributes.' *The Academy of Management Journal* **43**(6), 1045-1067.

Stigler, G. J. (1951). 'The division of labor is limited by the extent of the market.' *Journal of Political Economy* **59**(3), 185-193.

Stigler, G. J. (1961). 'The economics of information.' *The Journal of Political Economy* **69**(3), 213-225.

Strambach, S. (2001). 'Innovation processes and the role of knowledge-intensive business services (KIBS)', in. K. Koschatzky, M. Kulicke and A. Zenker (eds), *Innovation Networks. Concepts and Challenges in the European Perspective, Technology, Innovation and Policy*, Heidelberg /New York, Physica Verlag, 53-68.

Stump, R. L. and J. B. Heide (1996). 'Controlling supplier opportunism in industrial relationships.' *Journal of Marketing Research* **33**(4), 431-441.

Suarez, F. F. (2004). 'Battles for technological dominance: an integrative framework.' *Research Policy* **33**(2), 271-286.

Sutton, J. (1986). 'Non-cooperative bargaining theory: An introduction.' *The Review of Economic Studies* **53**(5), 709-724.

Tao, J., J. Daniele, E. Hummel, D. Goldheim and G. Slowinski (2005). 'Developing an effective strategy for managing intellectual assets.' *Research Technology Management* **48**(1), 50-58.

Taylor, C. T. and A. Silberston (1973). *The Economic Impact of the Patent System: A Study of the British Experience*. Cambridge, Cambridge University Press.

Teece, D. J. (1977). 'Technology transfer by international firms: The resource cost of transferring technological know-how.' *Economic Journal* **87**(346), 242-261.

Teece, D. J. (1981). 'The market for know-how and the efficient international transfer of technology.' *The ANNALS of the American Academy of Political and Social Science* **458**(November), 81-96.

Teece, D. J. (1986). 'Profiting from technological innovation: Implications for integration, collaboration, licensing and public policy.' *Research Policy* **15**(6), 285-305.

Teece, D. J. (1989). 'Inter-organizational requirements of the innovation process.' *Managerial and Decision Economics* **10**(Spring), 35-42.

Teece, D. J. (1998a). 'Capturing value from knowlege assets.' *California Management Review* **40**(3), 55-79.

Teece, D. J. (1998b). *Economic Performance and the Theory of the Firm*. Cheltenham, UK and Northampton, MA, USA, Edward Elgar Publishing.

Teece, D. J. (2000). 'Firm Capabilities and Economic Development: Implications for Newly Industrializing Economies', in. L. Kim and R. R. Nelson (eds), *Technology, Learning, and Innovation: Experiences of Newly Industrializing Economies*, Cambridge, Cambridge University Press, 105-129.

Teece, D. J. (2006). 'Reflections on "profiting from innovation".' *Research Policy* **35**(8), 1131-1146.

Teece, D. and G. Pisano (1994). 'The dynamic capabilities of firms: An introduction.' *Industrial and Corporate Change* **3**(3), 537-556.

Tekin-Koru, A. (2009). 'Technology transfers and optimal entry strategies for the multinational firm.' *Journal of International Trade & Economic Development* **18**(4), 553-574.

Thaler, R. (1980). 'Toward a positive theory of consumer choice.' *Journal of Economic Behavior and Organization* **1**(1), 39-60.

Thiele, C. (2003). Online versus Offline Beschaffungsverhandlungen. WHU - Otto Beisheim School of Management.

Thursby, J. G. and M. C. Thursby (2007). 'University licensing.' *Oxford Review of Economic Policy* **23**(4), 620-639.

Tietz, R. and C. Herstatt (2007). Virtuelle Communities als ein innovatives Instrument für Unternehmen: Eine explorative Fallstudienanalyse im Hobby- und Freizeitgüterbereich. *Institute for Technology and Innovation Management*. Hamburg, Hamburg University of Technology.

Tietze, F. (2004). Development of Intellectual Property Management: Cases from German and Swedish Biotechnology Companies. Göteborg, Chalmers tekniska högsk.: xi, 126.

Tietze, F. (2008). Technology Market Intermediaries to Facilitate External Technology Exploitation - The Case of IP Auctions. *Working Paper 55*. Hamburg University of Technology, Institute for Technology and Innovation Management.

Tietze, F. (2010). A Typology for Technology Market Intermediaries. *Working Paper 60*. Hamburg University of Technology, Institute for Technology and Innovation Management.

Tietze, F. and A. Barreto (2007). Intellectual Property Monetization: The Market and its Business Models. *Institute for Technology and Innovation Management*. Hamburg, Hamburg University of Technology.

Tietze, F. and C. Herstatt (2010). Technology Market Intermediaries and Innovation. *DRUID Summer Conference*. Imperial College London.

Tietze, F., Ervilia and C. Herstatt (2009). 'Exploring the relation of patent ownership and market success – Cases from the LCD Flat Panel Display Industry.' *International Journal of Technology Intelligence and Planning* **5**(1).

Tietze, F., C. Herstatt and S. Buse (2007). Externe Technologieverwertung: Maßnahmen zur Stärkung der Innovationskraft Hamburger KMU *Projekt Regionale Innovationsstrategien (RIS) Hamburg*. Hamburg, Hamburg University of Technology.

Tijssen, R. J. W. (2002). 'Science dependence of technologies: Evidence from inventions and their inventors.' *Research Policy* **31**(4), 509-526.

Tirole, J. (2008). *The Theory of Industrial Organization*. Cambridge, Mass, MIT Press.

Tobin, J. (1992). 'Financial Intermediaries', in. P. K. Newman, M. Milgate and J. Eatwell (eds), *The New Palgrave Dictionary of Money & Finance*, London, Macmillan, 834.

Trajtenberg, M. (1990). 'A penny for your quotes: Patent citations and the value of innovations.' *The RAND Journal of Economics* **21**(1), 172-187.

Troy, I. and R. Werle (2008). Uncertainty and the Market for Patents. *Max Planck Institute for the Study of Societies*. Cologne.

Tschirky, H. and S. Koruna (1998). *Technologie-Management: Idee und Praxis*. Zürich, Verlag Industrielle Organisation.

Tschirky, H., J.-P. Escher, D. Tokdemir and C. Belz (2000). 'Technology marketing: a new core competence of technology intensive enterprises.' *International Journal of Technology Management* **20**(3/4), 459-474.

Tushman, M. L. and P. Anderson (1986). 'Technological discontinuities and organizational environments.' *Administrative Science Quarterly* **31**(3), 439-465.

Tushman, M. L. and L. Rosenkopf (1992). 'Organizational determinants of technological change: Towards a sociology of technological evolution.' *Research in Organizational Behavior* **14**, 311-47.

University of the West of England. (2007). "Data analysis on-line learning programme." from http://hsc.uwe.ac.uk/dataanalysis/index.asp.

Utterback, J. M. (1996). *Mastering the Dynamics of Innovation*. Cambridge, Mass., Harvard Business Press.

Utterback, J. M. and F. F. Suarez (1993). 'Innovation, competition, and industry structure.' *Research Policy* **22**(1), 1-21.

Van de Ven, A. H. (1993). 'A community perspective on the emergence of innovations.' *Journal of Engineering and Technology Management* **10**(1-2), 23-51.

Van de Ven, A. H. (2000). 'On the nature, formation, and maintenance of relations among organizations.' *Academy of Management Review* **1**(4), 24-36.

Van de Vrande, V., C. Lemmens and W. Vanhaverbeke (2006). 'Choosing governance modes for external technology sourcing.' *R&D Management* **36**(3), 347-363.

van den Berg, G. J., J. C. Van Ours and M. P. Pradhan (2001). 'The declining price anomaly in Dutch Dutch rose auctions.' *American Economic Review* **91**(4), 1055-1062.

van Lente, H., M. Hekkert, R. Smits and B. A. S. van Waveren (2003). 'Roles of systemic intermediaries in transition processes.' *International Journal of Innovation Management* **7**(3), 247-280.

van Zeebroeck, N. (2007). Patents only live twice: A patent survival analysis in Europe. *Working Papers CEB*.

Várdy, F. (2010). Institutional Traps. *University of California Berkeley, Haas School of Business*.

Vickery, G. (1988). 'A survey of international technology licensing.' *STI Review* **4**(December), 7-49.

Vickrey, W. (1961). 'Counterspeculation, auctions, and competitive sealed tenders.' *Journal of Finance* **16**(1), 8-37.

Vincent, D. R. (1995). 'Bidding off the wall: Why reserve prices may be kept secret.' *Journal of Economic Theory* **65**(2), 575-584.

von Graevenitz, G., S. Wagner and D. Harhoff (2008). How Cost, Complexity and Technological Opportunity Affect the Rate of Patenting. Munich School of Management.

von Hippel, E. (1987). 'Cooperation between rivals: Informal know-how trading.' *Research Policy* **16**(6), 291-302.

von Hippel, E. (1988). *The Sources of Innovation*. Oxford, UK, Oxford University Press.

von Hippel, E. (1994). 'Sticky information and the locus of problem solving: Implications for innovation.' *Management Science* **40**(4), 429-439.

Wagner, S. (2006). Make-or-Buy Decisions in Patent Related Services. *University of Munich - Department of Business Administration*.

Walker, G. and D. Weber (1984). 'A transaction cost approach to make-or-buy decisions.' *Administrative Science Quarterly* **29**(3), 373-391.

Walker, G. and D. Weber (1987). 'Supplier competition, uncertainty, and make-or-buy decisions.' *Academy of Management Journal* **30**(3), 589-596.

West, J. and M. Bogers (2010). Contrasting Innovation Creation and Commercialization within Open, User and Cumulative Innovation. *Academy of Management Conference*. Montréal.

Williamson, O. E. (1975). *Markets and Hierarchies: Analysis and Antitrust Implications. A Study in the Economics of Internal Organization*. New York, NY, Free Press.

Williamson, O. E. (1979). 'Transaction-cost economics: The governance of contractual relations.' *Journal of Law and Economics* **22**(2), 233-261.

Williamson, O. E. (1981). 'The economics of organization: The transaction cost approach.' *American Journal of Sociology* **87**(3), 548-577.

Williamson, O. E. (1985). *The economic institutions of capitalism. Firms, markets, relational contracting*. New York, NY, Free Press.

Williamson, O. E. (1991). 'Comparative economic organization: The analysis of discrete structural alternatives.' *Administrative Science Quarterly* **36**(2), 269-296.

Williamson, O. E. (1998). 'The institutions of governance.' *The American Economic Review* **88**(2), 75-79.

Williamson, O. E. (1999). 'Strategy research: Governance and competence perspectives.' *Strategic Management Journal* **20**(12), 1087-1108.

Wilson, M. (1996). 'Asking Questions', in. R. Sapsford and V. Jupp (eds), *Data Collection and Analysis*, Thousand Oaks, Ca, Sage, 94-121.

Winkelmann, R. (2008). *Econometric analysis of count data*. Berlin, Springer.

WIPO. (2009). "International Patent Classification (IPC)." Retrieved 28.09.2009

Wuchty, S., B. F. Jones and B. Uzzi (2007). 'The increasing dominance of teams in production of knowledge.' *Science* **316**(5827), 1036.

Wurman, P. R., M. P. Wellman and W. E. Walsh (2001). 'A parametrization of the auction design space.' *Games and Economic Behavior* **35**(1-2), 304-338.

Xu, Q., J. Chen, Z. Xie, J. Liu, G. Zheng and Y. Wang (2007). 'Total innovation management: A novel paradigm of innovation management in the 21st century.' *The Journal of Technology Transfer* **32**(1), 9-25.

Yang, L. and K. E. Maskus (2009). 'Intellectual property rights, technology transfer and exports in developing countries.' *Journal of Development Economics* **90**(2), 231-236.

Yazdanparast, M. A., I. Manuj and S. M. Swartz (2010). 'Co-creating logistics value: A service dominant logic perspective.' *International Journal of Logistics Management* **21**(3), 375-403.

Yin, R. K. (2003). *Case Study Research: Design and Methods*. Beverly Hills, Sage.

Yin, R. K. and K. A. Heald (1975). 'Using the case survey method to analyze policy studies.' *Administrative Science Quarterly* **20**(3), 371-381.

Zander, U. and B. Kogut (1995). 'Knowledge and the speed of the transfer and imitation of organizational capabilities: An empirical test.' **6**(1), 76-92.

Zeebroeck, N. v., B. v. Pottelsberghe de la Potterie and W. Han (2005). Issues in measuring the degree of technological specialization with patent data. *Working Papers CEB*. Solvay Business School.

Zhang, A. (2006). 'Transaction governance structure: Theories, empirical studies, and instrument design.' *International Journal of Commerce and Management* **16**(2), 59-85.

Ziedonis, R. H. (2008). 'Intellectual Property and Innovation', in. S. Shane (eds), *Handbook of Technology and Innovation Management*, Chichester John Wiley & Sons, 295-333.

Zucker, L. G. and M. R. Darby (1996). Star scientists and institutional transformation: Patterns of invention and innovation in the formation of the biotechnology industry. *Proceedings of the National Academy of Sciences of the United States of America*. Irvine, CA. **93:** 12709-12716.

Index

Academic patents 116
Asset
  heterogeneous 92
  homogenous 92
  intangible 92
  tangible 92
Asset class 10
Auction 76
  3G UMTS radio spectrum 27
  absentee bids 86, 175
  art 27
  bidding procedure 84, 322, 330
  bidding room 86
  catalogue 38, 140, 152, 177, 207, 229, 236, 313
  choice determinant 83
  closing rules 86
  commission fees 87, 172, 181, 200, 210, 229, 298, 305, 315
  common value 92, 96, 222, 301, 322
  competitive bidding 85, 222, 298, 322
  design 29, 35, 83, 180, 204, 240, 295
  Dutch 84, 174
  English 84, 174, 263, 301, 325
  final bid 87, 172, 225, 305
  first-price 84, 180
  multi-unit 92
  open framework 85
  open outcry 86
  opening bid 87, 156, 167, 175, 240
  price logic 87
  pricing 28
  private value 92, 120, 239
  process 84, 168, 206
  prospective buyers 21, 35, 54, 59, 72, 110, 170, 188, 222, 294
  public 11, 27, 35, 42, 73, 85, 113, 186
  record retention policy 87
  reserve price 81, 87, 90, 169, 172, 180, 225, 291, 307, 323,
  sealed-bid 84, 180
  single-unit 92
  Vickrey 84, 180, 301
  wine 27, 76, 317
  winner's curse 35, 91, 222, 301, 322, 330
  winning bid 86
Auction firm 36, 210
Auction theory 35, 76, 81

Bargaining negotiation 27, 34, 77, 123, 192, 205, 299, 304
  bargaining costs 124
  bargaining power 78, 215
  buyer number 82
Blackberry 4
Blohm und Voss 40
Boundaries of the firm 11, 47, 98, 320
Business model 35, 73, 118

Clearinghouse 73
Coase Theorem 124
CommerceOne 42, 301, 315, 323
Community innovation survey 146
Comparative advantage 108
Competitive Technologies, Inc 155
Competitiveness 6, 11
Complexity 4, 16, 91, 102, 119, 189, 207, 210, 230, 249, 270, 294
Contract law 125, 135
Corporate invention 116

Dominant design 104, 119, 216
Dow Chemicals 8
DPMA 40, 151
Due diligence 72, 168, 186, 205, 223, 294, 302, 318, 325, 330
DVD media technology 3

Entrepreneurship 72
EPO 4, 27, 41, 99, 113, 147, 151, 292
Ethernet technology 3
European Patent Convention 118
Evergreening 6
External technology exploitation 21, 34, 47, 76, 185, 196, 206, 298, 304

Firm boundaries 6
Firm growth 6
Ford Corporation 155
Forward citation 235
Fraunhofer-Gesellschaft 164
Freedom-to-operate 4

Game theory 76
Governance costs 124, 134
Governance structure 125

Hold-up problem 5, 16

IBM 7, 165
IIP management 41, 128, 303, 316, 320, 326, 332
IIP manager 24, 73, 110, 122, 308
Independent inventors 22, 41, 73, 115, 154, 241, 259, 323
Inflexion Point 164
Information asymmetries 80
Information carrier 98
Infringement 4, 170, 188, 198, 316
Innovation 3, 7, 19, 107, 126, 231, 303
  cumulative 3, 11, 106, 119
  disruptive 116
  incentive 42, 63, 75, 94, 118, 219, 223, 299, 307, 325
  incremental 104, 234
  modularity 16
  multi-invention 16
  radical 107, 116, 313
  source 107
  strategy 69
Innovation management 69
Innovation policy 41
Innovation process 5, 7, 29, 47, 70, 106, 119, 234, 312, 318
  fuzzy front end 19
Innovation race 6
Innovation speed 6, 11, 101, 312, 320, 331
Innovation system 34
Institutional structure 9, 21
Intel 4
Intellectual Auctions GmbH 41, 139, 173
Intellectual Property Right 98, 170, 207, 327

Intellectual Ventures 299, 327
Invention 98, 107, 112, 118, 231, 237
Inventor 5
IP assembly problem 5, 16, 327
IP asset 4, 10, 21, 52, 61, 74, 92, 105, 132, 176
IP capitalism 48
IP portfolio 5, 185
IPC class 121, 151, 152, 329
IP-Com 4

Keep-or-sell 34, 47, 56, 122, 303, 321
Knowledge 16, 50, 73, 97
  disembodied 50
  stickyness 18
  tacit 18, 23, 52, 60, 97, 102, 118, 172, 207, 213, 295
Knowledge intensive business services 73
Knowledge transfer 180, 201, 205, 210, 211

Lead time 116, 331
License 5, 17, 27, 40, 52, 60, 76, 98, 111, 170, 209, 313
  cross licensing 6, 50
  grant-back agreements 52
  licensing revenues 8
  out-licensing 7
  royalty 4, 16, 121, 175, 191, 221, 324
Licensing contracts 123
Lifecycle 6, 47, 104, 111, 216, 233, 291
Lump sum payment 127, 173, 302, 324

Make-or-buy 34, 47, 122, 135, 303, 320
Market design 9
Market for lemons 63
Market imperfections 18, 65
Markets for technologies and ideas 5, 23, 34, 59, 331
  demand side 6, 18, 21, 71
  supply side 6, 18, 21, 68, 299, 312
Medical equipment 108
Monopoly 111, 126
Moral hazard 62
MPEG 3
Multinational firms 4

Negotiation 24, 58, 76, 129
Network 5, 26, 68, 106
Not-invented-here-syndrome 9
Novel, Inc. 42
Novo Nordisk A/S 8
NTP 4

Obstacles 35, 54, 318, 331
  information asymmetries 62, 67, 72, 208, 222
  opportunism 63
  transparency 28, 62, 80, 186, 195, 202, 210, 300, 314
Ocean Tomo 42, 139, 168
OECD 27
Only use here (OUH) syndrome 61
Open innovation 6, 11, 17, 47, 101, 118, 128, 134, 294, 303, 321

Paradigm 7, 119, 134, 303
Patent 3, 23, 40, 111, 155, 236, 266
  application 4, 72, 104, 237
  assignee 241
  blocking 6
  bundle 39, 111, 228, 237
  definition 99
  department 54, 116, 205

family 3, 100, 119, 232, 294, 313
fencing 6
holder 100
infringement 56, 218, 236
IPC class 39, 43, 114
jurisdiction 19, 39, 100
law firm 22, 74, 132, 155, 189, 191, 204, 299, 306, 326
litigation 4, 73, 80, 170, 204
management 196
management software 323
office 22, 151, 235
pool 50, 72
portfolio 42, 110, 120
premium 110, 296
prior art 108
priority date 174, 234
sytem 19, 100, 118, 121
troll 4
unused 117
valuation 10, 34, 173, 204, 239
value 111, 120
Patented technology 9, 101, 168, 261
PatVal 41, 109, 218, 239, 292, 328
Procter and Gamble 6
Propensity to patent 117

Research in Motion 4
Resource allocation 58, 61, 78, 332
Resource-based view 34, 97, 135
Robert Bosch GmbH 4

Siemens AG 155
Small and medium sized enterprise 22, 40, 116, 131, 183, 196, 241, 259
Standards 5, 170

Technology
acquisition 17, 51, 56, 57, 123
complexity 101, 212, 248
definition 97
deployment 51, 56, 59
exploitation channel 53
impact 105, 236, 248, 294
marketing 47, 51, 193
quality 108, 220, 237, 248, 294
transfer 50, 60
type 24
uncertainty 103, 248, 215, 251, 295
value 110, 257
Technology auction 11, 24, 34, 139, 254
Technology base 19, 47, 303
Technology Market Intermediary 7, 10, 26, 34, 41, 73, 91, 127, 134, 189, 208, 304, 314, 330
broker 18, 57, 66, 70, 127, 208, 326
consultants 69
non-practicing entities 4, 22, 316
patent troll 191, 201
Technology scouting 71
Technology stock 108
Technology transaction 7, 23
Technology transfer office 40, 112, 121, 154, 316, 323
Technology-based firm 23
TFT display technology 108
Transaction 9, 47, 71, 101, 127, 202, 311, 327
management 11
obstacles 9
spot market 127
traded asset 23, 55, 66, 132
transfer agent 21
transfer object 21

transferred medium 21
Transaction cost theory 100, 215, 294, 305
asset specificity 100, 128, 211
transaction frequency 129
Transaction costs 10, 59, 72, 122, 134,187, 203, 210, 222, 301
dynamic 123
Transaction governance structure 11, 24, 59, 73, 110, 125, 139, 168, 202, 210
bilateral 131
market governance 132
selection 130
trilateral 131
unilateral 130
Transaction management 9, 59
Transaction model 10, 23, 27, 41, 60, 122, 212, 305
Transaction process 23, 55, 71
Typology 48, 53, 66, 71, 126, 327, 331

Uncertainty 119
University 41, 71, 116, 121, 154, 166, 323, 326
US spectrum auctions 40
USPTO 4, 147, 151, 166, 235

White & Case LLP 40
Wifi technology 3
Willingness-to-exploit 6
Willingness-to-pay 91, 207, 295, 329
WIPO 151, 235
World Trade Organization 22

Yet2.com 27, 311